Future North

The changing Arctic is of broad political concern and is being studied across many fields. This book investigates ongoing changes in the Arctic from a landscape perspective. It examines settlements and territories of the Barents Sea Coast, Northern Norway, the Russian Kola Peninsula, Svalbard and Greenland from an interdisciplinary, design-based and future-oriented perspective.

The Future North project has travelled Arctic regions since 2012, mapped landscapes and settlements, documented stories and practices, and discussed possible futures with local actors. Reflecting the multidisciplinary nature of the project, the authors in this book look at political and economic strategies, urban development, land use strategies and local initiatives in specific locations that are subject to different forces of change.

This book explores current material conditions in the Arctic as effects of industrial and political agency and social initiatives. It provides a combined view on the built environment and urbanism, as well as the cultural and material landscapes of the Arctic. The chapters move beyond single-disciplinary perspectives on the Arctic, and engage with futures, cultural landscapes and communities in ways that build on both architectural and ethnographic participatory methods.

Janike Kampevold Larsen is Associate Professor at the Institute of Urbanism and Landscape, The Oslo School of Architecture and Design, Norway. With a background in literature and philosophy, she specializes in landscape theory and particularly the configuration and conceptualization of contemporary landscapes. She is project leader of the Future North project and the Landscape Journeys project before that. She is one of the article editors for the *Journal of Landscape Architecture* (JoLA).

Peter Hemmersam is Associate Professor at the Institute of Urbanism and Landscape, The Oslo School of Architecture and Design, Norway. He is an architect and received his PhD from the Aarhus School of Architecture in 2008. His main research deals with urban design and urban policy and focusses on liveability, enabling technologies, sustainability, community engagement and the public realm. He directs the Oslo Centre for Urban and Landscape Studies at AHO, and is a senior researcher in the Future North project.

Landscape Architecture: History – Culture – Theory – Practice
Series Editors: Ellen Braae and Henriette Steiner

This series forms a central outlet for research exploring new approaches and vocabularies in relation to this field. It offers both young and established researchers a much-needed platform for publishing their work in a way that allows for both traditional and interdisciplinary methodologies to come into play. Volumes focussing on the critical interpretation of Landscape Architecture from a historical, theoretical or philosophical perspective are particularly encouraged. Questions of ethics and sustainability are also emphasised.

For a full list of titles, please visit: www.routledge.com/Landscape-Architecture-History-Culture-Theory-Practice/book-series/ASHSER-1440

Future North
The Changing Arctic Landscapes
Edited by Janike Kampevold Larsen and Peter Hemmersam

Future North

The Changing Arctic Landscapes

Edited by Janike Kampevold Larsen and Peter Hemmersam

LONDON AND NEW YORK

First published 2018 by Routledge

2 Park Square, Milton Park, Abingdon, Oxfordshire OX14 4RN

52 Vanderbilt Avenue, New York, NY 10017

Routledge is an imprint of the Taylor & Francis Group, an informa business

First issued in paperback 2020

British Library Cataloguing-in-Publication Data
A catalogue record for this book is available from the British Library

Library of Congress Cataloging-in-Publication Data
A catalog record for this book has been requested

Typeset in Bembo
by Apex CoVantage, LLC

ISBN 13: 978-1-4724-8125-2 (hbk)
ISBN 13: 978-0-367-59237-0 (pbk)

Contents

Figures

Maps

Contributors

Susan Jayne Carruth was born in Scotland in 1980, and received her Bachelors and Masters in Architecture at Strathclyde University. She practiced in London for almost a decade, mostly with Piercy & Company, before relocating to Denmark in 2012 to begin a PhD at the Aarhus School of Architecture. In 2015 Susan defended her thesis entitled 'Infrastructural Urbanism that Learns from Place'. Using Greenland as a central case study, her research explored the spatial, aesthetic and civic dimensions of renewable energy infrastructures, contributing a praxis for embedding and situating technical infrastructures in place and culture. As Head of Research and Sustainability for White Arkitekter she aims to bridge between research and practice.

Elizabeth Ellsworth is an artist and a member of the Atomic Photographers Guild. She has published extensively on the subjects of media and public pedagogy and media and change. Elizabeth co-directs smudge studio. Their collaboration works across video, photography, graphic and web design and installation. Their process involves visual field research, aesthetic response and public pedagogy. As part of the collaboration, Elizabeth uses photography (digital and 35mm) and video to visualize or 'signal' invisible forces (natural and human-made) that shape daily life. Elizabeth is a guest researcher at the Future North project.

Aileen A. Espíritu is a researcher at the Barents Institute, The University of Tromsø – The Arctic university of Norway. She has ongoing research on sustainable development in the Arctic regions, notably its urban areas; region-building in the Arctic and the Barents Region; identity politics in indigenous and non-indigenous northern communities; the impact of industrialisation and post-industrialisation on mono-industry towns in the High North and the politics of community sustainability in Russia in comparative perspective. Espíritu is also a senior researcher in the Future North project that focussed on the knowledge production on new landscape typologies, include the everyday in the category of landscape, exploring new tools to articulate and narrate such perspectives. She is a co-editor in the international refereed journal *Barents Studies*.

William L. Fox is a writer whose work is a sustained inquiry into how human cognition transforms land into landscape. His numerous nonfiction books rely upon fieldwork with artists and scientists in extreme environments to provide the narratives through which he conducts his investigations. He has published poems, articles, reviews and essays in more than seventy magazines, has had fifteen collections of poetry published in three countries, and has written eleven nonfiction books about the relationships among art, cognition and landscape. He has also authored essays for numerous exhibition catalogues and artists' monographs. In the visual arts, Fox has exhibited text works in more than two dozen group and solo exhibitions in seven countries. In 2001–2002 he spent two-and-a-half months in the Antarctic with the National Science Foundation in the Antarctic Visiting Artists and Writers Program. He has also worked as a team member of the NASA Haughton-Mars

Project, which tests methods of exploring Mars on Devon Island in the Canadian High Arctic. He was a visiting scholar at the Getty Research Institute, the Clark Institute, the Australian National University and the National Museum of Australia. He has also twice been a Lannan Foundation writer-in-residence

Peter Hemmersam directs the Oslo Centre for Urban and Landscape Studies at the Oslo School of Architecture and Design. His background is in architecture and he is a former partner in the architectural practice Transform. He received his PhD from the Aarhus School of Architecture in 2008. He has previously directed the Institute of Urbanism and Landscape at AHO, and currently chairs the AHO Research Committee. His main research interest lies in the field of retail based urban design, Arctic urbanism and digital urbanism. He is a senior researcher in the Future North project and was Research Fellow at the Center for Art + Environment at the Nevada Museum of Art (2014–2016).

Morgan Ip's research is focused on ethnographic methods interrogating local context, participation and engagement in Arctic urban design and landscapes. He holds a Master of Architecture from Carleton University with a thesis centred on participatory co-design of a cultural healing centre in Cape Dorset. Ip also worked on two interdisciplinary International Polar Year projects focused on community responses to arctic changes. Prior to joining The Oslo School of Architecture and Design, he worked at Lateral Office, a renowned Toronto firm with extensive experience tackling architectural and cultural design challenges of the Arctic.

Jamie Kruse is an artist, designer and part-time lecturer at Parsons, The New School for Design (New York). In 2006 she co-founded smudge studio, with Elizabeth Ellsworth, based in Brooklyn, NY. Their multi-media practice seeks to invent aesthetic provocations that assist humans in feeling for themselves the reality of contemporary forces and scales of change (natural and human-made). She is the author of the Friends of the Pleistocene blog (fopnews.wordpress.com) and recently co-edited a collection of essays entitled, *Making The Geologic Now: Responses to Material Conditions of Contemporary Life* (2012). Jamie is a guest researcher at the Future North project.

Janike Kampevold Larsen has a background as a literary scholar, and now specialises in landscape theory, particularly the configuration and conceptualisation of contemporary landscapes. At the Oslo School of Architecture and Design, she is project leader of the Future North project and the Landscape Journeys project before that. She was Research Fellow at the Center for Art + Environment at the Nevada Museum of Art (2014–2016). She is currently the coordinator of the Tromsø Academy of Landscape and Territorial Studies.

Henry Mainsah is a Marie Curie scholar at the Centre for Interdisciplinary Methodologies, University of Warwick. He was previously Associate Professor at the Centre for Design Research, Oslo School of Architecture and Design. He holds a PhD in Media and Communication from the University of Oslo. His research connects the social sciences with design research, with particular focus on interdisciplinary research methods. A lot of his teaching has focused on qualitative research methods for students in media, design, and architecture. He has previously researched and taught a wide range of topics such as digital culture, design, identity, globalization, learning and youth culture. His research has been published in journals such as *European Journal of Cultural Studies, Computers and Composition* and *Journal of Media Innovations*.

Andrew Morrison leads design research projects on Communication and Interaction Design, collaborating on service, systems and product design. He works on design writing, fiction and criticism, and design and technology critiques and collaborates with the Institute of Urbanism and Landscape. He has coordinated the AHO PhD School, has supervised and examined widely is on the board of several international journals and the AHO Board. Prominent projects include YOUrban. Recent books are *Inside Multimodal Composition* (2010) and *Exploring Digital Design* (2010). Andrew is co-chair of the *Design + Power* NORDES 2017, a member of *Anticipation 2017 Conference* and leads the AHO 2017 research evaluation He is co-authoring books on doctoral design education and the network city, and his own on Futures and Design Studies.

Johan Schimanski is a Professor of comparative literature and Head of Research of his department at the University of Oslo, and also part-time research Professor of Cultural Encounters at the University of Eastern Finland. His research interests include borders, the Arctic, science fiction and Welsh literature. He has co-coordinated research projects within Arctic discourses, border aesthetics, Arctic modernities and within the EU FP7 project EUBORDERSCAPES. He has co-edited the volumes *Border Poetics De-Limited* (2007), *Arctic Discourses* (2010), *Reiser og ekspedisjoner i litteraturens Arktis* (2011) and *Border Aesthetics: Concepts and Intersections* (2017). Recent publications include 'Border Aesthetics and Cultural Distancing in the Norwegian-Russian Borderscape' (2015) and 'Seeing Disorientation: China Miéville's *The City & the City*' (2016); with Ulrike Spring he has written 'The Useless Artic: Exploiting Nature in the Arctic in the 1870s' (2015) and the monograph *Passagiere des Eises: Polarhelden und arktische diskurse 1874* (2015).

Kjerstin Uhre is an architect MNAL currently finishing her PhD project in Arctic landscapes at the Oslo School of Architecture and Design. Her thesis – 'The Perforated Landscape' – addresses ongoing and contested transitions of outfield landscapes in Sápmi. She holds a diploma from the Bergen School of Architecture and has studied Philosophical Aesthetics at the University in Bergen. Uhre has built, exhibited, taught, debated and published extensively, and she is Director of Dahl & Uhre architects in Tromsø. D&U has alongside with designing built projects – alone and in collective achievements – contributed to Nordic urbanism through conducting prize-rewarded urban landscape projects and winning entries in open architecture and idea competitions at regional and territorial scales in the Nordic Countries.

Acknowledgements

The research in this book was conducted in the Future North project (www.futurenorth.no), funded by Research Council Norway under the SAMKUL program. The project was located at the Oslo School of Architecture and Design and was a collaboration between the Institute of Urbanism and Landscape and the Institute of Design. The Barents Institute/UiT, the Arctic University of Norway was a partner in the project.

Project team:

  Janike Kampevold Larsen, researcher, project leader
  Peter Hemmersam, researcher
  Andrew Morrison, researcher
  Aileen A. Espíritu, researcher
  Henry Mainsah, researcher
  Kjerstin Uhre, PhD-fellow
  Ip, Morgan, PhD-fellow
  Ann-Sofi Rönnskog, PhD-fellow
  Kathleen John-Alder, guest researcher
  William L. Fox, guest researcher
  Elizabeth Ellsworth & Jamie Kruse (smudge studio), guest researchers
  Alessandra Ponte, guest researcher

We would like to thank Svein Harald Holmen and Vardø Restored for continued collaboration, and the Fritt Ord foundation for supporting our public seminar in Oslo.

Students at the Oslo School of Architecture and Design and the Tromsø Academy of Landscape and Territorial Studies have been extensively involved in studies of territorial drivers of change, cultural heritage and urban development.

We are grateful for contributions to this volume from Johan Schimanski og Susan Carruth, both allies of the project. Sarah Bonnemaison, Tom Nielsen and Brita Nøstvik have contributed valuable advice in the project period.

The Centre for Art + Environment at the Nevada Museum of Art and the International Centre for Northern Governance and Development at the University of Saskatchewan generously hosted researchers from the project, and researchers from the Kola Science Center of the Russian Academy of Sciences provided valuable insight into the social and environmental conditions of the Kola Peninsula.

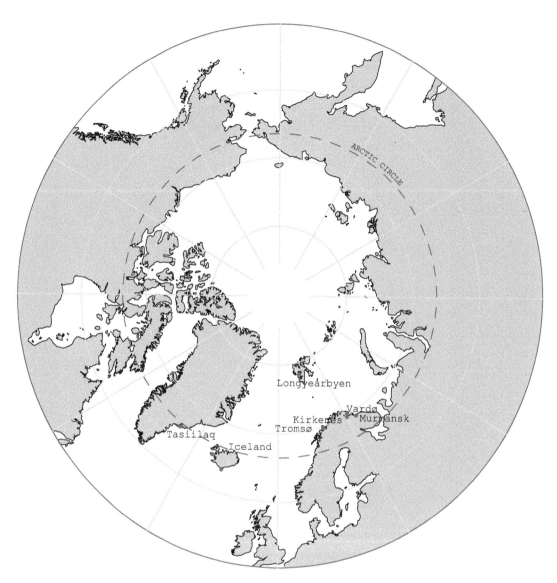

Map 1 The Future North project has traveled to various changing Arctic territories in order to study locations exposed to various current and future forms of landscape exploitation: mining, fishing, hunting and, more recently, tourism. The extreme events of these territories challenge our landscape perceptions. They are entering the global economic geography, and this calls for consciousness and care with regard to how we approach them (map: Eimear Tynan).

1 What is the future North?

Janike Kampevold Larsen and Peter Hemmersam

North is a cardinal point. It is a relative place, a point on a compass. Even within the geographies of the North – such as Norway (which literally means the "northern way") – there is always an even further north, different from where we live. What is the Future of this North? What will it look like? Who will live there and how will they sustain themselves? More important still: How do our actions today contribute to shaping what is also known as the Arctic – its communities, its landscapes, its industries – in the future? There is no one answer to these questions; the answers depend on a broad variety of changing landscape conditions, including climate change, sea-ice movements and animal migration patterns, evolving industries and shifting populations as well as increasing connectivity and tourism.

To most people, the Arctic, despite encircling the globe and spanning many countries and peoples, is a mythological topos. This perception is continuously being reproduced. The landscapes of the North have been described for centuries by explorers, painters, writers and scientists as wild, terrifying and, not least, precious. These landscapes have been portrayed as places of retreat and of terror, but also of bliss.[1] Discourses on the Arctic and the North have often rested upon the unknown, what had to be charted and conquered.[2] What was the fate of the last of Sir John Franklin's expeditions to the Northwest Passage, which departed England in 1845? How can one endure the polar night? Are there ice monsters on the North Pole? Or land underneath it? Does the ice move? Imagination has played a large role in these often fantastic discourses.

In the Arctic, landscapes and communities seem to be changing faster than anywhere else. The current massive interest in the Arctic pivots around two images: one of the Arctic as a 'canary' for climate change – an image supported by scientific measurements and observations – and one of industrial development based on resources and strategic potential. The second of these is partly possible due to climate change, as ice retreats and opens new territories for prospecting and extraction. However, although knowledge continues to expand, such images are still essentially formed at a distance. Landscapes are sensed remotely, and local communities are rarely considered to have any significant agency with regard to their futures. An abundance of literature reiterates the meta-narratives of a shifting climate, accelerating globalization and low ecological resilience.[3] Fearful discourses are echoed within the communities of the Arctic, where populations are confused about the future. Their livelihoods may be threatened, and they may even suffer from 'Solastalgia' – the feeling of not only loss of home, but loss of a healthy and meaningful relation to the landscape they inhabit due to the effects of environmental change.[4] At the same time, the populations of the Arctic find themselves in an emerging geopolitical center, as this diverse region has been increasingly exposed to speculation about resource extraction and global flows of trade and people. This also means that the *landscapes* of the Arctic – its waters, its coastlines, its fisheries, its sea floors and, not least, its terrestrial territories – have moved to the center of political and societal debate. In modern history, landscape has been something that existed at the margins of society or enclosed within it as parks or recreational areas. In the age of the Anthropocene, landscapes are central to any debate about urbanization, land use, environmental challenges, preservation issues and extractive industries.[5]

Landscapes of the North

In the Future North project, we have studied territories that are imbued with the promise of future potential. Traditionally they have been regarded as resting beyond the well-known world and have been particularly mythologized due to their remoteness.[6] To us, the circumpolar Arctic is a laboratory for investigating landscapes in the making – their materiality and appearance as well as their cultural layers. Specifically, we have looked into the landscapes of the European Arctic that face the Barents Sea. Due to its location, its generally ice-free waters and strategic importance, this region is the most accessible, populated and industrialized part of the Arctic. Here, three landscapes were chosen for their particular exposure to various forces of change, past, present and future: the Russian Kola Peninsula, the Barents Coast in northern Norway and the high Arctic archipelago of Svalbard (or Spitsbergen, as it is sometimes called). These very different arenas for research may not all be 'Arctic' – in the climatographic sense of being located north of the 10°C July isotherm which roughly corresponds to the dividing line between the snow forest of the taiga and the tree-less tundra – but our concern has been to study the Arctic as a *condition* where the relationship between landscape and people is charged by both climate and human agency in unique ways. These landscapes share an uncertain future. And there is an urgency to this future, given increasing and speculative interest from globalized industries and states. In trying to understand this condition, we have been inspired by Canadian geographer Louis-Edmond Hamelin, who demonstrated that perceptions of 'North' are dynamic and change over time. There are several parameters that contribute to defining the nordicity of a region, and settlements and places will become more or less 'Nordic' as climate, infrastructures, economies and perceptions shift.[7] During the duration of the Future North project, we have observed rapid shifts in such parameters, including shifting conditions for hunting and fishing, changes in the use of urban and landscape spaces due to temperature rise, rapid demographic fluctuations due to mineral prices, increasing geo-hazards and changes in infrastructural connectivity.

The future northern landscapes are subject to diverging interests and intentions from different groups – including pastoralists, industrialists, fishermen, preservationists and states – but also to the gaze of tourists. They are exposed to multiple perspectives: as landscapes, they are put to practical use, while at the same time providing immaterial and aesthetic experiences. As an example of how complex and multifaceted the embodied experience of landscapes can be, philosopher Jakob Meløe divided the landscape of Northern Norway into two types: the fishermen's coastal landscape and the reindeer herders' practice landscape.[8] The fisherman's implied use of tools and his ability to find port is – like the Sámi reindeer herder's reading of the seasons, the weather and the snow – the result of long experiences in a given landscape, and of working with it. While this understanding points to important dimensions of the relation between people and their environment, Meløe's description of the northern landscape has met criticism for being reductive. Geographer Michael Jones points out that it fails to include the agricultural landscape which is crucial to the traditional fisherman–farmer smallholdings in Northern Norway, the industrial landscape (including forestry and military installations) and, not least, the modern urban landscape where most of the population of Northern Norway live and work.[9] In addition to Jones' categories, we suggest that an updated operational division of northern landscapes consists of a 'viewing landscape', defined by the tourist, and an 'extraction landscape', defined by trade cycles and world mineral markets: one above ground and one below ground. One landscape governed by the gaze, one by the prices of petroleum and raw materials.

Engaging Arctic landscape research

The changing Arctic attracts researchers from many disciplines – climatology, geology, biology – but also from the social sciences, the humanities and the design disciplines, including architects and landscape architects, who study the interaction between material and cultural dimensions of territories. Arctic landscapes may be looked at from the perspectives of practical use or visual experience, or they may be regarded as consumption and production landscapes. In any case, negotiation of the still prevailing image of the Arctic as vast, pristine and sublime requires a move beyond the aesthetics of magnificence that derives from the romantic painterly tradition. This tradition gave rise to what has been a prevailing concept of landscape as a way of seeing, both by cultural geographers and others, until problematized by writers such as Denis Cosgrove.[10] Developing alternatives to the preoccupation with the isomorphic Arctic involves relating to forms of cultural geography that argue that human societies *produce* landscapes, and brings attention to the vernacular or everyday landscape.[11] However, we do not only seek to reproduce representational approaches of 'reading' cultural landscapes as a kind of text.[12] We also explore approaches leaning on object-oriented ontologies that more actively engage with the agency of materiality and highlight the equality of organic and non-organic landscape components (such as ground cover, gravel, rock and minerals, but also the materiality of settlements such as buildings, roads and people).[13] Being aware that vital matter directs us and influences our investigations, we continuously seek to demonstrate that materiality not only affects us as researchers but also drives industrial, geopolitical and aesthetical appreciation of the Arctic: It triggers the human desire to act upon and make use of landscapes.[14] Furthermore, searching for ways to appreciate the process of production of contemporary Arctic landscapes, we draw on perspectives posed by geographer Don Mitchell and others who offer insight into the continuous material process of landscape production through power and economic agency, which are evidently at play in the Arctic.[15]

In cultural landscape discourses, ideas of remote observation and aerial perspectives for the purpose of landscape reading, inspired by the likes of J.B. Jackson and perpetuated by landscape architect Ian McHarg, are still prevalent among cultural landscape researchers, architects and landscape architects.[16] While such approaches may be valuable, we find that in order to more closely address ongoing and future perspectives for cultural landscape change, it is important to seek out forms of observation that edge closer to the social dimensions of space. We are inspired by various alternative approaches to landscape and urban studies that highlight embedded, eye-level perception.[17] This resonates with authors such as Tim Ingold,[18] Guy Debord, Rebecca Solnit, Robert Smithson and Sarah Pink, who all in various ways explore the phenomenological dimension of walking as an approach to map cultural landscapes.[19] This relates also to Tim Ingold's notion of 'Taskscape', for example: interpreting landscapes as socially constructed by interconnected tasks – thus bringing time into the reading of landscapes and space.[20]

Near future Arctic

In addition to exploring the role of the material agency of landscapes, we have sought to highlight the potential of architectural and design-based speculation. Such speculation presents near future views or imaginaries that are neither dystopian nor utopian, but help us face the present.[21] We have experienced a need for engaging with the field of futures studies and its relations, especially to architectural design and landscape research. The transdisciplinary field of future studies has typically centered on strategic planning, but in recent years it has widened its interests and scope to include interrogations of knowledge domains and representations. This has included issues of emergence through narrativity and performativity. In our research on Arctic cultural landscapes, we have had to engage with what might be called 'Future Anthropocene', which is concerned with yet unseen results of humankind's largescale interventions on the earth's surface. The project speculates on these by experimenting with fictive personas, design fiction and 'fabulous' constructions of place and perspective that bring together concerns with architecture, design and futures and notions of antici-pation.[22] A prominent example of such modes of landscape inquiry within the project has involved the imaginary nuclear-powered narwhale, Narratta, a story-telling device that has served as a shared voice and persona, allowing us to write speculatively in relation to the places we study.[23] She comes out of the ice, sometimes launches into the air, catch glimpses of new terrains, then re-emerges in an entirely different place – or in the same place, transformed by what she witnessed.[24]

Studying landscapes from within

We try to bring attention to Arctic landscapes as *produced*, moving away from an idea of landscape as a way of *seeing*. However, it is obvious that landscape as both 'scene' and 'seen' is still important in the Arctic due to tourism, which is a projected future path for many local communities. When, for instance, traditional mining societies in the Arctic transform in response to global change and min-eral prices, both the geologic and the visual landscape remain important for the future.[25] In all prob-ability, Arctic landscapes will persist as hyper-networked tourist locations and places.[26] Sense of place is evident in these locations, but, leaning on Doreen Massey and others, we argue that they would not be able to subsist unless also part of a complex network of trade and transport within a global economy and society.[27] We also argue, with Lucy Lippard, that suggestions for local interventions and design practices in such locations should be "*place specific* – that is, include people and consider the network of social and environmental agencies in a given place".[28] By insisting on place specificity, one expands the local capacity for thinking about landscape change. We find this particularly apt in our approach to small Arctic communities that, as we have witnessed, fluctuate between positioning themselves in relation to external flows in the form of future industry and infrastructure and mobi-lizing local social resources.

In order to develop practices of interventions and design, and to explore ways of addressing future Arctics, we have adapted methods and concepts for addressing the complexity, emergence and change in these landscapes from different fields of knowledge. The project brought together a multidisciplinary group, consisting of landscape theorists, specialists in visual media, design, art and architecture, an ethnographer and social scientist as well as invited guest researchers. Many of us had not visited the different locations until we arrived there as a group. As experts in different academic fields, we would then start interpreting – both alone and by combining perspectives – and work-ing our ways into an intimacy with each other's expertise with the given territory and its various representations.

Architectural practice is about social exchange in negotiation with a material context. In this project, we have employed the apparatus of architecture to engage communities and landscapes. Adding to this, humanistic rhetorical forms of inquiry have been employed to conceptualize land-scapes and to look critically at how they are perceived and articulated by both locals and outsiders. Future and historic perspectives among local populations and stakeholders have been approached through ethnographic methods, including extended fieldwork, interviews, tours and photographic documentation, but also digital online participatory mapping and narrative and performative forms of inquiry.[29] Our approaches for "viewing the world from within" have also included experimental mapping of material, iconographic and narrative landscapes.[30] One prominent approach to inter-rogating the interaction of local materiality with the globalizing forces of change is found in the work of architect Raoul Bunschoten, which inspired us to develop a form of cultural mapping based on collectively walking transects through urban landscapes.[31] The transect activates encounters with what he calls 'Urban Flotsam', which for us includes every component of matter that we encountered as landscape observers, organic as well as inorganic, human as well as non-human. These Urban Transect Walks functioned as introductory framing devices that enabled our multidisciplinary group to develop initial concepts and perspectives on landscape reading as we walked, paused and talked.[32] Through this action, we attempt to appreciate landscapes as assemblies of materialities that are put to use – and are also acting back. As Nato Thompson says in his book *Experimental Geography*: "Not only do we make the world – the world makes us".[33]

Travel as research

Scientists and specialists have explored and represented the Arctic through expeditions and documentation for centuries. In encountering territories, Michael Bravo and Sverker Sörlin argue that the traditional field sciences perform a practice of "collecting, sketching, measuring, recording, classifying" as an "external order" imposed from an outside perspective of rationality.[34] Our research has involved repeatedly moving in, through and out of the landscapes of study, from Kirkenes to Murmansk on the Kola Peninsula, to Vardø and the Varanger Peninsula in Norway and to Longyearbyen on Svalbard. Cars, buses and even planes have performed as ambulating viewing platforms and landscape machines that allowed views to continually develop and connections to be made. Like the transect walk, these vehicles delimited and directed our access to the territories. In order to approach the territories as settlements and landscapes in ways that engage with forces and materialities, with social and cultural dimension, we have conceptualized travel as a research mode that productively exploits the tendency to always compare something new with what we already know. Travel has a double role; not only does it introduce encounters with the new and unseen, but it also reproduces the travelers' preconceived ideas, expectations and anticipations, often through preconfigured aesthetic appreciation. This double role means that the underlying ideas and configurations that enable such appreciation can be teased out of the experience, as argued by architectural theorist Jilly Traganou.[35] However, while we seek to appreciate landscapes beyond the aesthetic, we do not regard this dimension as irrelevant, but rather aspire to maintain a tension between the two functional levels.

We lean on the tradition of travel as a research format in architecture, as introduced in Robert Venturi, Denise Scott Brown and Steven Izenour's seminal book *Learning from Las Vegas*. Mapping and describing the gambling capital, the visiting research team attempted "to maintain an aura of objectivity and a tone of scholarly dispassion", attaining to scientific credibility by enforcing unbiased and repeatable protocols in the fieldwork they undertook.[36] In Martino Stierli's opinion, this attempt at neutral observation may be linked to the "radical, skeptical epoché" of Husserl's phenomenology, which involves a bracketing of the phenomena looked upon; through "phenomenological reduction", a phenomenon is delimited from its surrounding context to serve as an object to the intentional gaze.[37] However, as Rem Kolhaas stresses, what we are 'learning from' are situations, lands and cities where the structuring forces of globalization may be identified as the predominant driver of change.[38] These are complex situations of forces and drivers that are not easily 'reduced' to pure objects. In our travels, we found that we were indeed looking at situations reflecting globalization. For us, this has led to a two-fold strategy: Where the *astonishment* involved in epoché was found to be interesting in an initial phase of exploration, we argue that this needs to be extended through complex analysis of a broad range of actors and forces, as the chapters of this book demonstrate.

Affecting Arctic landscapes

Building intimacy with a landscape, as a materiality, place and community, means getting involved, returning repeatedly and working with communities on issues that emerge. Over the course of the project, we have conducted repeated visits to, as well as extended stays in, the landscapes of study, which has resulted in interaction and collaboration with individuals and local groups seeking to impact their communities.

Reflecting on action research as a mode of inquiry, we have sought insight into, impact on and the possibility of raising awareness of the complex interaction of social and material aspects of local landscapes and their exposure to forces of change.[39] Work in different territories has involved students of architecture and landscape architecture from the Tromsø Academy of Territorial Studies and the Oslo School of Architecture and Design working on landscape and urban analysis and design proposals. The interactions between specific locations and researchers/students have had impact through providing empowerment and inspiration for local inhabitants, revealing embedded physical and organizational structures in a community and suggesting new ways in which these may be reappropriated for new uses as landscape and urban change occurs.[40]

Vardø

- *Challenges:* Shrinking population and a perceived lack of capacity for future thinking.
- *On-going processes:* Community mobilization and a developing culture of appreciation, care and memory.

Located on an island east of the Norwegian mainland, Vardø was in the nineteenth and twentieth centuries an important hub for trade along the Barents Coast into Russia. During the last 30 years it has lost population as a result of a declining fishing industry. Today, nearby Kirkenes, close to the only land border crossing to Russia, has taken over the role of primary border town. With the construction of a new harbor, Vardø tried to position itself in expectation of future industries, hoping to become a supply base for offshore hydrocarbon extraction. Business never arrived, but this venture revealed that Vardø, and the nearby Varanger Peninsula, is a cultural landscape under pressure from extractive industries and their associated speculations and expectations. The local response to these pressures has for years entailed waiting for the petroleum industry, leaving a landscape that is effectively 'on hold'.[41] In this context of stagnation and 'wait and see', new actors have nevertheless emerged that explore alternative avenues of future thinking. As a project, we have, at an intimate level, been engaged in a long-term work to map cultural value and heritage in the city in collaboration with the local organization Vardø Restored, which was set up by local enthusiasts with a focus on the restoration of the pre-war historical buildings of town.[42] Working with this group of enthusiasts turned experts, and involving students of architecture and landscape architecture, we have been involved in a city-wide documentation of historical and heritage value, with the purpose of enabling the municipality to develop a strategic heritage plan. While such work is presumably about conservation, our shared ambition has been to explore how it actually informs strategic decisions and enables future thinking and urban planning. In this work, we have used both architectural and humanistic methods such as participatory mapping and interviews with 'living archives', business owners and collections of stories.[43] This has involved theoretically developing a cultural geography perspective on the potency of a multi-layered landscape involving a wide range of aspects of the town. Narrative inquiry through Narratta has enabled collective reflection within the research group, and performative acts by our associated artists/researchers in Smudge Studio have inserted Vardø into global narratives of anthropomorphic environmental change.[44]

Beyond knowledge-building through documentation and speculative and performative acts of thinking, one immediate impact through the project is the initiation of the explorative design project on The Grand Hotel Hub in Vardø. The initiative and collaboration with the owner of this former hotel is forming much quicker than we had anticipated, and has already resulted in a plan for gradual development and 'clients' in the form of innovators and designers who design the spaces while inhabiting them.[45]

The Kola Peninsula

- *Challenges*: Shrinking population; environmental degradation; policy restrictions on civil society.
- *On-going processes*: New petrochemical industrialization; emerging civil society; continued mining industry; increasing militarization.

In the twentieth century, the Barents Coast gained strategic importance as a mineral region and as a potentially ice-free sea route from the Soviet Union to the Atlantic. In particular, the Russian Kola Peninsula saw heavy industrialization and militarization in the Soviet period, swelling the population of the peninsula to well over one million (compared to the 75,000 inhabitants currently residing in the Norwegian border county of Finnmark). The Kola Peninsula is still today the most heavily populated part of the Arctic. Extensive mineral industry and military activity has developed the region's settlements to impressive scales, but current national and international shifts in economy and politics have rendered their future uncertain.

The resource territory of the peninsula extends over the border to include the Norwegian border town of Kirkenes, which functions as the diplomatic hub of the Barents Region. Like its Russian counterparts, this is also a mining community. It has a permanent Russian population and is

Figure 1.1 Teriberka on the Kola Peninsula was the designated land hub for the planned Shtokman gas field, which was mothballed only days before our arrival. Approaching the town, we saw what we imagined was the last digger being moved out of town.

Photo: Peter Hemmersam.

frequented by cross-border shoppers and visitors from Murmansk and other Kola towns.[46] The Kola Peninsula is an industrial but also a post-industrial landscape. In some places, the very evident environmental degradation leaves it with an almost post-apocalyptic appearance of ruination and landscape damage. In fact, we argue with philosopher Timothy Morton, the apocalypse has already happened here, and in an Anthropocene perspective, the apocalypse is us – enmeshed with non-living things in a 'Dark Ecology'.[47] Nevertheless, the more or less ruined industrial landscapes and cities of the Kola are open for speculation around possible futures, while at the same time, their vertical attachment to the minerals in the ground will in all probability endure.[48] Adding complexity to the territory, the Norwegian–Russian borderscape evidences a 'dark landscape' made up of infrastructures of both historical and contemporary politics and conflict, while also reflecting global interconnectedness as a 'hyperlandscape'.[49]

Longyearbyen/Svalbard

- *Challenges*: Deindustrialization; increased landscape hazards due to climate change.
- *On-going processes*: Urbanization and influx of tourists; emergence of urban planning.

The island territory of Svalbard, demarcating the northern edge of the Barents Sea, has a history as an energy landscape – first through whaling and hunting and later through coal mining, which has been the main industry here for over a century. While the territory is open to businesses from everywhere, today only Norway and Russia maintain settlements and activities in the archipelago. However, both are phasing out coal mining in favor of other industries: research and tourism. Far from being forgotten as a post-industrial landscape, Svalbard, and its 'capital' Longyearbyen, is increasingly seen as a central political and symbolic hub and laboratory for the future of the Arctic as a whole, which is evidenced by the role that the location, its architecture and its art attribute to it.[50]

In 2016 and 2017, avalanches took lives and buildings in Longyearbyen, and whole districts of the town now face evacuation. Screeds and escarpments are literally starting to come down, an effect of thawing permafrost that was anticipated, but not so soon and at such magnitude. These sudden results of climate change indicate that future planning in the Arctic is still dependent on variables that are highly uncertain and render the territory one of future speculation: The receding polar ice cap opens up for hydrocarbon prospecting and the possibility for new businesses associated with cross-polar shipping routes between Europe and Asia.

Working with students, we have mapped forces at work in this territory. Urban proposals were developed in response to the dramatically shifting urban situation of Longyearbyen, refined by interaction with the local planning authority and other stakeholders. These proposals were based on an insistence that the place is not a transitory camp or research outpost, but is transitioning towards an urban complexity comparable to any other town, with an urgent need for comprehensive and strategic urban planning. More than any other place in our research area, Longyearbyen and Svalbard (due to its transient population) demonstrate that places cannot be considered in isolation, but have to be seen as entangled in extended networks of economy, ecology, politics and people. This extends also to the landscape with its strong geologic character. While still beautiful and attractive to tourists, it is infused with cultural landscape layers of resource extraction and is manmade to a much greater extent than immediately evident. Today, even the glaciers are melting due to human actions – a condition paradoxically reinforced by visiting tourists, eager to witness climate change play out.

The presence of future in the North

Change has happened before our eyes as we traveled the Arctic, and much more quickly than we anticipated. The further north we ventured, the more precarious human habitation appeared and the more intense the speculation on futures became. We found that speculation on future ways of living is intimately linked with our initial hypothesis that the pressure posed by climate change, extraction industries and potentially new sea routes brings confusion, disagreement and contestation as to how to mobilize for the future. Effectively, many Arctic territories are subjected to the dynamics of a frontier perceived by many actors as being open for industrial exploitation and quick revenue, not dissimilar to the old American West.[51] Adapting a learning-from perspective, being on the ground in different places, engaging in discussions with a range of local actors, we realize that the Arctic in many ways represents a new landscape reality and dynamism. It is not only a territory that is open to increased activity; it represents a future territory and a future Anthropocene that is already charged by human infiltration. Consideration of the Arctic as an economical frontier is not entirely an outside view. The different territories and settlements in the Arctic are populated by very different groups and compositions of actors. Among these, from within and from without, imaginaries of the future evolve, converge and collide: "Arctic imaginaries, like the Arctic itself, are never settled".[52]

Arctic landscapes and settlement are often beautiful, but they have to be approached at the same level of complexity and interconnectedness as cities and territories elsewhere and in ways that supplement or counteract hegemonic discourses on landscapes and development in the region. Mediating such perspectives, as well as acting transdisciplinarily within communities and with individuals and groups, is critically important when engaging future thinking in the development of the future north. Through experimental, participative and action-oriented research modes that include architectural production, the Future North project demonstrates that narrative and performative future speculation works towards developing multi-views beyond any of the individual disciplinary fields involved. Doing this, we supplement other, more conventional cultural landscape perspectives. This is important in order to progress our understanding of Arctic landscapes as not only diverse, conflictual and narrated, but also as continually emerging and speculative hyperlandscapes that will materialize in multiple forms, yet unknown, in a future Anthropocene.

Notes

1 Johan Schimanski and Ulrike Spring point out that current images of the Arctic rest on older discourses, created mostly through expeditions such as those of Fridtjof Nansen and John Franklin, as well as the Austro-Hungarian Arctic Expedition (1872–1874). Johan Schimanski and Ulrike Spring, "A Black Rectangle Labelled 'Polar night': Imagining the Arctic After the Austro-Hungarian Expedition of 1872–1874," in *Arctic Discourses*, ed. Anka Ryall, Johan Schimanski, and Henning Howlid Wærp (Newcastle upon Tyne: Cambridge Scholars Publishing, 2010).

2 Jennifer Blessing, *True North* (New York: Deutsche Guggenheim, 2008).

3 E.g. Laurence Smith, *The New North: The World in 2050* (London: Profile, 2011); Philip E. Steinberg et al., *Contesting the Arctic: Politics and Imaginaries in the Circumpolar North* (London: I.B. Tauris, 2015); Peter Arbo et al., "Arctic Futures: Conceptualizations and Images of a Changing Arctic," *Polar Geography* 36, no. 3 (2013).

4 Glen Albrecht, "Solastalgia, a New Concept in Human Health and Identity," *Philosophy Activism Nature* 3 (2005).

5 E.g. Klaus Dodds and Mark Nuttall, *The Scramble for the Poles: The Geopolitics of the Arctic and Antarctic* (Cambridge: Polity Press, 2016); E. C. H. Keskitalo, *Climate Change and Globalization in the Arctic: An Integrated Approach to Vulnerability Assessment* (London: Earthscan Publications, 2008); Eric Paglia, *The Northward Course of the Anthropocene Transformation, Temporality and Telecoupling in a Time of Environmental Crisis* (PhD diss., Stockholm: Royal Institute of Technology, 2016), http://urn.kb.se/resolve?urn=urn:nbn:se:kth:diva-179139.

6 See Johan Schimanski's chapter in this volume on the mythologization of the Arctic: "Reading the future North."

7 Louis-Edmond Hamelin, *Discours du Nord* (Québec City: Université Laval, 2002).

8 Jakob Meløe, "The Two Landscapes of Northern Norway," *Inquiry* 31, no. 3 (1988).

9 Michael Jones, "The 'Two Landscapes' of North Norway and the 'Cultural Landscape' of the South," in *Nordic Landscapes. Region and Belonging on the Northern Edge of Europe*, ed. Michael Jones and Kenneth R. Olwig (Minneapolis: University of Minnesota Press, 2008).

10 Denis Cosgrove, *Social Formation and Symbolic Landscape* (Totowa, NJ: Barnes & Noble Books, 1985).

11 E.g. John Brinckerhoff Jackson, "Discovering the Vernacular Landscape," in *Human Geography: An Essential Anthology*, ed. John Agnew (Oxford: Blackwell, 1996).

12 Tim Cresswell, *Geographic Thought: A Critical Introduction* (Chichester: Wiley-Blackwell, 2013).

13 E.g. Nigel Clark, Doreen B. Massey, and Philip Sarre. *Material Geographies: A World in the Making* (London: Sage Publications, 2008); John Wylie, *Landscape* (London: Routledge, 2007); Ben Anderson and John Wylie J., "On Geography and Materiality," *Environment and Planning A* 41, no. 2 (2009); Sarah Whatmore, "Materialist Returns: Practicing Cultural Geography in and for a More-Than-Human World," *Cultural Geographies* 13 (2006).

14 Jane Bennett, *Vibrant Matter: A Political Ecology of Things* (Durham: Duke University Press, 2010).

15 Don Mitchell, "Landscape and Surplus Value: The Making of the Ordinary in Brentwood, CA," *Environment and Planning D: Society and Space* 12, no. 1 (1994); Kenneth Olwig and Don Mitchell, eds., *Justice, Power and the Political Landscape* (London: Routledge, 2009).

16 J. B. Jackson, "The Need for Being Versed in Country Things," *Landscape* 1 (1951); Ian L. McHarg, *Design with Nature* (New York: Doubleday, 1969); Mitchell Schwarzer, *Zoomscape: Architecture in Motion and Media* (Princeton Architectural Press, 2004); Nina E. Anker and Peder Anker, "Viewing the Earth from Without or From Within," *New Geographies* 4 (2011); Dolores Hayden, *A Field Guide to Sprawl* (New York: W W Norton & Company, 2004).

17 e.g. Tom Nielsen, *Formløs* (Århus: Arkitektskolens Forlag, 2001); Robert Venturi, Denise Scott Brown, and Steven Izenour, *Learning from Las Vegas* (Cambridge; Massachusetts Inst. of Technology, 1972).

18 Tim Ingold and Jo Lee Vergunst. *Ways of Walking: Ethnography and Practice on Foot* (London: Routledge, 2016).

19 Sarah Pink, "An Urban Tour: The Sensory Sociality of Ethnographic Place-Making," *Ethnography* 9, no. 2 (2008); Sarah Pink, *Doing Sensory Ethnography* (London: Sage, 2015); Kimberly Powell, "Making Sense of Place: Mapping as a Multisensory Research Method," *Qualitative Inquiry* 16, no. 7 (2010).

20 Tim Ingold, "The Temporality of the Landscape," *World Archaeology* 25, no. 2 (1993).

21 Ann Bergman, Jan Karlsson, and Jonas Axelsson, "Truth Claims and Explanatory Claims: An Ontological Typology of Futures Studies," *Futures* 42 (2010).

22 See Cynthia Selin et al., "Scenarios and Design: Scoping the Dialogue Space," *Futures* 74 (2015); Kirkka Heinonen and Elina Hiltunen, "Creative Foresight Space and the Futures Window: Using Visual Weak Signals to Enhance Anticipation and Innovation," *Futures* 44 (2012); Manuela Celi and Andrew Morrison, "Anticipation and Design Inquiry," in *Handbook of Anticipation*, ed. Roberto Poli (Vienna: Springer, 2017).

23 See Andrew Morrison's account in this volume of Narratta as design fiction: "Future north, nurture forth: design fiction, anticipation and Arctic futures."

24 Her reflections are recorded in our online project blog: www.futurenorth.no.

25 Dieter K. Müller, "Issues in Arctic Tourism," in *The New Arctic*, ed. Birgitta Evengård, Joan Nymand Larsen, and Øyvind Paasche (Berlin: Springer, 2013).

26 See Janike Kampevold Larsen's chapter in this book.

27 Doreen B Massey, *Space, Place, and Gender* (Minneapolis: University of Minnesota Press, 1994); Doreen Massey, *For Place* (London: Sage Publishing, 2005). See also Susan J. Carruth's chapter in this book, "Place as progressive optic: reflecting on conceptualisations of place through a study of Greenlandic infrastructures."

28 Lucy Lippard, *The Lure of the Local: Senses of Place in a Multicentered Society* (New York: New Press, 1998); Janike Kampevold Larsen and Peter Hemmersam, "Landscapes on Hold: The Norwegian and Russian Barents Sea Coast in the New North," in *Critical Norths, Space, Nature, Theory*, ed. Sarah J. Ray and Kevin Maier (Fairbanks: University of Alaska Press, 2017).

29 In his chapter in this book, "Hyperlandscape: the Norwegian-Russian borderlands," Morgan Ip reflects on the potential for community involvement in different cities, while also conceptualizing and revealing their differences through locative media as both a research and a planning tool. In her chapter titled "The perforated landscape," Kjerstin Uhre reports on her fieldwork in the contested landscapes of Northern Norway.

30 Anker and Anker, "Viewing the Earth from Without or from Within."

31 Raoul Bunschoten, Hélène Binet, and Takuro Hoshino, *Urban Flotsam: Stirring the City* (Rotterdam: 010 Publishers, 2001); Peter Hemmersam, Andrew Morrison, and Jonny Aspen, "Serendipity and the Urban Transect Walk Reflections on Design and Cultural Mapping in Arctic Cities," in *Proceedings of Cumulus Hong Kong 2016 Open Design for E-very-thing* (Hong Kong: Cumulus, 2017).

32 See William L. Fox's account of transect walks in his chapter, "Branding ice: contemporary public art in the Arctic." On transect walks, see also Lisa Diedrich, Gini Lee and Ellen Braae, "The Transect as a Method for Mapping and Narrating Water Landscapes: Humboldt's Open Works and Transareal Travelling," *NANO* November (2014), www.nanocrit.com/issues/6-2014/transect-method-mapping-narrating-water-landscapes-humboldts-open-works-transareal-travelling/.

33 Nato Thompson, *Experimental Geography* (New York: Melville House, 2008), 15.

34 Michael Bravo and Sverker Sörlin. *Narrating the Arctic: A Cultural History of Nordic Scientific Practices* (Canton: Science History Publications: 2002), 18.

35 Jilly Traganou, *Travel, Space, Architecture* (Farnham: Ashgate: 2009), 25.

36 Robert, Scott Brown, and Izenour, *Learning from Las Vegas*; Aron Vinegar, Michael J. Golec, eds., *Relearning from Las Vegas* (Minneapolis: University of Minnesota Press, 2009).

37 Martino Stierli, *Las Vegas in the Rearview Mirror: The City in Theory, Photography, and Film* (Los Angeles: Getty Publications, 2013).

38 Rem Koolhaas, "Harvard Project on the City, 'Lagos'," in *Mutations*, ed. Stefano Boeri, Francine Fort, and Rem Koolhaas (Barcelona: Actar, 2000).

39 Gerald I. Susman and Roger D. Evered, "An Assessment of the Scientific Merits of Action Research," *Administrative Science Quarterly* 23, no. 4 (1978); Sara Kindon, Rachel Pain, and Mike Kesby, "Participatory Action Research: Origins, Approaches and Methods," in *Participatory Action Research Approaches and Methods: Connecting People, Participation and Place*, ed. Sara Kindon, Rachel Pain, and Mike Kesby (London: Routledge, 2007).

40 Specifically, this happens through on-site exhibitions in Longyearbyen and Vardø and a series of pamphlets that present site readings and proposals.

41 Hemmersam and Kampevold Larsen, "Landscapes on Hold."

42 It is now associated with the local branch of the regional Sørvaranger Museum. See www.vardorestored.com.

43 Henry Mainsah's article in this volume, "Visual and sensory methods of knowing place: the case of Vardø," unfolds various ethnographic approaches taken to developing place knowledge in Vardø. For an example of architectural methods, see Morgan Ip's account of the development of the digital mapping platform MyBarents.com that allows for community interaction, enabling future thinking in urban planning in cities across the national borders of the region

44 See the account of Smudge Studio's performative inhabitation in various locations: Elizabeth Ellsworth and Jamie Kruse, "Inhabiting change," and Andrew Morrison's unpacking of modes of collaborative design fiction for speculative inquiry, "Future north, nurture forth" (both in this volume).

45 For a presentation of the hotel: http://vardorestored.com/en/project/grand-hotel/.

46 See Aileen A. Espíritu's account of the spectacularization of Kirkenes as a diplomatic outpost, "Spectacular Speculation: Arctic futures in transition," and Morgan Ip's account of the cultural landscape and 'dark infrastructure' of the border region, "HyperLandscape" (both in this volume).

47 Timothy Morton, *Ecology Without Nature: Rethinking Environmental Aesthetics* (Cambridge, MA: Harvard University Press, 2009). See also Janike K. Larsen's chapter in this book, "The landscapes of the new north."

48 See Peter Hemmersam's account of layers of urban landscapes in "Ruins and monuments of the Kola cities," in this volume.

49 See Ip, "Hyperlandscape" in this book.

50 It is a site of spectacularization and a display window for national policies. See Espíritu, "Spectacular speculation" (in this volume). See also William L. Fox's account of the emergence of a 'brandscape' through public art in Svalbard in this volume, "Branding ice: contemporary public art in the Arctic."

51 The American West has become "heavily dependent on tourism, along with military, nuclear, and extractive industries". Lucy Lippard, *Undermining: A Wild Ride Through Land Use, Politics, and Art in the Changing West* (New York: The New Press, 2013), 94.

52 Steinberg, *Contesting the Arctic*, 9.

Bibliography

Albrecht, Glenn. "Solastalgia, a New Concept in Human Health and Identity." *Philosophy Activism Nature* 3 (2005): 41–4.

Anderson, Ben, and John Wylie. "On Geography and Materiality." *Environment and Planning A* 41, no. 2 (2009): 318–35.

Anker, Nina E., and Peder Anker. "Viewing the Earth from Without or from Within." *New Geographies* 4 (2011): 89–94.

Arbo, Peter, Audun Iversen, Maaike Knol, Toril Ringholm, and Gunnar Sar. "Arctic Futures: Conceptualizations and Images of a Changing Arctic." *Polar Geography* 36, no. 3 (2013): 163–82.

Bennett, Jane. *Vibrant Matter: A Political Ecology of Things*. Durham: Duke University Press, 2010.

Bergman, Ann, Jan Karlsson, and Jonas Axelsson. "Truth Claims and Explanatory Claims: An Ontological Typology of Futures Studies." *Futures* 42 (2010): 857–65.

Blessing, Jennifer. *True North*. Berlin: Deutsche Guggenheim, 2008.

Bravo, Michael, and Sverker Sörlin. *Narrating the Arctic: A Cultural History of Nordic Scientific Practices*. Canton: Science History Publications, 2002.

Bunschoten, Raoul, Hélène Binet, and Takuro Hoshino. *Urban Flotsam: Stirring the City*. Rotterdam: 010 Publishers, 2001.

Celi, Manuela and Andrew Morrison. "Anticipation and Design Inquiry." In *Handbook of Anticipation*, edited by Roberto Poli. Vienna: Springer, 2017.

Clark, Nigel, Doreen B. Massey, and Philip Sarre. *Material Geographies: A World in the Making*. London: Sage Publications, 2008.

Cosgrove, Denis. *Social Formation and Symbolic Landscape*. Totowa, NJ: Barnes & Noble Books, 1985.

Cresswell, Tim. *Geographic Thought: A Critical Introduction*. Chichester: Wiley-Blackwell, 2013.

Diedrich, Lisa, Gini Lee, and Ellen Braae. "The Transect as a Method for Mapping and Narrating Water Landscapes: Humboldt's Open Works and Transareal Travelling." *NANO* November (2014). www.nanocrit.com/issues/6-2014/transect-method-mapping-narrating-water-landscapes-humboldts-open-works-transareal-travelling/.

Dodds, Klaus, and Mark Nuttall. *The Scramble for the Poles: The Geopolitics of the Arctic and Antarctic*. Cambridge: Polity Press, 2016.

Hamelin, Louis-Edmond. *Discours du Nord*. Québec City: Université Laval, 2002.

Hayden, Dolores. *A Field Guide to Sprawl*. New York: WW Norton & Company, 2004.

Heinonen, Kirkka, and Elina Hiltunen. "Creative Foresight Space and the Futures Window: Using Visual Weak Signals to Enhance Anticipation and Innovation." *Futures* 44 (2012): 248–56.

Hemmersam, Peter, Andrew Morrison, and Jonny Aspen. "Serendipity and the Urban Transect Walk: Reflections on Design and Cultural Mapping in Arctic Cities." In *Proceedings of Cumulus Hong Kong 2016 Open Design for E-very-Thing*. Hong Kong: Cumulus, 2017.

Ingold, Tim. "The Temporality of the Landscape." *World Archaeology* 25, no. 2 (1993): 152–74.

Ingold, Tim, and Jo Lee Vergunst. *Ways of Walking: Ethnography and Practice on Foot*. London: Routledge, 2016.

Jackson, J. B. "Discovering the Vernacular Landscape." In *Human Geography: An Essential Anthology*, edited by John Agnew, 316–28. Oxford: Blackwell, 1996.

Jackson, J. B. "The Need for Being Versed in Country Things." *Landscape* 1, no. 1 (1951): 1–5.

Jones, Michael. "The 'Two Landscapes' of North Norway and the 'Cultural Landscape' of the South." In *Nordic Landscapes. Region and Belonging on the Northern Edge of Europe*, edited by Michael Jones and Kenneth R. Olwig, 283–99. Minneapolis: University of Minnesota Press, 2008.

Keskitalo, E., and Carina H. *Climate Change and Globalization in the Arctic: An Integrated Approach to Vulnerability Assessment*. London: Earthscan Publications, 2008.

Kindon, Sara, Rachel Pain, and Mike Kesby. "Participatory Action Research: Origins, Approaches and Methods." In *Participatory Action Research Approaches and Methods: Connecting People, Participation and Place*, edited by Sara Kindon, Rachel Pain, and Mike Kesby, 9–18. London: Routledge. 2007.

Koolhaas, Rem. "Harvard Project on the City, 'Lagos'." In *Mutations*, edited by Stefano Boeri, Francine Fort, and Rem Koolhaas, 652–719. Barcelona: Actar, 2000.

Larsen, Janike Kampevold, and Peter Hemmersam. "Landscapes on Hold – The Norwegian and Russian Barents Sea Coast in the New North." In *Critical Norths: Space, Theory, Nature*, edited by Kevin Maier and Sarah J. Ray, Fairbanks: University of Alaska Press, 2017.

Lippard, Lucy. *The Lure of the Local: Senses of Place in a Multicentered Society*. New York: New Press, 1998.

Lippard, Lucy. *Undermining: A Wild Ride Through Land Use, Politics, and Art in the Changing West*. New York: The New Press, 2013.

MacCannel, Dean. *The Tourist: A New Theory of the Leisure Class*. Oakland: University of California Press, 2013.

Massey, Doreen B. *Space, Place, and Gender*. Minneapolis: University of Minnesota Press, 1994.

Massey, Doreen B. *For Place*. London: Sage Publishing, 2005.

McHarg, Ian L. *Design with Nature*. New York: Doubleday, 1969.

Meløe, Jakob. "The Two Landscapes of Northern Norway." *Inquiry* 31, no. 3 (1988): 387–401.

Mitchell, Don. "Landscape and Surplus Value: The Making of the Ordinary in Brentwood, CA." *Environment and Planning D: Society and Space* 12, no. 1 (1994): 7–30.

Morton, Timothy. *Ecology Without Nature: Rethinking Environmental Aesthetics*. Cambridge: Harvard University Press, 2009.

Müller, Dieter K. "Issues in Arctic Tourism." In *The New Arctic*, edited by Birgitta Evengård, Joan Nymand Larsen, and Øyvind Paasche, 147–58. Berlin: Springer, 2013.

Nielsen, Tom. *Formløs*. Århus: Arkitektskolens Forlag, 2001.

Olwig, Kenneth, and Don Mitchell, eds. *Justice, Power and the Political Landscape*. London: Routledge, 2009.

Paglia, Eric. "The Northward Course of the Anthropocene Transformation, Temporality and Telecoupling in a Time of Environmental Crisis." PhD diss. Stockholm: Royal Institute of Technology, 2016. http://urn.kb.se/resolve?urn=urn:nbn:se:kth:diva-179139.

Pink, Sarah. "An Urban Tour: The Sensory Sociality of Ethnographic Place-Making." *Ethnography* 9, no. 2 (2008): 175–96.

Pink, Sarah. *Doing Sensory Ethnography*. London: Sage, 2015.

Powell, Kimberly. "Making Sense of Place: Mapping as a Multisensory Research Method. *Qualitative Inquiry* 16, no. 7 (2010): 539–55.

Ryall, Anka, Johan Schimanski, and Henning Howlid Wærp. *Arctic discourses*. Newcastle upon Tyne: Cambridge Scholars Publishing, 2010.

Schimanski, Johan, and Ulrike Spring. "A Black Rectangle Labelled 'Polar night': Imagining the Arctic After the Austro-Hungarian Expedition of 1872–1874." In *Arctic Discourses*, edited by Anka Ryall, Johan Schimanski, and Henning Howlid Wærp, 19–43. Newcastle upon Tyne: Cambridge Scholars Publishing, 2010.

Schwarzer, Mitchell. *Zoomscape: Architecture in Motion and Media*. Princeton Architectural Press, 2004.

Selin, Cynthia, Lucy Kimbell, Rafael Ramirez, and Yasser Bhatti. "Scenarios and Design: Scoping the Dialogue Space." *Futures* 74 (2015): 4–17.

Smith, Laurence. *The New North: The World in 2050*. London: Profile, 2011.

Steinberg, Philip E., Jeremy Tasch, Hannes Gerhardt, Adam Keul, and Elizabeth Nyman. *Contesting the Arctic: Politics and Imaginaries in the Circumpolar North*. London: I.B. Tauris, 2015.

Stierli, Martino. *Las Vegas in the Rearview Mirror: The City in Theory, Photography, and Film*. Los Angeles: Getty Publications, 2013.

Susman, Gerald I., and Roger D. Evered. "An Assessment of the Scientific Merits of Action Research." *Administrative Science Quarterly* 23, no. 4 (1978): 582–683.

Thompson, Nato. *Experimental Geography*. New York: Melville House, 2008.

Traganou, Jilly. *Travel, Space, Architecture*. Farnham: Ashgate, 2009.

Venturi, Robert, Denise Scott Brown, and Steven Izenour. *Learning from Las Vegas*. Cambridge: Massachusetts Institute of Technology, 1972.

Vinegar, Aron, and Michael J. Golec. *Relearning from Las Vegas*. Minneapolis: University of Minnesota Press, 2009.

Whatmore, Sarah. "Materialist Returns: Practicing Cultural Geography in and for a More-Than-Human World." *Cultural Geographies* 13 (2006): 600–9.

Wylie, John. *Landscape*. London: Routledge, 2007.

2   Reading the future North

Johan Schimanski

'A very interesting sound', the mining director says, 'it sounds like science fiction.'

(Kjerstin Uhre, pp. 156)[1]

The future North must involve an act of reading in order to salvage discourses and views that are blotted out by the rhetoric of Arctic futures and Arctic frontiers. At the symposium "Constructions of North: Landscape, Imagery, Society" in Tromsø in 2014, William L. Fox asked us to use art to make us aware of spaces we do not see.[2] The irony of the occasion was not lost on his audience, for the symposium took place as a side event to the policy/industry/science conference named "Arctic Frontiers". His demand was inaudible to most of the delegates at that conference, a highly profiled yearly meeting which Aileen Espíritu analyses in this volume as an example of the "spectacular speculations" now besetting the Arctic.

I will use here an art form, as a historical producer of speculative future landscapes and science fiction, as a prism to understand the future north and some of the ideas coming out of the Future North project, which arranged that side event. Science fiction is intimately related to the concept of landscape: as science fiction author and theorist Samuel R. Delany has said, "the episteme was always the secondary hero of the s-f novel – in exactly the same way that the landscape was always the primary one".[3]

The political and democratic agenda Fox set out necessitates, I argue, 1. a paradoxical *temporality* of reading, already hinted at in the title of his talk, "Backwards Landscape the Reading" (a title to be read backwards), and 2. a paradoxical *epistemology* of reading, in which one reads for the unreadable. In their contributions to the present volume, Fox and Espíritu suggest that art can resist the commodification of the landscape, but also be co-opted by that commodification. As Fox and as Susan Jayne Carruth point out in this volume, and as Finn-Arne Jørgensen has suggested elsewhere, the North is now a site of consumption.[4] My argument is that only by reading with an awareness of different temporalities and discourses, along with attention to that which resists reading, can one see and create a future north without commodifying the landscape, or to use the term introduced by Kjerstin Uhre in this volume, without "perforating" it.

The rhetoric of the Arctic as a "future" in the stock trading sense as a form of speculation seems to build on the investment promise the Arctic holds as a repository of natural resources (be they oil, gas, minerals or biological material) and shipping routes. Rhetorically however, as A.C.H. Keskitalo has shown, this use of the word "future" is stuck in the past, building on the frontier discourses of the golden age of Arctic exploration and heroism formulated in a nineteenth-century colonial context.[5] As she shows, this is an unhistoricized rhetoric: it does not acknowledge its debt to the past, or indeed to other temporalities, such as images of the Arctic as a pristine or primitive place, situated before or outside of the progress of time.[6] Work within the "Arctic Modernities" research project based at UiT The Arctic University of Norway has shown how modernity's version of the Artic not only builds on the primordial as constitutive other, but, in line with modernist aesthetics, also creates the modern through a return to archetypal forms.[7] The future North builds on a coevalness of the Arctic modernities and primitivities in which the past must be allowed to haunt the future.

The North is the future. The North is science fiction. Many recurrent European images of the Arctic – as empty, unknown, inaccessible, dangerous, possibly exploitable, often catastrophic – have their origin in a period when the poles were seen as the final unexplored territories of the world. The temporality of the not-yet-explored is neatly summed up in the tension between Vilhjalmur Stefansson's two book titles from between the wars, *Unsolved Mysteries of the Arctic* and *The Northward Course of Empire*.[8] The standard Arctic chronotope was implicitly one of a distant, marginal and future space, a space which today has become increasingly close, global and contemporary. Distance, danger and the unknown made the Arctic susceptible to aesthetic categories such as the uncanny, the grotesque and the sublime, all related to the historical development of a literary genre, science fiction. In the mythological topography of antiquity and renaissance sources, the Arctic was either a cold and dark place or a place where one might reach paradisiacal islands in a warm and open sea hidden behind the ice; a fantastic paradigm which is echoed not only in John Cleve Symmes' nineteenth-century theories of the Arctic as a portal to a hollow earth, the volcanic visions of Jules Verne or Stefansson's friendly Arctic, but also in contemporary dreams of resource extraction, cruise tourism and cuddly polar bears.[9] In the golden age of Arctic exploration in the nineteenth century, the category of the sublime – danger enjoyed from a distance – underpinned the popular heroism of explorers entering into close proximity with such a deadly landscape. The fact that the Arctic was unknown, along with the metaphoric qualities of the ice as a preserver through time, deepened this sense of the sublime, allowing fantasies of discovering there societies or beings existing outside of time, or from primordial times, to flourish.

The parallels and connections between the history of the Arctic and the development of science fiction are striking. For each of the stereotypical images of the Arctic, science fiction has had its particular response. The Arctic is inaccessible, thus the fantasies of advanced technologies of mobility and habitation there. It is unknown, thus science fiction provides opportunities for an epistemological vision of a future in which it is known. It is strange, thus allowing science fiction to introduce aliens into the Arctic landscape, or create alien planets based on the Arctic. It is dangerous, thus some of these aliens or indeed the products of the biotechnical experiments of modernity may inhabit the Arctic as monsters. It is empty, thus it is a space to be filled with colonial fantasies of utopian settlements. It is potentially exploitable, thus the production of science fiction–like scenarios for how it may be used.

The North is a place which might be landscaped and designed, thus the utilization of "design fiction" and a "poetics of anticipation" in the Future North project, detailed in Andrew Morrison's contribution to this book.

The science fictional qualities of the Arctic have resulted in a line of central works of science fiction in which the Arctic or Arctic-like environments figure. Often claimed to be the first work of science fiction, *Frankenstein; or, The Modern Prometheus* by Mary Shelley (1818) is partly set in the Arctic, and, in it, the Arctic becomes a symbol of both the hubris of modernity and isolation from humanity.[10] Several of the works by which Jules Verne established his reputation a writer of scientific romances are set in the Arctic, and also in the Antarctic (the latter giving rise to a whole tradition of itself of Antarctic science fiction and horror, including works by authors such as Edgar Allen Poe and H.P. Lovecraft).[11] A recurrent topos in the study of fictional utopias is that such societies – which must always be set 'somewhere else' – are increasingly divided from us in time rather than space, as the globe gradually possesses fewer and fewer geographical spaces in which a utopia could be hidden from us. It is the Arctic, along with the Antarctic, which provides some of the last such spaces, for example in the various "arctopias" examined by Heidi Hansson.[12] The Arctic as a hiding place of superbeings, monsters and artefacts appears in superhero comics (Superman has his "Fortress of Solitude" in the Arctic, as did proto-superhero Doc Savage before him) and in various thrillers based around finding alien or prehistoric beings and things frozen in the ice.[13]

More recently, the Arctic has gone from being a science fictional future to being a science fictional past, with various alternative history novels being published in which the Arctic is allowed a future in some alternative past. Such narratives stand in stark contrast to our image of the Arctic today as a fully explored landscape which clearly does not contain any hidden fortresses or societies. An example would be Michael Chabon's *The Yiddish Policemen's Union* (2007), set in a past future in which European Jews have been provided with land in Alaska to settle, avoiding the holocaust.[14] Another would be Henning Howlid Wærp's similarly heterotopian global warming satire *En øyreise* (*An island journey*, 2015) set in a Franz Josef Land which has been leased to the Netherlands and landscaped into a tropical island tourism paradise ("Nieuwe Curacao"); the Dutch colonial past appears here within a simulacrum.[15]

Other past futures (or future pasts) belong to variants of steampunk, a popular sub-genre of science fiction originating in cyberpunk movement of the 1980s, in which the future is imagined as Victorian in style, attitude and technology. Steampunk has produced various offshoots which all anachronistically fulfil the desire for a retro version of future. Steampunk elements figure strongly in alternative histories such as Phillip Pullman's *Northern Lights* (1995) and Jean-Christophe Valtat's decadent *Aurorarama* (2010).[16] Kate Elliott, like Pullman, mixes steampunk with fantasy motifs in what she calls "icepunk"; her "Spiritwalker" novels are set in a fictional Afro-Celtic nineteenth century during an belated ice age.[17] In the French context (Valtat, while he writes his Arctic novels in English, is French), an anachronistic, Vernean image of the Arctic had already been established in Jacques Tardi's early graphic novel, *Le Démon des glaces* (1974).[18]

The anachronisms of steampunk exaggerate the underlying tendency in many of the works mentioned here: that the futurity of science fiction may be mixed with an appeal to the primordial and the idyllic. The uncanny hauntings in the Arctic and Antarctic works of Shelley and Lovecraft make other levels of the polar landscapes visible, bringing together the atavistic and the futuristic. The Arctopian tradition combines utopian tendencies to the "pastoral" and the "industrial", imagining a past or timeless world and progress towards the future in the same Arctic setting.[19] These mixed temporalities confuse the imperialist topos of the Arctic existing only in the future because it was yet to be discovered. Only lately have we seen a return of the Arctic future in fiction, as colonial discourses of a static Arctic have slowly been substituted by ecological discourses of a dynamic Arctic which is changing, possibly in a radical way, through the interaction with humans (post-humanism and the concept of the Anthropocene are both predicated on a dynamic, ecological approach to nature as no longer a static object). The tendency here has been to globalize the Arctic; many examples are provided through the "cli-fi" (science fictional treatment of climate change) which Hansson sees as a continuation of a utopian tradition.[20] One example is Kim Stanley Robinson's *Fifty Degrees Below* (2005), in which the primordial is introduced into a possible future as the landscape of Washington, D.C., is transformed into a polar waste and the protagonist actively seeks out a paleolithic lifestyle in a rewilded city.[21] However, a globalized polar winter is also present in for example the dystopian graphic novel created by Jacques Lob and Jean-Marc Rochette, *Le Transperceneige* (1982–83; translated as *Snowpiercer*); the film adaptation by Bong Joon-ho ends with a specifically Arctic image, a polar bear, signifying hope for the future.[22] In Valtat's *Luminous Chaos* (2013), the sequel to *Aurorarama*, the action takes place in Paris, which in dealing with increasingly colder winters requires the assistance of the denizens of the Arctic metropolis of New Venice.[23]

These fictional examples of Arctic futures – and Arctic futures will always be fictions – negotiate the Arctic through the chronotopic visions they present. One should not however make the mistake of thinking that science fiction primarily is there to make the future visible; science fiction is as much about the present as the future. Sometimes it will provide future perspectives through a process of extrapolation, but primarily science fiction tends to combine a form of what theoretician Darko Suvin has called "cognitive estrangement"[24] with an intertextual construction of landscapes. According to Suvin, science fiction proper will always be based around a pseudo-scientific premise (a "novum") involving both imagined and imaginable historical changes which help estrange ourselves and see ourselves anew.[25] In a sense then, science fiction about the Arctic is not about the future north, but about the present north. Suvin has furthermore been criticized for providing a definition of science fiction, which is unworkable with many works in the genre, i.e. those in which the novums have been conventionalized and no longer have any estranging force.[26] Damien Broderick has suggested that an important part of many science fiction texts is the "mega-text", a constantly developing pool of standard motifs (robot, alien, spaceship etc.), which can be reused in new works of science fiction.[27] This means that science fiction, and Arctic science fiction, can both involve a form of speculative estrangement and collages of intertexts regulated by the constant production of new subgenres, combining different temporalities. Such bricolages or "mash-ups" (as Elliott calls her "Spiritwalker" novels) may include disparate (and as we have seen, anachronistic) elements.[28] They bear formal similarities to the "heterogeneit[ies] of style", the "entanglement[s]" and "melange[s] of archaic and modern elements" that Hanna Eglinger has found in modern constructions of Arctic primitivism of the late nineteenth and early twentieth centuries.[29] Like Morrison's design fictions in this book, science fiction in general has a potential for relationality and "liquidity" beyond banal perceptions of science fiction as extrapolation or even Suvin's more sophisticated, but basically structuralist theory of science fiction as cognitive estrangement.

Seeing the future Arctic landscape through the lens of science fiction is thus to pose the question whether Arctic futures are as much about the present as about the future, and whether Arctic landscapes must be read as "mash-ups", disruptions and entanglements of nature and modernity rather than as white spaces to be opened up by the march of progress. Indeed, presentations by Janike Kampevold Larsen and Silje Figenschou Thorsen at symposia within the "Future North" project have focused on themes such as the industrial (and not the natural) sublime in the Arctic and the way in which mining enterprises have cut through the Arctic,[30] or the bricolage involved in idiosyncratic technical solutions to everyday problems created by Sámi inhabitants of the North Calotte.[31]

It is in this sense that reading the future northern landscape is like reading science fiction. In both cases, you read at different levels of depth. There is a surface reading in which you image that you are seeing a future, and a deeper reading where you become aware of a tangled multiplicity of temporalities. Looking at the landscape, you imagine what is underneath the landscape. You attempt to make the invisible visible. When you build on imaginations of the future in any political decision, you are partaking in the visible or the sensible, even though that vision may be a fictional one, taking place in the mind's eye. In Jacques Rancière's phrase, the political is a "partage du sensible", a "distribution/division/sharing of the visible";[32] but that which is sensible does not have to be real and must indeed include that which is unreal. From one perspective, representations, fictions and imaginaries exist on the same level as the realities they represent. This entanglement of levels is integral to the idea of *scape* as originally developed by Arjun Appadurai:[33] a flexible skein of threads (or of the Vardø fishnets Larsen showed photographs of during her talk at the "Future North" session in 2015) spreading through imaginary, symbolic and real, physical landscapes.

Landscapes carry within them many different discourses and temporalities: to any two layers, more may be added. At the "Future North" session in 2015, the participants were presented with various multi-layer models of the future north and the northern landscape. Peter Hemmersam, referencing the archaeologist Sefryn Penrose, suggested that the Artic landscape involves four Ps: Politics, People, Pleasure and Profit (he develops on these in this book).[34] Larsen spoke of four different landscapes:

the viewing landscape, the cultural landscape, the vernacular landscape, the practice landscape, along with a fifth, the extraction landscape.[35] Different landscapes can bring into play a sense of place embedded within the landscape, but also, as Carruth argues in this volume, counteract the power dimension implicit in the landscaping gaze. Fox brought other layers into play: the artscape and the brandscape, with the possibility that the artscape can both disrupt and become a brandscape (he develops on this relationship in his contribution here). Such layerings also appear throughout this volume, for example in the contribution by Henry Mainsah, and related conceptualizations such as the "hybrid" or "palimpsestual" landscape have also been used as analytical tools for understanding the northern landscape elsewhere.[36] Going back to previous work in Arctic discourses analysis and imagology, the geographer Louis-Edmond Hamelin's theory of multifaceted nordicities (cultural nordicity, seasonal nordicity, winteriness, altitudinality, slidicity, iciness etc.) provides a precursor to such models.[37]

I suggest that the future north will only be viable if we allow landscapes to become entangled by nordicities, the invisible, artworks, migration, vernaculars, heritage, sense of place, representations and materiality. Recently, border studies has used the term "landscape" in a metaphorical sense in understanding borders as being entangled in "borderscapes" made up of many different markers and performances of bordering in different spaces and on different levels: parallel to the verb "to landscape", one now speaks of active processes of "borderscaping".[38] Designing the brandscape of future north (and future west/future east) may be an act of resistant "borderscaping" if it can create new strategies of co-narration and collaborative/participatory mapping presented by Mainsah in this volume: new mash-ups, necessitating new (and to follow Mainsah, multisensory) ways of reading the landscape. Collaborations in other geographical contexts may suggest models of hybrid agency: the ongoing "antiAtlas des frontières" project on the borders to the global north, in the Mediterranean or between Mexico and the USA, has initiated and mapped various art-science projects.[39] In the northern context (a borderscape in many senses), one could imagine art/science, but also art/tourism, tourism/art and travel/planning collaborations (just as in the border context there exist borderguard-poets).[40] Indeed, Fox lists several such hybrid projects in the North in this volume, and Elizabeth Ellsworth and Jamie Kruse of smudge studio detail how they invent a hybrid, collaborative voice for their projects within Future North. The northern landscape invites collaboration between different disciplines or ways of looking at the world, along with the creation of future hybrid, cyborg agencies such as Narratta the narwhal, a design fiction imagined by Morrison as part of the "Future North" project and described in his contribution to this book, where he also introduces the term "co-narration" I am using here.[41] Recently, at an "Arctic Day" arranged by the University of Oslo in February 2016, Sverker Sörlin suggested in his introductory keynote that what he called the "future of Arctic futures" lies in elaboration of the concept of "co-habitation" in the Arctic (in the scientific context implying a need for collaboration between the natural science and the humanities).[42]

It is at this point of the cultural argument that the industrial sublime threatens to disturb such visions. We may well talk about combining art and science, but "industry-art" collaboration and co-narration is already well-established in the symbolic aestheticization of the urban landscape in northern company towns.[43] Following Morgan Ip, writing in this volume, I would like to utilize the adjective "dark" here. Photographs from the "Future North" field trips suggest that some of the major disruptions of the landscape created by mining operations in sparsely habited areas (i.e. areas with few constituencies who can make audible complaints) are not only reminders of what Timothy Morton calls "dark ecology",[44] but also works of what one might call "dark landscape art". The industrial sublime is not an effect of the artworks brought in to decorate or make interventions in mining communities, but of the mining and processing operations themselves. In northern borderscapes, the sublime may easily turn into the grotesque.[45] The major challenge for strategies of resistant co-narration – apparent at the present time in the art of my contrapuntal example, the Afro-European borderscape – is how to appropriate hegemonic techniques of surveillance, mapping, imaging, exploring, ethnography, scenario-building and reading without oneself being appropriated by the industrial-military complex. In her contribution to this volume, Uhre points to the potential of mapping as both an instrument of power and an instrument of liberation. How do we define the difference between maps produced by artist-researchers (for example in the "Barents Urban Survey 2009" project) and those maps produced as future scenarios by mining and drilling companies?[46] Is reading the future north an innocent pastime? Is research, is mapping, an innocent pastime? Mapping is a highly controversial practice, central as it has been to the imperial project. So what happens when it is used by counter-hegemonic forces, for example within critical cartography, the artistic and indigenous counter-mappings (and counter-prospecting) described by Uhre here, or within interactive community GIS projects such as mybarents.com, described in Ip's contribution?[47] Art may, as Holger Pötzsch argues in an article on the "X-Border Art Biennial" (the northernmost art biennial in the world), "de-habitualize established frames for perception and cognition", but will also rely on and reinforce "economic and political interests"; it is inherently ambiguous.[48]

To conclude these reflections on the future northern landscape, on science fictional Arctic futures, and the potentials and ambiguities of "sharing the sensible" in acts of co-narration, I would like to make a call – ironically inspired by industrial landscapes which invite categorization as sublime – to focus our thinking on points of unreadability. The sublime, since it refers to that which we perceive as greater than ourselves, has often been associated with an inability to express and by implication to interpret sublime objects.[49] To be able to say something about the future north, or to make marginalized voices audible and minority subjects visible, it is crucial to reflect upon those instances in landscapes and in science fictional futures which resist reading and which stage the epistemological sublime. Such junctures, readily available in the North, are not to be limited to those objects apparent in the post-industrial northern landscape that Janike Kampevold Larsen in her presentation of Future North field trips described as "non-Euclidean".[50] They may also have discursive dimensions produced in the northern spectacles and hyperbuilding analyzed by Espíritu in this volume, or on the social media platforms landscaping the North described by Ip in his contribution here as hyperobjects. They demand an understanding of the northern landscape (indeed, of the northern geology of the Anthropocene) not as a static exploitable, but as moving and changeable beyond the limits of design, as Ellsworth and Kruse of smudge studio argue in this volume. Elements of the unreadable in the landscape is easily banished in neo-imperial imageries of an explorable and exploitable Arctic, but will be encountered in the complex redistributions of the sensible, in the intertextual bricolages of science fiction and the entanglements of different layers of northernness which I have suggested here.

Such unreadability invites a discussion of the embodiment and embeddedness of the sensing apparatus in ourselves, in the natural landscape as well as in the industrial landscape. When we read the landscape, even when we read the future, we cannot play the "god trick" implicit in traditional conceptions of science and governance.[51] We are – as readers – situated in a world of reading, in which each ecological being reads its *Umwelt*, in which reading, as philosophical hermeneutics implies, is an aspect of being. We are not privileged readers reading unreading landscapes. As Uhre suggests in this volume, we must immerse ourselves, which is to invert reading, to let the landscape read us – or even ignore and unread us, as it chooses. Only by positioning ourselves in hybrid subject positions as human, animal and cybernetic readers of the North may we read without perforating the landscape and create without commodifying.

Acknowledgements

I would like to thank the Future North project for providing a framework for this discussion, Ulrike Spring for her incisive comments, and also Aniara – studentenes science fiction forening (the student science fiction society in Oslo) for the opportunity for sharing and developing some ideas about

Arctic science fiction at one of their meetings. The research for this article took place as part of the Arctic Modernities research project (Research Council of Norway, POLARPROG 226030).

Notes

1 From field notes quoted by Kjerstin Uhre in her contribution to this volume.
2 Fox, William L. "Backwards Landscape the Reading." Arctic Frontiers 2014 side event: Constructions of North: Landscape, Imagery, Society, 23 January, Tromsø: 70° arkitektur/NAF/University of Tromsø/AHO, 2014.
3 Samuel R. Delany, *Triton* (New York: Bantam, 1977), 333.
4 Finn-Arne Jørgensen, "The Networked North: Thinking About the Past, Present, and Future of Environmental Histories in the North," in *Northscapes: History, Technology, and the Making of Northern Environments*, ed. Dolly Jørgensen and Sverker Sörlin (Vancouver: UBC Press, 2013), 278.
5 E. C. H. Keskitalo, *Constructing 'the Arctic': Discourses of International Region-Building* (Rovaniemi: Lapin yliopisto, 2002), 303–7.
6 Ibid., 306–7.
7 E.g. Hanna Eglinger, "Nomadic, Ecstatic, Magic: Arctic Primitivism in Scandinavia around 1900," *Acta Borealia* 33, no. 2 (2016): 190–91; Heidi Hansson, "Arctopias: The Arctic as No Place and New Place in Fiction," in *The New Arctic*, ed. Birgitta Evengård, Joan Nymand Larsen, and Øyvind Paasche (Cham: Springer, 2015); Anka Ryall, "Introduction," *Acta Borealia* 33, no. 2 (2016). See also http://uit.no/forskning/arcmod.
8 Vilhjalmur Stefansson, *Unsolved Mysteries of the Arctic* (New York: Palgrave Macmillan, 1938) and *The Northward Course of Empire* (New York: Harcourt, Brace and Company, 1922).
9 Jules Verne, *The Extraordinary Journeys: The Adventures of Captain Hatteras*, trans. William Butcher (Oxford: Oxford University Press, 2005); Vilhjalmur Stefansson, *The Friendly Arctic: The Story of Five Years in Polar Regions* (New York: Palgrave Macmillan, 1921).
10 Mary Wollstonecraft Shelley, *Frankenstein, or The Modern Prometheus*, ed. Maurice Hindle (London: Penguin, 1992).
11 Cf. Elizabeth Leane, *Antarctica in Fiction: Imaginative Narratives of the Far South* (Cambridge: Cambridge University Press, 2012). It should be noted that works such as Shelley's *Frankenstein* and Verne's *Captain Hatteras* were not actually set in the future in relation to the time of publication, but in fictional recent pasts; they both however build on the implication of potential future journeys to the North Pole.
12 Hansson, "Arctopias."
13 E.g. James Rollins, *Ice Hunt* (New York: William Morrow, 2003).
14 Michael Chabon, *The Yiddish Policemen's Union* (New York: HarperCollins, 2008).
15 Henning Howlid Wærp, *En øyreise* (Tromsø: MARGbok, 2015). A declaration of interest is worth noting here, since Wærp is a previous colleague and project collaborator of the author.
16 Philip Pullman, *Northern Lights* (London: Scholastic, 2002); Jean-Christophe Valtat, *Aurorarama* (Brooklyn, NY: Melville House, 2010).
17 Kate Elliott, *Cold Magic* (New York: Orbit, 2010); *Cold Fire* (New York: Orbit, 2011); *Cold Steel* (New York: Orbit, 2013). For "icepunk," see *Cold Magic*, kindle ed. (London: Hachette Digital, 2010), 581.
18 Tardi, *Le Démon des glaces: Une aventure de Jérome Plumier* (Paris: Casterman, 2014).
19 Northrop Frye, "Varieties of Literary Utopias," in *Utopias and Utopian Thought*, ed. Frank E. Manuel (London: Souvenir Press/Condor, 1973). For images of a timeless Arctic, see Henning Howlid Wærp, "Den arktiske pastoralen – Fridtjof Nansen, Knud Rasmussen, Helge Ingstad," *Norsk Litterær Årbok* (2012).
20 Hansson, "Arctopias," 75.
21 Kim Stanley Robinson, *Fifty Degrees Below* (New York: Bantam Dell, 2007).
22 Jacques Lob, Jean-Marc Rochette, and Benjamin Legrand, *Transperceneige: Intègrale* (Paris: Casterman, 2014); Bong Joon-ho, *Snowpiercer* (South Korea/Czech Republic, 2013).
23 Jean-Christophe Valtat, *Luminous Chaos* (New York: Melville House, 2013).
24 Darko Suvin, "On the Poetics of the Science Fiction Genre," *College English* 34, no. 3 (1972).
25 *Metamorphoses of Science Fiction: On the Poetics and History of a Literary Genre* (New Haven: Yale UP, 1979), 63–84; "On the Poetics of the Science Fiction Genre".
26 E.g. Roger Luckhurst, "The Many Deaths of Science Fiction: A Polemic," *Science-Fiction Studies* 21 (1994).
27 Damien Broderick, *Reading by Starlight: Postmodern Science Fiction* (London: Routledge, 1995), 59–61.
28 Elliott, *Cold Magic*, 581.
29 Eglinger, "Nomadic, Ecstatic, Magic," 12, 7, 22 respectively; see also 15.
30 Larsen, Janicke Kampevold. "The New Landscapes of North." Arctic Frontiers 2015 side event: Future North, 23 January, Tromsø: Future North, 2015.

31 See also http://indigenuityproject.com. Thoresen, Silje Figenschou. "The Indigenuity Project, Improvised Design in Sápmi." Arctic Frontiers 2014 side event: Future North, 22 January, Tromsø: 70° arkitektur/NAF/University of Tromsø/AHO, 2014.

32 Jacques Rancière, *The Politics of Aesthetics: The Distribution of the Sensible*, trans. Gabriel Rockhill (London: Continuum, 2004).

33 Arjun Appadurai, "Disjuncture and Difference in the Global Cultural Economy," *Theory, Culture & Society* 7, nos. 2–3 (1990).

34 Hemmersam, Peter. "Arctic Urbanism: Kola Mining Towns." Arctic Frontiers 2015 side event: Future North, 23 January, Tromsø: Future North, 2015.

35 Larsen, "The New Landscapes of North."

36 E.g. Jørgensen, "The Networked North," 277; Nadir Kinossian and Urban Wråkberg, "Palimpsests," in *Border Aesthetics: Concepts and Intersections*, eds. Johan Schimanski and Stephen F. Wolfe (New York: Berghahn, 2017).

37 Cf. Daniel Chartier, "Towards a Grammar of the Idea of North: Nordicity, Winterity," *Nordlit*, no. 22 (2007): 39–41. See also Hemmersam in this book.

38 Anne-Laure Amilhat Szary, "Latin American Borders on the Lookout: Recreating Borders through Art in the Mercosul," in *Placing the Border in Everyday Life*, ed. Reece Jones and Cory Johnson (Farnham: Ashgate, 2014); Chiara Brambilla, "Exploring the Critical Potential of the Borderscapes Concept," *Geopolitics* 20, no. 1 (2015); Chiara Brambilla et al., "Introduction: Thinking, Mapping, Acting and Living Borders Under Contemporary Globalisation," in *Borderscaping: Imaginations and Practices of Border Making*, ed. Chiara Brambilla, et al. (Aldershot: Ashgate, 2015); Elena dell'Agnese and Anne-Laure Amilhat Szary, "Borderscapes: From Border Landscapes to Border Aesthetics," *Geopolitics* 20, no. 1 (2015); Prem Kumar Rajaram and Carl Grundy-Warr, "Introduction," in *Borderscapes: Hidden Geographies and Politics at Territory's Edge*, ed. Prem Kumar Rajaram and Carl Grundy-Warr (Minneapolis: University of Minnesota Press, 2007); Johan Schimanski, "Border Aesthetics and Cultural Distancing in the Norwegian-Russian Borderscape," *Geopolitics* 20, no. 1 (2015).

39 Cédric Parizot et al., "The AntiAtlas of Borders: A Manifesto," *Journal of Borderlands Studies* 29, no. 4 (2014). See also antiatlas.net.

40 Mari Ristolainen, "Virtual Frontiers: Russian Border Guard Poems Online," in *Globalization and Borders: Cultural, Political and Regional Aspects of the Finnish and Russian Borders*, ed. Minna Jokela (Imatra: Finnish Border Guard, 2015).

41 http://www.oculs.no/people/narratta/.

42 Sörlin, Sverker: "Framtiden för 'Arktiska framtider'." Arktisk dag seminar, 15 February 2016, Oslo: Universitetet i Oslo, 2016.

43 Julia Gerlach and Nadir Kinossian, "Cultural Landscape of the Arctic: 'Recycling' of Soviet Imagery in the Russian Settlement of Barentsburg, Svalbard (Norway)," *Polar Geography* 39, no. 1 (2016).

44 Timothy Morton, *The Ecological Thought* (Cambridge, MA: Harvard University Press, 2010). See also the introduction to the present book. For an art/research/travel project inspired by Morton in a similar Arctic context, see http://darkecology.net.

45 Schimanski, "Border Aesthetics and Cultural Distancing in the Norwegian-Russian Borderscape," 42.

46 Espen Røyseland and Øystein Rø, eds., *Northern Experiments: The Barents Urban Survey 2009* (Oslo: 0047 Press, 2009).

47 For an introduction to critical cartography, see Jeremy W. Crampton and John Krygier, "An Introduction to Critical Cartography," *ACME: An International E-Journal for Critical Geographies* 4, no. 1 (2006). See also Uhre's contribution to the present volume.

48 Holger Pötzsch, "Art Across Borders: Dislocating Artistic and Curatorial Practices in the Barents Euro-Arctic Region," *Journal of Borderlands Studies* 30, no. 1 (2015): 122.

49 Jean-François Lyotard, "Answering the Question: What Is Postmodernism?" trans. Régis Durand, in *The Postmodern Condition: A Report on Knowledge* (Manchester: Manchester University Press, 1984), 77–8.

50 Larsen, "The New Landscapes of North."

51 Donna Haraway, "Situated Knowledges: The Science Question in Feminism and the Privilege of Partial Perspective," *Feminist Studies* 14, no. 3 (1988).

Bibliography

Amilhat Szary, Anne-Laure. "Latin American Borders on the Lookout: Recreating Borders Through Art in the Mercosul." In *Placing the Border in Everyday Life*, edited by Reece Jones and Cory Johnson, 346–78. Farnham: Ashgate, 2014.

Appadurai, Arjun. "Disjuncture and Difference in the Global Cultural Economy." *Theory, Culture & Society* 7, nos. 2–3 (1990): 295–310.

Bong Joon-ho. *Snowpiercer*. 126 mins. South Korea/Czech Republic, 2013.

Brambilla, Chiara. "Exploring the Critical Potential of the Borderscapes Concept." *Geopolitics* 20, no. 1 (2015): 14–34.

Brambilla, Chiara, Jussi Laine, James W. Scott, and Gianluca Bocchi. "Introduction: Thinking, Mapping, Acting and Living Borders under Contemporary Globalisation." In *Borderscaping: Imaginations and Practices of Border Making*, edited by Chiara Brambilla, Jussi Laine, James W. Scott and Gianluca Bocchi, 1–9. Aldershot: Ashgate, 2015.

Broderick, Damien. *Reading by Starlight: Postmodern Science Fiction*. London: Routledge, 1995.

Chabon, Michael. *The Yiddish Policemen's Union*. New York: HarperCollins, 2008.

Chartier, Daniel. "Towards a Grammar of the Idea of North: Nordicity, Winterity." *Nordlit*, no. 22 (2007): 35–47.

Crampton, Jeremy W., and John Krygier. "An Introduction to Critical Cartography." *ACME: An International E-Journal for Critical Geographies* 4, no. 1 (2006): 11–33.

Delany, Samuel R. *Triton*. New York: Bantam, 1977.

dell'Agnese, Elena, and Anne-Laure Amilhat Szary. "Borderscapes: From Border Landscapes to Border Aesthetics." *Geopolitics* 20, no. 1 (2015): 4–13.

Derrida, Jacques, and Derek Attridge. "'This Strange Institution Called Literature': An Interview with Jacques Derrida." Translated by Geoffrey Bennington and Rachel Bowlby. In *Acts of Literature*, edited by Derek Attridge, 33–75. New York: Routledge, 1992.

Eglinger, Hanna. "Nomadic, Ecstatic, Magic: Arctic Primitivism in Scandinavia Around 1900." *Acta Borealia* 33, no. 2 (2016): 189–214.

Elliott, Kate. *Cold Fire*. New York: Orbit, 2011.

———. *Cold Magic*. New York: Orbit, 2010.

———. *Cold Magic*. Kindle ed. London: Hachette Digital, 2010.

———. *Cold Steel*. New York: Orbit, 2013.

Fox, William L. "Backwards Landscape the Reading." Arctic Frontiers 2014 side event: Constructions of North: Landscape, Imagery, Society, 23 January, Tromsø: 70° arkitektur/NAF/University of Tromsø/AHO, 2014.

Frye, Northrop. "Varieties of Literary Utopias." In *Utopias and Utopian Thought*, edited by Frank E. Manuel, 25–49. London: Souvenir Press/Condor, 1973.

Gerlach, Julia, and Nadir Kinossian. "Cultural Landscape of the Arctic: 'Recycling' of Soviet Imagery in the Russian Settlement of Barentsburg, Svalbard (Norway)." *Polar Geography* 39, no. 1 (2016): 1–19.

Hansson, Heidi. "Arctopias: The Arctic as No Place and New Place in Fiction." In *The New Arctic*, edited by Birgitta Evengård, Joan Nymand Larsen and Øyvind Paasche, 69–77. Cham: Springer, 2015.

Haraway, Donna. "Situated Knowledges: The Science Question in Feminism and the Privilege of Partial Perspective." *Feminist Studies* 14, no. 3 (1988): 575–99.

Hemmersam, Peter. "Arctic Urbanism: Kola Mining Towns." Arctic Frontiers 2015 side event: Future North, 23 January, Tromsø: Future North, 2015.

Jørgensen, Finn-Arne. "The Networked North: Thinking about the Past, Present, and Future of Environmental Histories in the North." In *Northscapes: History, Technology, and the Making of Northern Environments*, edited by Dolly Jørgensen and Sverker Sörlin, 268–79. Vancouver: UBC Press, 2013.

Keskitalo, E. C. H. *Constructing 'the Arctic': Discourses of International Region-Building*. Rovaniemi: Lapin yliopisto, 2002.

Kinossian, Nadir, and Urban Wråkberg. "Palimpsests." In *Border Aesthetics: Concepts and Intersections*, edited by Johan Schimanski and Stephen F. Wolfe, 90–110. New York: Berghahn, 2017.

Larsen, Janicke Kampevold. "The New Landscapes of North." Arctic Frontiers 2015 side event: Future North, 23 January, Tromsø: Future North, 2015.

Leane, Elizabeth. *Antarctica in Fiction: Imaginative Narratives of the Far South*. Cambridge: Cambridge University Press, 2012.

Lob, Jacques, Jean-Marc Rochette, and Benjamin Legrand. *Transperceneige: Intégrale*. Paris: Casterman, 2014.

Luckhurst, Roger. "The Many Deaths of Science Fiction: A Polemic." *Science-Fiction Studies* 21 (1994): 35–50.

Lyotard, Jean-François. "Answering the Question: What Is Postmodernism?" Translated by Régis Durand. In *The Postmodern Condition: A Report on Knowledge*, 71–82. Manchester: Manchester University Press 1984.

Morton, Timothy. *The Ecological Thought*. Cambridge, MA: Harvard University Press, 2010.

Parizot, Cédric, Anne Laure Amilhat Szary, Gabriel Popescu, Isabelle Arvers, Thomas Cantens, Jean Cristofol, Nicola Mai, Joana Moll, and Antoine Vion. "The antiAtlas of Borders, A Manifesto." *Journal of Borderlands Studies* 29, no. 4 (2014): 503–12.

Pullman, Philip. *Northern Lights*. London: Scholastic, 2002.

Pötzsch, Holger. "Art Across Borders: Dislocating Artistic and Curatorial Practices in the Barents Euro-Arctic Region." *Journal of Borderlands Studies* 30, no. 1 (2015): 111–25.

Rajaram, Prem Kumar, and Carl Grundy-Warr. "Introduction." In *Borderscapes: Hidden Geographies and Politics at Territory's Edge*, edited by Prem Kumar Rajaram and Carl Grundy-Warr, ix–xl. Minneapolis: University of Minnesota Press, 2007.

Rancière, Jacques. *The Politics of Aesthetics: The Distribution of the Sensible.* Translated by Gabriel Rockhill. London: Continuum, 2004.

Ristolainen, Mari. "Virtual Frontiers: Russian Border Guard Poems Online." In *Globalization and Borders: Cultural, Political and Regional Aspects of the Finnish and Russian Borders*, edited by Minna Jokela, 14–16. Imatra: Finnish Border Guard, 2015.

Robinson, Kim Stanley. *Fifty Degrees Below.* New York: Bantam Dell, 2007.

Rollins, James. *Ice Hunt.* New York: William Morrow, 2003.

Røyseland, Espen, and Øystein Rø, eds. *Northern Experiments: The Barents Urban Survey 2009.* Oslo: 0047 Press, 2009.

Ryall, Anka. "Introduction." *Acta Borealia* 33, no. 2 (2016): 119–22.

Schimanski, Johan. "Border Aesthetics and Cultural Distancing in the Norwegian-Russian Borderscape." *Geopolitics* 20, no. 1 (2015): 35–55.

Shelley, Mary Wollstonecraft. *Frankenstein, or The Modern Prometheus.* Edited by Maurice Hindle London: Penguin, 1992.

Sörlin, Sverker: "Framtiden för 'Arktiska framtider'." Arktisk dag seminar, 15 February 2016, Oslo: Universitetet i Oslo, 2016.

Stefansson, Vilhjalmur. *The Friendly Arctic: The Story of Five Years in Polar Regions.* New York: Macmillan, 1921.

———. *The Northward Course of Empire.* New York: Harcourt, Brace and Company, 1922.

———. *Unsolved Mysteries of the Arctic.* New York: Palgrave Macmillan, 1938.

Suvin, Darko. *Metamorphoses of Science Fiction: On the Poetics and History of a Literary Genre.* New Haven: Yale UP, 1979.

———. "On the Poetics of the Science Fiction Genre." *College English* 34, no. 3 (1972): 372–82.

Tardi. *Le Démon des glaces: Une aventure de Jérome Plumier.* Paris: Casterman, 2014.

Thoresen, Silje Figenschou. "The Indigenuity Project, Improvised Design in Sápmi." Arctic Frontiers 2014 side event: Future North, 22 January, Tromsø: 70° arkitektur/NAF/University of Tromsø/AHO, 2014.

Valtat, Jean-Christophe. *Aurorarama.* Brooklyn, NY: Melville House, 2010.

———. *Luminous Chaos.* New York: Melville House, 2013.

Verne, Jules. *The Extraordinary Journeys: The Adventures of Captain Hatteras.* Translated by William Butcher. Oxford: Oxford University Press, 2005.

Wærp, Henning Howlid. "Den Arktiske Pastoralen – Fridtjof Nansen, Knud Rasmussen, Helge Ingstad." *Norsk Litterær Årbok* (2012): 197–216.

———. *En øyreise.* Tromsø: MARGbok, 2015.

3 Spectacular speculation

Arctic futures in transition

Aileen A. Espíritu

Arctic landscapes evoke visions of pristine, often snow-covered whiteness, teeming with wildlife and with only a sparse, often indigenous, human population. Vast empty spaces left to explorers, travellers and adventurers also come to mind. While true in some aspects, these perceptions are patently false in others. Recent interest in the Arctic, because of climate change and perceived scarcity and high demand for energy resources and other potentially lucrative commodities, have transformed these perceptions of an untouched Arctic. In fact, *the Arctic is industrialised and urbanised*. Private and public, local and international interests would like to open this land- and sea-scape to even more industrialisation by exploiting its natural resources and its potential as a significant source of oil and gas extraction, ocean highways of shipping routes and industrial transshipment hubs.

The transformation of perception is driven by a political economy discourse based on an anticipated future Arctic founded on intensified industrial development in order to meet national and international demands for energy, industrial commodities and heavy industry revenue. While Northern Europe is indeed industrialised, I argue that this heightened discourse of intensified industrial development is one founded on spectacular speculation, in which industrial and economic speculation has been elevated to the sphere of spectacle – a spectacle that in itself generates commerce and economy. The Arctic today is beset by future dreaming. Evidenced by an examination of the conversations around Arctic strategies and futures and three key event-generating places in the Arctic, I attempt to give an analysis of how the creation of spectacle and the speculative drives the political economy of the land- and sea-scape of the Norwegian High North. My touchpoint is a critical analysis of image, text, spectacle and their reinforcing performance in the making and remaking of the Arctic seen through Guy Debord's assertion that 'The spectacle manifests itself as an enormous positivity, out of reach and beyond dispute.'[1]

The backdrop

Images and image-making of the Arctic has not changed dramatically since nineteenth- and early twentieth-century scientists and adventurers attempted to explore and master it by heroically traversing it on ships, dog sledges and skis. Commonly regarded as pristine and wild, unforgivingly cold and snow white, the Arctic and/or the High North captures one's imagination of places untouched by human activity and industry. In Norway, explorers such as Fridtjof Nansen, Roald Amundsen, Otto Sverdrup and others were made heroes of exploration by the emerging independent Norwegian nation-state. We may argue that images of conquering the Arctic was very much within the framework of asserting Norwegian sovereignty over territory to its North, intertwined with its decidedly masculine nation-building efforts. Indeed, nationalism and Arctic exploration went hand-in-hand in the history of the Norwegian state- and nation-building. This was already evident when Fridtjof Nansen and Otto Sverdrup returned to the Norwegian capital in triumph in September 1896 from the first Fram expedition.[2] Notably, Norwegians were emigrating to North America by the tens of thousands because of the economic depression in the Nordic countries in that period, and yet Norwegian politicians and businessmen funded and encouraged costly expeditions to conquer both the Arctic and Antarctic. Ironically, these two events may have been symbiotic occurrences. The emerging Norwegian nation needed to shore up its heroic greatness in the face of insurmountable economic challenges, especially since a significant part of the population lived in penury.[3]

Of methodologies and subject positions

My research and observations for this article have been ongoing since I moved to Northern Norway from Western Canada in 2007. Already ensconced in research on the North and Arctic, the grammar of and vocabularies about the North were neither foreign nor unheard of in my immediate academic milieu. What was new was the robustness and energy behind the discourse and overarching pronouncements on the High North/Arctic that defined the politics in Northern Norway from the mid-2000s until the change from the Labour-led coalition government to a right-leaning coalition led by the Right (Høyre) Party beginning in the Autumn of 2013. Living in Northern Norway, one is exposed to the pervasive and omnipresent conversations and media coverage of the *Nordområdene* (the High North areas), and their inextricable link to oil and gas, and industrial development. It was clear that the Norwegian High North was and is undeniably industrialised and continuing to industrialise. These were familiar discourses that I had already been exposed to while embarking on an academic career that focused on indigenous peoples in Siberia, Russia and their relationship to oil and gas resource extraction and while researching and teaching on northern studies and research at a northern-focused university in Western Canada.[4] In other words, the concentrated focus on the High North and later, Arctic, in Northern Norway was part of the trajectory of my research, teaching and administrative work. Thus, my subject position both enhances and complicates how I see and analyse the spectacular speculation in the High North through the creation and promotion of events and conferences, not least because I have been an active and willing participant in these spectacles myself by traveling, looking and speculating on the future of Arctic landscapes. Moreover, as an employee at The Barents Institute at the UiT, The Arctic University of Norway, which is located in the High North in Kirkenes, I have benefitted from Norway's High North Strategy.

The spectacular

Guy Debord's *Society of the Spectacle* was first published in French in 1967, and translated into English in 1977. Within the Situationist thinkers, Debord's Marxist analysis decried society's alienation as it was overwhelmed by mass media and the ubiquitous celebration of celebrity. Debord himself summed up his entire argument in thesis 1 of the book: 'The whole life of those societies in which modern conditions of production prevail presents itself as an immense accumulation of spectacles. All that once was directly lived has become mere representation.'[5] Debord's additional 221 theses deepen this argument and emphasise the human condition of alienation and separation from what is actual, as he argues in thesis 8. Richard L Kaplan avers that, in *Society of the Spectacle*, Debord is profoundly critical of our capitalist culture that is founded on the atomised masses determined by 'isolation, fantasy, ideological blindness, manipulation.'[6] Lost in this spectacle, Debord argues, are the revolutionary masses and the utopian civilisation that it has the potential to create.

As Kaplan and others have claimed, however, Debord, like Marx, ignores the power of individual choice and what Michel Foucault would call governmentality. Wendy Larner observes that Foucault's post-structuralist construction of governmentality 'argues that while neoliberalism may mean less government, it does not follow that there is less governance.'[7] Further, while the state and government may involve itself less in the everyday decisions of individuals, what takes its place 'involves forms of governance that encourage both institutions and individuals to conform to the norms of the market.'[8] This is much more nuanced than Debord's conjecture that society is merely passively awe-struck by spectacle and celebrity. While society 'conforms to the norms of the market,' I argue that it does so willingly because of the promises of its rewards and future riches. More pointedly, communities participate in the creation of spectacular speculation because it hopes to benefit from the promised development, whether economically, socially, politically or culturally. Therefore, the eliciting awe and belief that the spectacle promises are given, accepted, sometimes criticised and always evaluated by the audience to which it is directed. Whether or not the decisions made in response to the spectacles' promises are transformed into good policy, economy or cultural products may only be determined much later, if at all. Thus, spectacles may also be seen as great gambles while presenting possibilities.

Besides Guy Debord's *The Society of the Spectacle*, this chapter also draws inspiration from Aihwa Ong's cogent intervention on 'Hyperbuilding: Spectacle, Speculation, and the Hyperspace of Sovereignty,'[9] of Asian cities. While Ong focusses on 'hyperbuilding' of physical buildings and spaces, I focus on the 'hyperbuilding' of events engineered to attract innovation, entrepreneurship and investments in urbanising communities in the High North. In effect, the staged spectacles in the Arctic are the equivalent of the hyperbuilding in Asia as the Arctic compete for attention and investments. I enquire and analyse how the High North has been enframed in spectacle-making through the staging of spectacular events by interrogating the attention given to this apparently remote region because of its potential for the riches that have yet to be explored and exploited.

Like the explorers of the past, present-day explorations rely on state-of-the-art knowledge and technology in order for resources to be identified and mapped. Unlike the explorers of the past, present-day explorations and mapping of the High North aim to use the latest knowledge and technology to place reputed and found commodities on the world market for monetary gain and profit, not merely for the sake of exploration, curiosity and personal grandeur. The spectacularisation of three cities in the Norwegian Arctic illustrates these representations of an industrialised and industrialising Arctic: Tromsø, Kirkenes and Longyearbyen, Norway.

Spectacularising the high North and the Arctic

With great fanfare, the Norwegian government introduced its first High North policy in 2005[10] followed by a detailed strategy in 2006.[11] Ostensibly, the speech was to introduce a tranche of funding for the High North, called *Barents 2020*, reserving monies for research and development projects in and on the Barents Region. The setting was carefully selected – taking place at the University of Tromsø, the world's northernmost University (changed to UiT, The Arctic University of Norway in 2013),[12] and presented by the decidedly charismatic Minister of Foreign Affairs at the time, Jonas Gahr Støre. Støre delivered his speech entitled 'A sea of opportunities – A sound policy for the High North' to a packed auditorium eager to hear and see the strategies for the North of the newly minted coalition government led by the Labour Party.[13] Støre and his speech were greeted with great anticipation and excitement, and arguably, was a forerunner to other spectacles that would follow in the form of the annual *Arctic Frontiers Conference*, even located in the very auditorium where the High North policy was first formally launched.

Støre's choice of using the term 'High North' instead of Arctic was strategic in itself, even though the three northern counties in Norway are situated either at or above the Arctic Circle. This policy distinction was (and still is today) to acknowledge that Norway has its own distinct Arctic sea- and land-scape – the Arctic Ocean and the Svalbard archipelago. Moreover, in the mid-2000s, the term 'High North' was mostly to mean the Barents Region and the Barents Sea. Indeed, Støre used Arctic not as a place but as a descriptor in his 2005 speech – in Norwegian 'arktisk'. Nevertheless, Norway's Stoltenberg II government's introduction of the High North through its *Barents 2020* funding programme (transformed in 2014 into Arktis (Arctic) 2030 by the Conservative-led coalition government) did not confine its reach just to the Norwegian High North, but importantly to Norway's neighbor to the East, Russia and other countries in the region including the other Arctic countries and, significantly, the largest economic market for Norway, the European Union.

Støre's speech not only gave the High North a central pride of place in the Norwegian state's policy directions, it also established the region within a transnational, even globalised, paradigm reaching beyond but rooted in its locality. It was for all intents and purposes a 'reterritorialization of the metropole,' placing the Arctic at the centre.[15] I argue that this instance of reterritorialisation can be read, at least in limited fashion, as the central state promoting what Ong describes as 'sovereign exception' inviting actors in the High North to be creative and innovative about how they would like to develop their territory with the caveat that it would have to be within the frame of a continued nation- and state-building envisioned by a unitary Norwegian polity. The vision of northern development was and is limited after all. The bulk of the High North Strategy was to come a year later in late 2006. The Stoltenberg II's High North/Arctic policy was dominated by the need for the region to industrialise by exploiting its potential energy resources that speculatively may be found in the depths of the Barents Sea. The background to this invitation to develop is the accepted and

Figure 3.1 Then and now images of Jonas Gahr Støre making speeches about the High North at the University of Tromsø, The Arctic University of Norway

Photo: Ynge Olsen Sæbe and Torgrim Rath Olsen, *Nordlys*.[14]

well-developed trope that the Northern Norwegian municipalities should and must become more self-sufficient and that subsidies and taxation benefits from the central government should be minimised or halted altogether.[16] Indeed, it was an invitation for the High North to create its own future sovereignty based on industrial and resource development, with culture, research and education and tourism rounding out such region-building.

By 2009, as the Norwegian's Chairship of the Arctic Council was approaching an end, it was becoming more common to use High North and Arctic as interchangeable terms or as terms used together. This is significant because we see more attention and resources now being placed in the research and development of the Arctic as distinct from just the Barents Region. Moreover, as mentioned previously, the current Norwegian government in recognising this has maintained research and project funding in the High North/Arctic but has renamed the funding programme from Barents 2020 to Arktis 2030. 'By merging the two schemes, Barents 2020 and Arctic Cooperation, to one new scheme, Arctic 2030 now covers the entire circumpolar Arctic.'[17] Such a shift in semantics was coincident with the political and economic sanctions levelled at Russia for its annexation of Crimea and its instigation of war in Eastern Ukraine. Covering the entire circumpolar Arctic ensured that Norway could justifiably spread its focus to the other Arctic states, not just Russia, Sweden and Finland, that is, not just the Barents Region.

Confident sovereignty

> The interactions between exception, spectacle, and speculation create conditions for hyperbuilding as both the practice and the product of world-aspiring urban innovations.[18]

My use of Ong's 'Hyperbuilding' imagery serves as an analogy and a descriptor to the kind of hyper-development discourse and practice that has ensued since Støre's 2005 address at the University of Tromsø, now the Arctic University of Norway. While the High North has not been engaged in the kind of urbanisation and construction of sky-scrapers ubiquitous in the emerging Asian economies that Ong analyses, I argue that the forms and nuances of the hyperdevelopment in the Norwegian Arctic is analogous. In other words, the hyperbuilding logic Ong identifies in Asia is consonant with the hyperdevelopment logic we find in the sparsely populated and relatively remote, yet urbanising, High North and Arctic that I focus on in this chapter. These logics are exemplified indeed in very similar ways in Asia and the Arctic, though manifested differently because of the great dissimilarities in political systems, population size, proximity to markets and economic resources.

In the High North, hyperbuilding was first based on hypermapping and speculation of what resources could possibly be underneath the land and sea surface of the Barents Region. In 2009, with the hype about peak oil and the need to find other sources of revenue in the eventuality of oil and gas being depleted, the Norwegian state granted funding of 105 million Norwegian crowns (NOK) to the exploration and mapping of minerals, ores and metals in Northern Norway over a four-year period.[19] That figure would be augmented by another 10 million NOK at the end of that period to include southern Norway. And in the autumn of 2015, another 173 million NOK were committed by the conservative coalition government to mapping what energy fossil fuel resources there may be underneath the Barents Sea.[20] The purpose of such mapping was not only for the sake of knowing what was in the ground in Northern Norway, but also to make known to developers and entrepreneurs that the High North was open for business; investment both from domestic and international investors were welcome. Imaginings and speculations about the potential and proven knowledge of what resources exists below sea and land were directed towards expectations of grand industries, wealth creation founded on resource exploitation and possible utopian futures.

Thus, hypermapping would be the foundation on which spectacular events (read: discourse) and speculation would be designed. Over the last decade, before fracking in the United States and the 2014–15 freefall of fossil fuel prices, as oil and gas were being depleted off the shores of southern Norway and in other parts of the world, the oft-cited metric was that the Arctic holds as much as 22 per cent of the world's remaining fossil fuel resources.[21] Such discourse fanned the flames of a so-called 'resource grab' that largely came from those outside the region, whereas those who live in the Arctic have adhered to peaceful cooperation rather than hot competition, as in Jonas Gahr Støre's now popularised adage 'High North, Low Tensions.' One exception was in 2007 when the colourful Polar explorer turned parliamentarian Artur Chilingarov theatrically planted a Russian flag at the seabed of the North Pole and announced to the world that 'the Arctic is Russian.'[22] The Canadian government offered an overheated response declaring that 'this isn't the 15th century. You can't go around the world and just plant flags and say: "We're claiming this territory"'.'[23] And more recently, amidst the renewed tensions between Russia and the 'West' because of Russia's renewed aggressions, the bombastic and politically sanctioned Deputy Prime Minister of Russia Dmitry Rogozin unexpectedly landed on Svalbard without informing the Norwegian government. While on Svalbard, he defiantly tweeted that 'the Arctic was Russia's Mecca,' while posting photographs of himself and his entourage with a Russian flag on the North Pole. Nevertheless, High North and low tensions (especially vis-à-vis Norway and Russia) were the buzz words repeated by policy-makers and the media, and at times attributed to the key champion of the Norwegian High North, Jonas Gahr Støre.[24]

Hyperbuilding through discourse, spectacle and speculation

Writing on hyperbuilding in the emerging economic powers in Asia, Ong

> proposes a theory of sovereign exception in shaping urban spectacles for political and economic ends. Asian cities and governments are neither merely the passive substrate on which capital erects and constructs itself, nor are they being reconfigured in a way that can be easily understood in terms on an implicit scale of 'more' or 'less' sovereignty. In emerging Asian countries, the rule of exception variously negotiates the dual demands of inter-city rivalries on the one hand, and the spectacle of confident sovereignty on the other.[25]

How do the rules of sovereign exception manifest themselves in the High North/Arctic? As I have argued previously, while hyperbuilding in Asia has been characterised by the impulse to build bigger, elaborate and signature structures of steel and mortar, in the European High North/Arctic the physical structures are replaced by discourse, events, conferences, futurism and speculation, underlined by a competition for scarce social, economic and political resources. Manifesting in the launching of conferences and cultural events, three cities in the Norwegian High North and Arctic have, to varying degrees, engaged in such hyperdevelopment and hyperbuilding strategies. They do so in order to exert their sovereign exception to draw attention to their localities in order to attract visitors, tourists and, importantly, investors. I will discuss the cases of Tromsø, Kirkenes and Longyearbyen each in turn and compare how such strategies have met with varying degrees of success in fomenting actual sustainable economic and/or political and social change. In all three cases the hype around the High North and the Arctic have given rise to unrealised speculative expectations that have had social and economic implications for the local residents in these cities.

> At the root of the spectacle lies that oldest of all social divisions of labor, the specialization of power. The specialized role played by the spectacle is that of spokesman for all other activities, a

sort of diplomatic representative of hierarchical society at its own court, and the source of the only discourse which that society allows itself to hear.[26]

Debord's assessment of spectacles falls within a cynical Marxist reading of the theatrical performance of hyperdevelopment and hyperbuilding, but, at least on the visual surface, there is a kernel of truth to his analysis. Paramount in the aims of the purveyors of Arctic spectacles is to call attention to this exoticised place, often rendered as pristine, tranquil and sparsely developed empty space in the mind's eye of those who have never set foot North of the 59th parallel, let alone the Arctic Circle at just over 66°. The discourse of sovereign exception from Northerners however is that the Arctic is industrialised and open for more industrialising. We Northerners spend much energy trying to convince Southerners that we live much like they do, have international airports just as they do, have a café culture as they do, have access to and can create art, culture and world class performances just like they do. We also aver that governance of the High North should be done by the stakeholders and residents who live in the region. And most adamantly from those with big voices and prominent business interests in the community, much noise has been and is made about economic development opportunities in the High North, supported by ice-free ports and, as in the case of Finnmark and Svalbard, favourable tax regimes for investment.[27] (Particularly for Svalbard, however, the caveat to intensified industrial development – apart from coal mining, is the stringent Norwegian policies on environmental conservation laws and regulations on the archipelago.)

Most prominently and intertwined with the impulse towards industrial development is that the High North/Arctic aims to realise its voice in the larger context of nation-state. Historically seen as a resource outpost for the Norwegian south, colonial attitudes can still be discerned from our Southern counterparts. Centrist decisions vis-à-vis the High North/Arctic are most acutely demonstrated with regards to business development and resource extraction. The state's decision to allow the dumping of tailings in environmentally sensitive fjords, the approval of mining in regions of traditional nature use by local indigenous peoples as Kjerstin Uhre argues elsewhere in this book and the granting of licenses to domestic and international companies searching for hydrocarbon fields in the Barents Sea smacks of what some Northerners regard as a kind of colonialism.[28] The attempt to reverse such discourses and policies have been reflected in the politics of resource extraction as local leaders of mining towns have attempted to change national policies pertaining to benefits to their municipalities. Having reached a certain momentum over the last five years before commodity prices plummeted, an association of mineral mayors and the Association of Outlying Municipalities (Utmarkskommunenes Sammenslutning) joined forces in order to change laws and legislations. These activities were particularly espoused by Sør-Varanger's previous mayor, Cecilie Hansen, attempting to ensure more revenue-sharing from the iron-ore mine in her municipality.[29] The Sámi Parliament past president, Aili Keskitalo, and other Sámi stakeholders have also expressed their concern that the North should be developed sustainably and in consideration of Sámi and local communities. Such voices have been diminished in the speculation and hopes for grand projects.

Grand land and sea projects have been defined by discourse, with conferences and events as the vehicle of transmission of such industrial development dreams of the future, such as the building of logistics of roads and a railroad connecting Northern Finland and Kirkenes, oil and gas extraction from the Arctic and Barents seas, mining, the construction of industrial ports to service freight ships to Asia, the transshipment of goods from Asia to Europe and back and so on. The ambitions of such future projects are meant to expand the High North/Arctic space of industry and production connecting the Arctic more directly to Asia, Europe and the rest of the Circumpolar North. Marked by intensified globalisation, wealth creation in this remote part of the world is distinguished by disproportionate development, such as Ong's East Asian cities. I aver that the Norwegian High North and especially Arctic Russia are both characterised by unequal development and of standards of living. As can be witnessed in all of the Circumpolar Arctic, and especially outside of the Nordic states, stark inequalities in standards of living vary from city to country village in the Danish, Canadian, Alaskan, and Russian Arctics. Therefore, perhaps the Arctic hype is thus an even more important vehicle for regional and economic development, and capital investment. As hype, Arctic resource development also promises economic and financial independence from the centres of power, in keeping with neoliberal policies followed by all Arctic 8 countries since the early 1990s.

Tromsø: *Arctic Frontiers*

> Future climatic and ecological impacts on Arctic ecosystems and human populations will have significant economic, political, and social implications on Arctic nations. It is thus important to balance the use of Arctic ecosystems with their conservation. Arctic Frontiers in Tromsø will unite stakeholders to facilitate the process of defining priorities for development and research of Arctic regions with a pan-Arctic perspective. . . . Arctic Frontiers Tromsø 2007 is the first conference of what is planned as an annual event in Tromsø, Norway. Tromsø is the gateway to the Arctic and has been a centre of Norwegian trade with Russia since its very beginning.[30]

These lofty words underscore where and how the *Arctic Frontiers Conference* was trying to position itself when it began in 2007. Enveloped in this positioning was and is the Arctic city of Tromsø and its development as a central place in the hyperbuilding of the city as an international and cosmopolitan urban space. Though this description emphasises science, research and sustainability, it also aims to address priorities for development in the entire Arctic region. The discourse would change from academic to business development as the *Arctic Frontiers* continued its annual exposition. Indeed, the 2011 *Arctic Frontiers Conference*'s tone was much different:

> New opportunities create new challenges. This principle is central to how we must explore, develop, and manage the Arctic. Today the Arctic is an emerging energy and mineral province, with the extraction of natural resources projected to increase dramatically in the coming years to decades. New industrial activities, a changing business community, and demographic dynamics will alter the established social structures in several regions of the Arctic.[31]

Since the first *Arctic Frontiers*, the discourse on the development of the Arctic regions has been rooted in the anticipation of climate change, that is, climate warming and the melting of sea ice. The paradoxes imbued in the identity of *Arctic Frontiers* as a place to discuss science, ecosystem balance and conservation on the one hand and of the speculation of enormous energy extraction and transshipment of goods made possible by less sea ice on the other hand cannot be ignored. And yet, in the transformation of this signature event from 2007 until present day allows such paradoxes. These contradictions were already set in the DNA of *Arctic Frontiers*, however, as the conference organisers, savvy in their event-creation, obtained sponsorship from the very beginning from large private energy companies and public institutions keen to attract investors. Even then, though the event was ostensibly a conference to bring scientific academic conversations about the Arctic to the foreground, the creation of spectacularity was pre-destined. The 2007 information circular produced by the event organiser Akvaplan-niva, a scientific research institute based in Tromsø, boasted that the conference would be streamed live, that a French television team filming a documentary about the Arctic was at the conference, a German radio station would be broadcasting from the conference venue all week and that 30 journalists were accredited to report the event.[32] The making of a spectacle was irretrievably set.

As a result, the *Arctic Frontiers* organisers are then compelled to better the conference each year. They have done so by inviting and including state leaders and or their representatives. For example, in the lead-up to the 2015 *Arctic Frontiers*, they declared that

> The program contains two Prime Ministers and a number of other Norwegian and foreign ministers in the company of Prince Albert II of Monaco and the Sámi Parliament President Aili Keskitalo. . . . Prime Minister Erna Solberg will come. So will her colleague the Finnish Prime Minister Alexander Stubb.[33]

For a conference organised by a research institute, located at an academic institution – the University of Tromsø, The Arctic University of Norway – and attempting to advance state-of-the-art knowledge on the Arctic, it seems out of place to have such luminaries attending and making speeches. One could read this as a shift in focus and strategy, but arguably it is but an intensification of the same impulse to take a chance on event spectacles, that somehow the wager is that it will attract more attention, more influence and more prestige to Tromsø. All of this is tinged with the city's perceived competition with other cities in Norway and globally, insisting that it is as great or can develop to be as great as other world-class cities, whether in the global North or in the global South. Notably, the 2017 *Arctic Frontiers* policy section was moderated by the well-known and highly respected BBC HARDtalk interviewer and commentator, Stephen Sackur.[35]

Figure 3.2 In the picture are Anne Husebekk, the University Rector, Prince Albert II of Monaco and Salve Dahle, Director of Akvaplan-niva, the creator and main organiser of *Arctic Frontiers*[34]

Photo: Karine Nigar Aarskog/UiT, The Arctic University of Norway.

Figure 3.3 BBC HARDtalk's Stephen Sackur (far left) in panel discussion with the Prime Minister of Norway, Erna
Solberg (third from the right) and also world-renowned professor of economics, Jeffrey Sachs (second
from the left)

Photo: Alberto Grohovaz/Arctic Frontiers 2017, Policy, Monday 23 January.

Undergirding this reach for attention is a sense of running to keep up. The stakes are high with
arguments based on the certainty that success in proving greatness and capability would be followed
by capital investments, resource exploitation and those willing to move and to work in the Arctic.
Leaders and policy makers understand that attracting investors and people has the potential to lead
to economic stimulation, revitalisation and recovery, and thus to employment and its concomitant
higher standards of living. The fulfillment of such hopes would solve problems of economic inequal-
ity, in Norway's case, between the North and the South, between metropole and hinterland.

Kirkenes: the annual *Kirkenes conference*

Over the last two decades, Kirkenes has very successfully positioned itself as the unofficial capital of
the Barents Region. Historically a mining town located 14 km from the Russian border, featuring
the Norwegian Barents Secretariat and the Kirkenes Business Association, and its open-for-business
industrial mindset, it has not been difficult to draw attention to this small border town. Also on offer in
Kirkenes is the curatorial company, *Pikene på broen* (*Girls on the bridge*, named after the Edvard Munch
painting), whose main annual event is the staging of the *Barents Spektakel*. Significantly, the Norwegian
Barents Secretariat is an arm of the Norwegian Ministry of Foreign Affairs and was established to
boost Norwegian-Russian bilateral cooperation by funding cross-border cooperation projects.

The main event in Kirkenes, comparatively along the same lines as the *Arctic Frontiers* in Tromsø, is the *Kirkenes Conference*. Staged by the Kirkenes Business Association since 2008, it has purposely coincided with the more artistic and indisputably more theatrical and curated *Barents Spektakel*. The *Kirkenes Conference* has the drive to define the political and economic legitimacy for Sør-Varanger as the centre of the Barents Region. The launch of the first conference in 2008 had the timely themes of the Shtokmann off-shore gas development and the Pomor Zone. The now mothballed Shtokmann development project was led by Russia's *Gazprom* and buttressed by Norway's semi-private oil and gas company, *Statoil* and France's *Total*. The *Pomor Zone*, an idea by bureaucrats in the Norwegian Ministry of Foreign Affairs, was meant to ease the economic cooperation and exchange in the Norwegian-Russian borderland.

> New this year is that the Sør-Varanger municipality, the Barents Secretariat and the Kirkenes Business Association are joining forces to create a conference that aims to set new aspects of the High North development on the agenda. The conference is also part of the Barents Days 2008, of which the Ministry of Foreign Affairs is an important supporter.[36]

As with the *Arctic Frontiers Conference* in Tromsø, the *Kirkenes Conference* has attracted top leaders to speak and make pronouncements regarding this sparsely populated place on the outer edges of

Figure 3.4 Minister of Foreign Affairs, Børge Brende at the *Kirkenes Conference* 10 February 2016
Photo: the author.

Norway and Europe, yet central to the Barents Region and Norwegian-Russian relations. It is a place determined by its geography, allowing it to have a privileged geopolitical position.

The attention focussed on Kirkenes is founded on the Barents region-building imperative that began in 1993, which was ratified by the *Kirkenes Declaration*. The Barents Region celebrated its twentieth anniversary in 2013. Advanced by such institutions as the Norwegian and International Barents Secretariats, these organisations have made Kirkenes a meeting place for Russian, Norwegian, Finnish and Swedish politicians to emphasise the good relations wrought by the Barents Region initiative.[37] Thus, political spectacles in the border city of Kirkenes were already established by the time the first *Kirkenes Conference* was launched in 2008. Distinguishing the *Kirkenes Conference* from the *Arctic Frontiers* was that it has unabashedly been focussed on economic and industrial development driven by the government's High North policies and by private business interests in the Sør-Varanger region and beyond. It was not coincidental that the first *Kirkenes Conference* took place at the start of the Arctic hype centred on the planned Shtokmann gas fields development just 600 kilometres off the shores of Kirkenes. Coincidentally, the iron-ore mine situated in Bjørnevatn, with ore to be transported five kilometres to the port of Kirkenes, was just about to re-open after having been closed since 1996.

Figure 3.5 "Dronning Sonja på plass i Kirkenes" [Queen Sonja in place in Kirkenes], NRK, 2 February 2011

Photo: NRK/Morten Ruud.

With the focus on high politics and big business development, the *Kirkenes Conference* has included luminaries in its all-important participants list that typically number just under 300 delegates. In past years, the conference has been regaled by Her Majesty Queen Sonja of Norway, who was in Kirkenes to open the 2011 *Barents Spektakel*. Her speech opening the *Spektakel* referenced the overall concern of the *Kirkenes Conference* and the theme of that year's arts festival.

> 2010 was a landmark year for Norway and for the border region in the north. The delimitation agreement for the border in the Barents Sea allows a wider and better cooperation between Norway and Russia on the future development of the region. The theme for this year's festival is 'Mind the Map!' The map is turned upside down! – Peripheries become the center![38]

The pageant of Queen Sonja's attendance to introduce the *Barents Spektakel* was an expected spectacle in itself, but her appearance at the *Kirkenes Conference*[39] just hours before underscored the spectacularity of the scene as the conference participants dutifully followed the protocols of having one of the nation's monarchs processing into the meeting hall. Two years later, her son, HRH Crown Prince Haakon would also attend the *Kirkenes Conference* and open the city's annual arts festival.[40]

Longyearbyen: Arctic exploration, nation building, sovereignty

That Queen Sonja opened the art gallery in Longyearbyen, Kunsthall Svalbard (Art Gallery Svalbard), in 2015 was meant to create a spectacle in itself in this town of approximately 2,100 people and for the entire Norwegian nation.

> The government strongly believes in such a venue. Art Gallery Svalbard can serve as a catalyst for regional, national and international cultural institutions and businesses can find their way to Svalbard.[41]

Also attending the opening of the Kunsthall Svalbard were members of the US Embassy in Norway, characterising the visit as 'spectacular' on their social media Facebook page.[42] Indeed, Svalbard, and its main Norwegian city, Longyearbyen, captures the imagination. Located at 78 degrees North, it is easy to be convinced of its spectacularity in the magical Polar light of summer or the deep indigo nights of winter. For Norwegians, it is a place that somehow is definitive of their national identity as explorers and also as adventurers, trappers, map-makers and, in the modern period, its heroic resource industrialists, these foregrounding identities wrapped up in nationalism.[43]

Since the downgrading of the coal mine and the lay-offs of hundreds of miners, local residents fear the inevitability of a diminished population, fostering ideas of innovation and business development around Svalbard and Longyearbyen's strengths, namely art, tourism and research. Programmes to combat such an eventuality welcome spectacular ideas in order to stem such a tide of outmigration and to attract newcomers. In the case of the Svalbard archipelago, such ideas tend to come from those from outside, for example the launching of *Artica Svalbard* in 2017, which is significantly funded through the NGO Fritt Ord (Free Word) Foundation, the Norwegian Ministry of Culture and a major Norwegian bank's cultural industry fund. *Artica Svalbard's* main funding focus, thus far, will be an artists' residency that aims to bring the world to Longyearbyen and Svalbard. As outlined by *High North News*, one of the major aims of *Artica Svalbard* is to 'have an international and arctic orientation through public arrangements, to contribute to the focus on and discuss the meaning of the High North/Arctic.'[44]

Up until very recently Longyearbyen was a company town built by the iconic (at least for Norwegians) coal mining company, *Store Norske Spitsbergen Kulkompani* (The Great Norwegian Spitsbergen

Figure 3.6 Longyearbyen with the cable car junction in the distance, 24 August 2016
Photo: the author.

Coal Company) – shortened to *Store Norske*. Indeed, *Store Norske*, which may be literally translated as Great Norwegian, was the foundation of Norwegian presence on Svalbard. While Norway has held 'full and undivided sovereignty over Svalbard' since the Svalbard Treaty was signed in 1920, signatories of the treaty are allowed to undertake commercial activities on the archipelago as long as they heed Norwegian regulations and laws.[45] Significantly, sighting the Svalbard Treaty's provision of business development access, the presence of first the Soviet Union then Russia in the coal-mining towns of the now closed Pyramiden and the still operational Barentsburg flies in the face of Norway's nation-building myth of sovereignty over the Svalbard archipelago.

What differentiates the spectacular spectacles staged in Longyearbyen is not just their smaller size, but also their focus. With tourism and research now seen as the pillars of the economy deemed to buttress the downgrading of the coal mining industry on Svalbard, spectacles have been less about massive industrial development, though that is not out of the question. Longyearbyen has also not yet established one signature event that brings together policy-makers, investors, academics and stakeholders. What Longyearbyen has, however, is more in line with Ong's idea of hyperbuilding exemplified by the arachnid so far disused cable car junction; and the focus of my discussion here, the construction of the Svalbard Global Seed Vault. Managed by the Norwegian Directorate of Public Construction and Property (Statsbygg), it was serendipitously built with only local contractors because the specialised competencies for digging into the permafrost mountain and creating straight walls could already be found on Svalbard. With its front jutting out majestically from the mountain across from the Longyearbyen airport,

> it was decided that the visible exterior parts of the facility should be given an exceptional design – to adapt and integrate it into the magnificent natural surroundings, but also to give visual expression to the special significance of the facility's content and purpose.[46]

Although one has to search for the structure as it juts out of the mountain, the seed bank is designed to conserve seeds from all over the world, serving to brand Norway as an environmental superpower. It has garnered intense media attention. Moreover, as a symbol of hyperbuilding, it is also in accordance with Svalbard's all-important resource management ethic as elucidated by William L. Fox in chapter 10 of this book.

We may attribute the limitations on discourses of massive industrialisation to the stringent regulations that the Governor (Sysselmannen) wields over the archipelago, with strict adherence to the Svalbard Treaty and the reification of values of environmental conservation. Yet, Svalbard lays claim to being spectacular simply by the virtue of its geography and its history. While the politics of sovereign exception has been well-developed on Svalbard, illustrated by the power of the governorship of Svalbard, what is less well developed is the economic sovereignty of the archipelago and places of habitation therein. In the years to come, Svalbard and its main city, Longyearbyen, will depend on the Norwegian state to lend legitimacy to its society and political economy. The question is whether

Figure 3.7 The Global Seed Vault

Photo: Jaro Hollan/Statsbygg.

such legitimacy would be helped by an extended or new kind of spectacularisation. The very real threat of climate change, the impulse for people to live and stay in Longyearbyen and global economic trade of natural resources and manufactured goods will determine the answer to how the polity on Svalbard and its management will answer this question.

A conclusion: extravagant boosterism to events as tourism

Speaking at the 2014 *Kirkenes Conference*, Jan Tore Sanner, Minister of Local Government and Modernisation, claimed that 'some think expectations are daunting. I find it stimulating. . . . On behalf of the North, we need expectations. Expectations that can propel us forward.'[47] As I have argued here previously, the conspicuous architectural and infrastructure hyperbuilding and hyperdevelopment in Asian cities is mimicked by a similar hype in the Arctic. This phenomenon of hype and speculative futures in the Arctic has even been periodised by scholars: 2008–12 as one possible period because of the anticipations around the Shtokmann gas field in the Barents Sea and the attendant meanings attached to it – that if that field was opened up for production, others would follow precipitously, bringing wealth and prosperity. And 2008–14 as the other proposed period of hype in the Arctic, deflated by one of the Arctic 8 countries, Russia and the West's aforementioned sanctions against it.[48] Despite the analysis on the creation of hype and its deflation because of geopolitical concerns outside the Arctic and High North, it continues in events such as the *Arctic Frontiers Conferences*, the *Kirkenes Conference* and celebrations of the natural and human-made spectacularity of Longyearbyen and Svalbard with its attempts to develop research and tourism into job- and profit-making enterprises. Arctic and High North events have also expanded with Russia's *International Arctic Forum* in Arkhangelsk, inaugurated in 2010 and, like *Arctic Frontiers*, attended by the country's leaders, including President Vladimir Putin and Foreign Minister Sergei Lavrov. Following were Iceland's Arctic Circle event established three years ago to rival and compete with the *Arctic Frontiers*, then the lesser-known though no less spectacular the annual *The High North Dialogue Conference* at Nord University in Bodø, Norway, beginning in 2007 and more recently the organisation of the first University of the Arctic scientific symposium, staged in St Petersburg in August 2016. While these events cannot reasonably be called mega-events, their aims and ambitions do not stray far from such goals for the communities that arrange them. As Hiller argues,

> The showcase argument points out that mega-events are spectacles that can best be understood as either instruments of hegemonic power or public relations' ventures far removed from the realities of urban problems and challenges. Whatever the motivation, there is increased awareness that the mega-event can also be a vehicle for some form of urban transformation.[49]

Concomitant with and beyond urban transformation, organising signature events in the Arctic has the power of shifting the axis from centre to periphery, and portending the speculative promises of a sustainable industrial future.

While the cities of Tromsø, Kirkenes and Longyearbyen have vastly varying population sizes, with 73,296, 3,498 and approximately 2,100, respectively, each can be defined within frameworks of urban, urbanising or urban forms.[50] Thus, the staging of the theatres of spectacles not only aim to attract visitors and thus spending and commerce, spectacles have the multiple duties of suspending reality, imagining economic prosperity or daring to dream of a future utopia attracting people and international ideas. Though it would be misleading to assert that these events and spectacles are all-inclusive and are without their critics. The *Arctic Frontiers* policy segment has become highly securitised because of the noted politicians who attend, such as the Prime Minister of Norway, her counterparts, some of her various Ministers and dignitaries such as the Crown Prince of Monaco.

Figure 3.8 The *Arctic Frontiers* securitised

Photo: the author, 25 January 2016.

Critical voices have also reared because of the very expensive conference fees for the *Arctic Frontiers* as well as the *Kirkenes Conference*.

The use of the country's monarchy, whether in this place in the Arctic, or other places in the High North, is meant to draw interest from the media and the public. Spectacles for such places in peripheral geographical spaces make them not so peripheral. Spectacles are performed in order to create identities of belonging to the community and the larger nation-state, and also to establish a unique pride of place – a sovereign exception, in order to attract visitors – tourists, investors and potential residents. In the High North and Arctic, the dramaturgy of spectacle is linked to economic development and regional interests. One can also extend the description of such spectacles as events tourism, with participants and certainly speakers, coming from outside of the cities I have discussed in this chapter.

All of this spectacle-making and speculative discourse incontrovertibly must be critically analysed. Mega-events scholar Maurice Roche argues that often the staging of 'prestige,' signature or mega-events are the purview of elites within the community, and that they tend to be 'autocratic' in their execution. He further argues that the community is essentially excluded from how event planning is decided and fulfilled. Roche writes: 'There is typically little democratic community input, and decisions are largely determined by the will and power of urban political leaderships and/or other relevant and powerful urban elite groups (such as business and cultural elites).'[51] To be sure, such an analysis fits with the examples we have been discussing in this chapter, calling into question the sustainability of this type of hyperdevelopment because of its basis in speculation, its autocratic character and its disengagement from the local population and community at large. As the Arctic faces the certain future of climate change coincident with hyperindustrialisation – speculative and real – such elite projects as signature spectacular events will lose their power because they tend to alienate rather than include those living quotidian lives in the High North/Arctic. In order for the spectacular to retain its purpose, it too must change to be more responsive to the enormous sea-change occurring in the Arctic and the needs of the largely urban populations who live here. Now happening before our very eyes because of climate change and the opportunities and challenges it fosters for people who live here, such monumental changes in a future Arctic will not only mean transformed land- and sea-scapes in rapid order. It also ensures that spectacles must change to become more relevant to the needs of the populations that spectacle-makers intend to impress. The question, however, is whether this moment is when the spectacle's connection to the everyday and mundane would diminish their spectacularity. Considering the real and imminent threat of the transformations we face, such connections to the quotidian and the ordinary is necessary for a future Arctic, and thus the relinquishment of spectacular speculations.

Notes

1 Guy Debord, "Thesis 12," in *The Society of the Spectacle,* trans. Donald Nicholson-Smith (New York: Zone Books, 1994).

2 See "The First Fram Expedition (1893–1896)," *The Fram Museum,* accessed October 30, 2015, www.fram museum.no/Polar-Expedition/Expedition-1.aspx.

3 Erling Røed Larsen, "The Norwegian Economy 1900–2000: From Rags to Riches," *Economic Survey* 4 (2001).

4 I was academic faculty teaching and doing research at the University of Northern British Columbia, Prince George, British Columbia, Canada from January 1995 to July 2009, with a leave of absence between July 2007 to July 2009 to take a position at The Barents Institute in Kirkenes, Norway.

5 Debord, "Thesis 1."

6 Richard L Kaplan, "Between Mass Society and Revolutionary Praxis: The Contradictions of Guy Debord's *Society of the Spectacle,*" *European Journal of Cultural Studies* 15, no. 4 (2012): 458.

7 Wendy Larner, "Neo-Liberalism: Policy, Ideology, Governmentality," *Studies in Political Economy* 63 (2000): 12, 5.25.

8 Ibid.

9 Aihwa Ong, "Hyperbuilding: Spectacle, Speculation, and the Hyperspace of Sovereignty," in *Worlding Cities: Asian Experiments and the Art of Being Global,* ed. Ananya Roy and Aihwa Ong (Oxford: Wiley-Blackwell, 2011).

10 Jonas Gahr Støre, "Et hav av muligheter – en ansvarlig politikk for nordområdene" [An ocean of possibilities – a sound policy for the High North], November 10, 2005, accessed December 17, 2015, www.regjeringen.no/no/aktuelt/et-hav-av-muligheter–en-ansvarlig-poli/id273194/. See Inger Præsteng Thuen Og Torgrim Rath Olsen, "Ser du hvor lang tid som har gått mellom disse to bildene?" [Do you see how much time has passed between these two pictures?], *Nordlys,* November 10, 2015, www.nordlys.no/politikk/nordomradene/ap-politisk-parti/ser-du-hvor-lang-tid-som-har-gatt-mellom-disse-to-bildene/s/5-34-284840.

11 The Norwegian Ministry of Foreign Affairs, *The Norwegian Government's High North Strategy* (Oslo and Tromsø, 2006). The document was signed by the Prime Minister at the time, Jens Stoltenberg.

12 The idea to change the name of The University of Tromsø to The Arctic University of Norway came from the Ministry of Education, thus the change was political, and reflects the annexation of smaller institutions of higher learning in the Norwegian High North.

13 Støre, "Et hav av muligheter."

14 Thuen and Olsen, ""Ser du hvor lang tid som har gått mellom disse to bildene?"

15 Emanuela Guano, "Spectacles of Modernity: Transnational Imagination and Local Hegemonies in Neoliberal Buenos Aires," *Cultural Anthropology* 17, no. 2 (2002).

16 Subsidies and tax privileges have been reconfigured or taken away for Troms and Nordland in order to comply to EU regulations.

17 "Ved å slå sammen de to tilskuddsordningene Barents 2020 og Arktisk samarbeid til én ny ordning Arktis 2030 omfattes nå hele det sirkumpolære Arktis." The Norwegian Government's Ministry of Foreign Affairs, "Arktis 2030," last modified December 19, 2014, www.regjeringen.no/no/aktuelt/arktis-2030/id2356599.

18 Ong, "Hyperbuilding," 207.

19 Prime Minister's Office, "Nye byggesteiner i nord – Neste trinn i Regjeringens nordområdestrategi," last modified March 12, 2009, www.regjeringen.no/globalassets/upload/ud/vedlegg/nordomradene/byggesteiner_nord090323_2.pdf.

20 Norwegian Government, "173 millioner til kartlegging av våre nordlige havområder," news release, October 10, 2015, www.regjeringen.no/no/aktuelt/173-millioner-til-kartlegging-av-vare-nordlige-havomrader/id2457010.

21 U.S. Geological Survey, "90 Billion Barrels of Oil and 1,670 Trillion Cubic Feet of Natural Gas Assessed in the Arctic," July 23, 2008, accessed December 20, 2015, www.usgs.gov/newsroom/article.asp?ID=1980#.VnjzH5MrKRs.

22 Artur Chilingarov quoted in Tom Parfitt, "Russia Plants Flag on North Pole Seabed," *The Guardian*, August 2, 2007, accessed February 29, 2016, www.theguardian.com/world/2007/aug/02/russia.arctic.

23 Canadian Foreign Minister Peter Mckay quoted in ibid.

24 See Richard C. Powell and Klaus Dodds, *Polar Geopolitics? Knowledges, Resources and Legal Regimes* (Cheltenham: Edward Elgar Publishing, 2014), 136.

25 Ong, "Hyperbuilding," 2.

26 Debord, "Thesis 23."

27 Petter Hojem, *Mining in the Nordic Countries: A Comparative Review of Legislation and Taxation*, The Nordic Council of Ministers Publication (Copenhagen: TemaNord, 2015), 542.

28 Skjalg Fjellheim, "Finnmark som koloni" [Finnmark as a colony], *Nordlys*, March 3, 2016. http://nordnorsk debatt.no/article/finnmark-koloni.

29 Kaspar Fuglesang and Erik Lieungh, "Ordførere i Finnmark: – Gruvedrifta må bidra mer til kommunene," *NRK*, February 27, 2012, accessed November 30, 2015, www.nrk.no/troms/drar-til-finnmark-for-gruveinnspill-1.8013507.

30 "The Unlimited Arctic," the 2007 Arctic Frontiers Conference, accessed December 13, 2015, www.arctic frontiers.com/arctic-frontiers/archive/2007-the-unlimited-arctic/270-policy.

31 "Arctic Tipping Points," the 2011 Arctic Frontiers Conference Tromsø, accessed December 13, 2015, www.arcticfrontiers.com/arctic-frontiers/archive/45-2011-arctic-tipping-points/265-arctic-frontiers-2011.

32 Arctic Frontiers, "Storinnrykk til nordområdekonferansen Arctic Frontiers," news release, 2007.

33 "Programmet inneholder to statsministre og en rekke andre norske og utenlandske statsråder i selskap med fyrst Albert II av Monaco og sametingspresident Aili Keskitalo. . . . Statsminister Erna Solberg kommer. Det samme gjør hennes finske statsministerkollega Alexander Stubb." Ragnhild Gustad, "Rekordmange høyprofilerte på Arctic Frontiers" [A record number of VIPs at the Arctic Frontiers], *Nordlys* December 16, 2014, accessed December 13, 2015, www.nordlys.no/forskning/politikk/rekordmange-hoyprofilerte-pa-arctic-frontiers/s/5-34-49638.

34 This photograph appears in Karine Nigar Aarskog, "Behov for arktisk samarbeid" [Need for Arctic Cooperation], accessed March 8, 2017, https://uit.no/nyheter/artikkel?p_document_id=402070.

35 See Stephen Sackur, "HARDtalk," www.bbc.co.uk/programmes/profiles/3jnG4P3WMvmGk370qtJ3TJt/stephen-sackur and accessed January 24, 2017, www.arcticfrontiers.com/speaker/stephen-sackur/.

36 "Nytt av året er at Sør-Varanger kommune, Barentssekretariatet og Kirkenes Næringshage går sammen om å lage en konferanse som har til hensikt å sette nye sider av nordområdeutviklingen på dagsorden. Konferansen er også en del av Barentsdagene 2008, der Utenriksdepartementet er en viktig støttespiller." Linda Beate Randall, "Velkommen til Kirkeneskonferansen 2008" [Welcome to the 2008 Kirkenes Conference], Kirkenes Conference 2008, accessed December 1, 2015, https://sor-varanger.custompublish.com/getfile.php/615192.652.vxcrurqtsv/Kirkeneskonferansen.pdf.

37 The Norwegian Barents Secretariat was established in 1993 and focusses on the bilateral relations between Russia and Norway. The International Barents Secretariat was established in 2008 to support the bureaucratic work

of the Barents Euro-Arctic Council that includes the Northern regions of NW Russia, Norway, Finland and Sweden.

38 "2010 ble et merkeår for Norge og for grenseregionen her i nord. Delelinjeavtale for grensen i Barentshavet åpner for et bredere og bedre samarbeid mellom Norge og Russland om fremtidig utvikling i regionen. Tema for årets festival er 'Mind the Map!'. Kartet snus på hodet! – Periferi blir sentrum!" Her Majesty the Queen of Norway, "Barents Spektakel 2011: Åpningstale" [Barents Spectacle 2011: Opening Speech] *The Royal House of Norway,* December 13, 2015, www.royalcourt.no/tale.html?tid=89574&sek=28409&scope=27248.

39 "Dronning Sonja på plass i Kirkenes" [Queen Sonja in place in Kirkenes], *NRK,* February 2, 2011, accessed March 29, 2016, www.nrk.no/troms/dronning-sonja-pa-plass-i-kirkenes-1.7490301.

40 Amund Trellevik, "Hektisk Spektakel-døgn for kronprins Haakon," *iFinnmark,* February 3, 2013, www.ifinn mark.no/nyheter/hektisk-spektakel-dogn-for-kronprins-haakon/s/1-30002-6478838

41 "Regjeringen har stor tro på en slik visningsarena. Kunsthall Svalbard kan fungere som en døråpner for at region-ale, nasjonale og internasjonale kulturinstitusjoner og bedrifter kan finne veien til Svalbard." Minister of Culture Thorhild Widvey quoted in Torill Ustad Stav, "Dronningen åpner kunstmuseum på Svalbard," *NRK* Febru-ary 5, 2015, accessed December 20, 2015, www.nrk.no/troms/dronningen-apner-kunstmuseum-pa-svalbard-1.12191148.

42 See U.S. Embassy Facebook pages, Images of "Spectacular Svalbard," accessed October 13, 2015, www.facebook. com/media/set/?set=a.10152727861667123.1073741846.20650012122&type=3.

43 See Otto Ulseth, *Røde Robert Hermansen: redningsmann og syndebukk* (Oslo: Nordli 2014); and Michael Bravo and Sverker Sörlin, *Narrating the Arctic: A Cultural History of Nordic Scientific Practices* (Cambridge: Science History Publications, 2002).

44 "Ny stiftelse skal fremme kunst og kulturvirksomehet på Svalbard" [A new foundation shall advance art and cultural activities on Svalbard] *High North News,* November 21, 2016, accessed March 1, 2017, www.high northnews.com/ny-stiftelse-skal-fremme-kunst-og-kulturvirksomhet-pa-svalbard; Ministry of Culture, "The establishment of the foundation artica Svalbard," News release, November 17, 2016, www.frittord.no/aktuelt/ etablering-av-stiftelsen-artica-svalbard.

45 Svalbard Treaty, February 9, 1920, accessed December 13, 2015, www.sysselmannen.no/en/Toppmeny/ About-Svalbard/Laws-and-regulations/Svalbard-Treaty.

46 Statsbygg, "Svalbard globale frøhvelv Longyearbyen Nybygg" [Svalbard Global Seed Vault Longyearbyen New construction], Ferdigmelding nr. 671/2008, Project No.11098, accessed March 8, 2017, www.statsbygg.no/files/ publikasjoner/ferdigmeldinger/671_SvalbardFrohvelv.pdf.

47 'Noen synes forventninger er skremmende. Jeg syns det er stimulerende. . . . På vegne av nordområdene trenger vi forventninger. Forventninger som kan drive oss videre.' Jan Tore Sanner, Minister of Local Government and Modernisation, "Nordområdene og ny regjering Kirkenes-konferansen 2014" [The High North and new gov-ernance Kirkenes Conference 2014] Speech, January 4, 2014, accessed December 13, 2015, www.regjeringen. no/no/aktuelt/kirkenes-konferansen-2014/id750406/.

48 Monica Tennberg, "Barents Region in a Flux I: Local Communities, Sustainable Development and Neoliberali-sation' and 'Barents Region in a Flux II: Neoliberal Regionalism at Work' lectures for the GENI/MNGD Field School in Northern Norway, March 18, 2016.

49 Harry H. Hiller, "Mega-Events, Urban Boosterism and Growth Strategies: An Analysis of the Objectives and Legitimations of the Cape Town 2004 Olympic Bid," *International Journal of Urban and Regional Research* 24, no. 2 (2000).

50 "Norwegian Statistics Bureau: Tromsø," accessed February 15, 2016, www.ssb.no/245913/folkemengde-og-kvartalsvise-befolkningsendringar.heile-landet-fylke-og-kommunar; Kirkenes; Ibid, accessed February 15, 2016, www.ssb.no/186162/urban-settlements.population-and-area-by-municipality.1-january-2013; Governor of Svalbard [Sysselmannen], *Information for Foreign Citizens in Longyearbyen,* www.sysselmannen.no/Documents/ Sysselmannen_dok/Trykksaker/Brosjyre%20for%20utenlandske%20borgere%20i%20Longyearbyen,%202016/ ENGELSK%20-Informasjon%20til%20utenlandske%20statsborgere%20ENGELSK.pdf.

51 Maurice Roche, "Mega-Events and Urban Policy," *Annals of Tourism Research* 21 (1994).

Bibliography

"Ny stiftelse skal fremme kunst og kulturvirksomehet på Svalbard" [A new foundation shall advance art and cultural activities on Svalbard]. *High North News,* November 21, 2016, Accessed March 1, 2017, www.highnorthnews.com/ ny-stiftelse-skal-fremme-kunst-og-kulturvirksomhet-pa-svalbard.

Arctic Frontiers. "Arctic Tipping Points," *The 2011 Arctic Frontiers Conference Tromsø.* Accessed December 13, 2015. www.arcticfrontiers.com/arctic-frontiers/archive/45-2011-arctic-tipping-points/265-arctic-frontiers-2011.

Arctic Frontiers "The Unlimited Arctic," *The 2007 Arctic Frontiers Conference Tromsø*. Accessed December 13, 2015. www.arcticfrontiers.com/arctic-frontiers/archive/2007-the-unlimited-arctic/270-policy.

Bell, Elizabeth. *Theories of Performance*. London: Sage Publications, 2008.

Debord, Guy. *The Society of the Spectacle*. First published in 1967. Translation by Donald Nicholson Smith. New York: Zone Books, 1994.

Governemnt of Norway. "173 millioner til kartlegging av våre nordlige havområder." Release Nr: 8007, October 2015. Accessed December 19, 2015. www.regjeringen.no/no/aktuelt/173-millioner-til-kartlegging-av-vare-nordlige-havomrader/id2457010.

Governor of Svalbard [Sysselmannen]. *Information for Foreign Citizens in Longyearbyen*. www.sysselmannen.no/ Documents/Sysselmannen_dok/Trykksaker/Brosjyre%20for%20utenlandske%20borgere%20i%20Longyear byen,%202016/ENGELSK%20-Informasjon%20til%20utenlandske%20statsborgere%20ENGELSK.pdf.

Guano, Emanuela. "Spectacles of Modernity: Transnational Imagination and Local Hegemonies in Neoliberal Buenos Aires." *Cultural Anthropology* 17, no. 2 (2002): 181–209.

Hiller, Harry H. "Mega-Events, Urban Boosterism and Growth Strategies: An Analysis of the Objectives and Legitimations of the Cape Town 2004 Olympic Bid." *International Journal of Urban and Regional Research* 24, no. 2 (2000): 439–458.

Kaplan, Richard L. "Between Mass Society and Revolutionary Praxis: The Contradictions of Guy Debord's Society of the Spectacle." *European Journal of Cultural Studies* 15, no. 4 (2012): 457–478.

Larner, Wendy. "Neo-Liberalism: Policy, Ideology, Governmentality." *Studies in Political Economy* 63, Autumn (2000): 12, 5.25.

Ministry of Culture, "Etablering av stiftelsen Artica Svalbard." *News Release*, November 17, 2016, www.frittord.no/ aktuelt/etablering-av-stiftelsen-artica-svalbard.

Mitchell, Lawrence E. *The Speculation Economy: How Finance Triumphed Over Industry*. San Francisco: Berrett-Koehler Publishers, 2008.

Norwegian Statistics Bureau. "Folkemengde og kvartalsvise befolkningsendringar. Heile landet, fylke og kommunar" [Population and quarterly population changes. All country, county and municipalities]. Accessed February 15, 2016. www.ssb.no/245913/folkemengde-og-kvartalsvise-befolkningsendringar.heile-landet-fylke-og-kommunar.

Ong, Aihwa. "Hyperbuilding: Spectacle, Speculation, and the Hyperspace of Sovereignty." In *Worlding Cities: Asian Experiments and the Art of Being Global*, edited by Ananya Roy And Aihwa Ong. Oxford: Wiley-Blackwell, 2011.

Prime Minister's Office. "Rekordsatsing på nordområdene." www.regjeringen.no/no/aktuelt/nord/id615757/. Last modified October 15, 2015.

Prime Minister's Office. "Nye byggesteiner i nord – Neste trinn i Regjeringens nordområdestrategi." Last modified March 12, 2009. www.regjeringen.no/globalassets/upload/ud/vedlegg/nordomradene/byggesteiner_nord090323_2.pdf.

Roche, Maurice. "Mega-Events and Urban Policy," *Annals of Tourism Research* 21 (1994): 1–19.

Tennberg, Monica. "Barents 'Region in a Flux I: Local Communities, Sustainable Development and Neoliberalisation' and 'Barents region in a Flux II: Neoliberal Regionalism at Work'" lectures for the GENI/MNGD Field School in Northern Norway, March 18, 2016.

The Fram Museum. "The First Fram Expedition (1893–1896)." Accessed October 30, 2015. www.frammuseum.no/ Polar-Expedition/Expedition-1.aspx.

Thuen, Inger Præsteng and Torgrim Rath Olsen. "Ser du hvor lang tid som har gått mellom disse to bildene?" *Nordlys*. November 10, 2015. <www.nordlys.no/politikk/nordomradene/ap-politisk-parti/ser-du-hvor-lang-tid-som-har-gatt-mellom-disse-to-bildene/s/5-34-284840.

U.S. Geological Survey. "90 Billion Barrels of Oil and 1,670 Trillion Cubic Feet of Natural Gas Assessed in the Arctic." July 23, 2008. Accessed December 20, 2015. www.usgs.gov/newsroom/article.asp?ID=1980#.VnjzH5MrKRs.

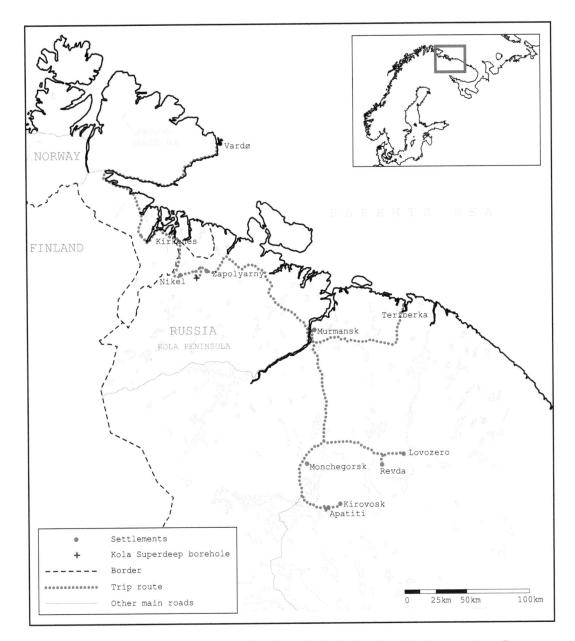

Map 2 The urban planning on the Kola Peninsula reflects a desire for the minerals in the ground as well as geopolitical ambitions. We visited various locations of mineral mining, processing and shipping. Walking the urban landscapes, we encountered postindustrial and destroyed landscapes that forced reflections on ecology. As a demonstrative Anthropocene landscape, the Kola Peninsula performs as an extremity of human exploitative drives (map: Eimear Tynan).

4 Ruins and monuments of the Kola cities

Peter Hemmersam

There are many definitions of the Arctic, from climatographic and botanical delineations to social and economic categorisations. Now also considered a political region, its future is increasingly becoming a topic for world affairs, particularly when it comes to unfolding geo-politics, expanding resource extraction and increased focus on security and climate change.[1] Global competition has led to structural crises in Arctic communities, and in many places new economies such as research, education or tourism are assumed to become increasingly important.[2] The local communities of the region are often considered to be in need of social and economic development. They are frequently seen as marginal, and the various effects of climate change are viewed as a direct threat to their lifestyles and livelihood. Prevailing discourses on the Arctic oscillate between global change and local conditions, but the role, design and structure of settlements and cities themselves receive little scrutiny.[3] They are, for instance, often simply framed as spaces for the transitory workforce involved in resource extractions in nearby territories, rather than thriving and diversified urban societies.

The *Megatrends* report on Arctic development by the Nordic Council of Ministers highlights urbanisation as a significant on-going social and economic process.[4] Rural areas and smaller settlements are depopulating in favour of nearby cities, a development coupled with other trends, such as out-migration of older people and women, lower birth rates and continued dependency on resource extraction and subsidies from national states. However, urbanisation in Arctic communities is not only an economic and demographic phenomenon, but also reflects cultural evolutions, changing values and lifestyle choices.[5] Thus, the future of Arctic cities may follow very different trajectories. For instance, questions of identity and the quest of indigenous populations for self-determination (as we witness in Greenland and Nunavut) may inform the design and layout of cities in ways that differ from the plans representing the projected territorial policies and concerns of central governments.

Future conflicts and issues of change and development will play out in Arctic cities in different ways, not only in entirely new settlements built from scratch, but also in existing cities. While some communities will experience an influx of new residents, others will shrink as 'fly-in-fly-out' resource extraction operations become more widespread. This means that existing settlements will have to be re-imagined and restructured, while new kinds of urban designs for accommodating a transitory work force will also need to be developed. As tourism grows and becomes an important economic factor in many Arctic regions, urban planning will have to negotiate between facilitating mass tourism and maintaining the flavour of wilderness sought among Arctic tourists. An increased focus on the Arctic as a 'research landscape', as is the case in Svalbard and parts of Greenland, also poses particular urban design challenges, as the actual object of research is the undisturbed landscape itself – not the cities and the settlements in which the researchers reside and the research infrastructure is located.

The Future North project studies the material as well as social agency in the creation of future landscapes along the Norwegian–Russian Arctic Barents Coast. Through reading the transforming urban landscapes across this part of the Arctic, a basis is established from which to reconsider what is already there and to launch reconceptualised landscapes onto which future urban change can be imagined. Using photography and forms of urban mapping, the interdisciplinary project team traced the urban landscapes of the Kola Peninsula, which, along with other regions of the Soviet Arctic, was the scene of systematic urbanisation after the Soviet revolution.[6] This region saw the development of key concepts of Artic urbanism that have influenced later developments in other parts of the Arctic.

Building cities on Kola

The Kola Peninsula is the most populated part of the Arctic and also one of the most urbanised regions in Russia, with more than 92 per cent of the population residing in cities.[7] In the 1980s and early 1990s, it had over a million inhabitants.[8] The cities of the peninsula are clustered around the rail and road link between St. Petersburg and Murmansk in the western part of the peninsula. The fabric of these cities, almost without exception, was constructed after the Soviet Revolution, and particularly after the Second World War; the latter saw damage to the existing cities as a result of military confrontation over the mineral-rich and strategic region. The development of the region required large-scale planning and economic effort on behalf of the Soviet state, and initially the construction of infrastructure and cities included extensive use of forced labour.[9]

Militarisation of the Russian–Norwegian border along with the strategic location of the Northern Fleet in and around the all-year ice-free harbour of Murmansk, which provides open access to the Atlantic, has left a significant imprint on the territory. The key driver of urban development, however, was the extensive, large-scale extraction of minerals in the region, which is a direct result of Soviet central economic planning. The planned new cities were the products and properties of different economic organisations within the state system, including the military–industrial complex and the military itself. They served various national strategies, and the existence of the Kola cities can to a large extent be explained by their designated roles in supplying strategic mineral resources to the national industrial and agricultural production; their location reflects rational balancing of access to minerals, energy and transport. They were built with particular production purposes in mind, a fact that is reflected in the rational layout of the cities. Cities like Apatity and Nikel are even named after the minerals they extract and process.

The logic of urban development in the Kola differs from the market economic logic that has dominated Western European city development in the twentieth century.[10] Soviet central economic planning meant that industrial and urban development happened on a large scale, following a utilitarian and scientific rationale. At the same time, Arctic city building provided the possibility to demonstrate the superiority of the Soviet system over nature through the construction of ideal communities in adverse conditions. According to this logic, the Soviet industrial cities in the North were explicitly constructed as 'workers' paradises' in order to attract a workforce motivated to 'Build Communism'.[11]

The Kola cities were well-developed examples of twentieth-century utopian modernist urban planning, in which the inhabitants' welfare was achieved through systematic provision of essential goods and services, healthy dwellings, cultural and educational facilities and an urban layout that ensured air and light as well as efficient circulation of people and goods. Characteristic of this model is the zoning that separates the core urban functions of living, working and free-time activities, the industrially produced standardised housing blocks and the pre-devised provision of communal and service functions. The cities had prescribed population size targets and included various versions of the model pedestrian residential districts called 'microrayons'; these were reproduced in socialist countries around the world with little or no adaptation to climate, topography or local cultures, including for instance the indigenous population and other groups inhabiting the Kola Peninsula.

The Kola cities today

The population of Murmansk Oblast that constitutes the Kola Peninsula has shrunk by around 30 per cent over the last two decades. The cost of energy, travel and food in the region has increased from the early 1990s, while the social welfare programs have been drastically reduced, resulting in out-migration.[12] The Soviet era buildings that make up most of the urban fabric are mostly still either in continued use or have lost their original function and been transformed for other purposes. This means that the Soviet-era planning and urban design footprint is well-preserved, and more recent additions and alterations to buildings and urban areas are easily identifiable.

The core cluster of cities on Kola are the result of Soviet urban planning, but differ when it comes to their role in the industrial and infrastructural system of the territory. Murmansk is a rail, road and shipping hub, as well as a diversified economic and administrative centre, with a population of some 300,000. The other cities are predominantly mono-industrial settlements with populations ranging from 10,000 to 55,000. These cities all feature forbidding industrial complexes on their outskirts – often as big as the residential districts – and in the nickel processing cities of Nikel and Monchegorsk, environmental damage is visually evident in the surrounding landscape. Cities like Zapolyarny and Olenegorsk have operational mines, but in Revda and the former fishing settlement of Teriberka industries have closed down, with severe depopulation as a result. The Kola Peninsula also has several closed military towns and installations such as Severomorsk, home of the Russian Northern Fleet, just north of Murmansk. Other settlements with a special status include Poljarnje Zori, which houses the region's nuclear power plant, and Lovozero in the centre of the peninsula, the centre of the indigenous Sami population and the reindeer herding industry.

The Future North team's travel to the Kola Cities revealed the monumental scale of the Soviet era urban spaces: urban squares and boulevards – such as the over 60-metre wide Prospekt Metallurgov in Monchegorsk. Other prominent features include the broad variety of public monuments and war memorials, such as the 35-metre high 'Alyosha' monument to the Defenders of the Soviet Arctic during the Great Patriotic War that rises over Murmansk. The cities all have impressive cultural palaces and green urban parks. In fact, urban vegetation and tree-lined streets are found throughout Kola, and have obviously been a planning priority in order to mitigate the harsh climate. The majority of the urban fabric is made up of residential neighbourhoods, dominated by standardised multi-storey apartment buildings that reflect the various development phases of the Soviet construction industry. The landscapes of these districts are easily accessible by foot, and there is still very little actual privatisation of land. All the cities feature what might be called parallel or 'mirror' cities in the form of extensive garage districts where the male population has traditionally escaped domestic settings. Due to the Soviet planning ideology the cities lack purpose-built city centre retail structures with street frontages, and these have often been added later as low temporary structures or 'kiosks'.[13] In Murmansk we also found several recent, centrally located, purpose-built shopping centres.

Buildings and urban infrastructure outside the immediate urban centres are dilapidated and in many cases in dire need of repair. Private initiative and ownership is evident in, for instance, the newer windows in many apartment blocks, but the facades themselves seem to be under communal (or no-one's) ownership and care. An economy of scarcity is also observable in the re-use of materials in the new or refurbished constructions, such as military steel road panels used for fences and gates, repurposed pre-cast concrete slabs used for garage construction or even the repurposing of entire Soviet-era housing blocks for new uses, such as the Radisson Poliarnie Zori Hotel in Murmansk in which the project team stayed.

During our visits, we identified new types of spaces and functions that reflected the changes in politics, economy and lifestyle over the last couple of decades. These include the upgraded central

Figure 4.1 The new church in Nikel
Photo: the author.

retail district in Murmansk, and the emergence of a car-based city at the fringes of the older urban structure – with no facilitation for pedestrians. Here, we found both expected programs, such as big-box shopping complexes, but also surprising functions such as a ski slope, 'new–old' traditional style wooden churches and even a full monastery in the Murmansk urban fringe – complete with palisade.

Based on our observations, it became clear that the Kola cities share many characteristics, and that their urban landscapes displayed many similar features. For analytical purposes, these urban landscapes can be read as layers. Not the historical layers of archaeology or the structural urban layers of experience and memory of urban morphology, but rather as elements of the cities – new and old – that were repeated over the region and could be attributed to particular 'modes' and imaginaries of urban landscape production.[14] These range from historic, ideological Soviet layers of monumentality, to the industrial modernist logic of the standardised urban modular housing units, and to the contemporary commercial aesthetics that reveal the changes in economy and consumer preferences.

Due to the clear original structure of the urban planning and design and the relative absence of new construction, reading the distinct layers of Soviet and post-Soviet development was remarkably easy on Kola. Identifying and reading these layers of the urban landscape revealed to us that the Kola cities have a distinct history that is very different from other cities we had visited in the Arctic – and that their overlapping urban landscape layers not only revealed their development history, but also had consequences for how they could develop in the future.

A visual urban landscape typology

In order to read the urban landscapes of the Kola cities, an analytical visual approach was adopted, inspired by methods of landscape characterisation developed in landscape history.[15] This approach belongs to the tradition of cultural landscape reading, which is again grounded in a notion that human societies *produce* landscapes and that cultures can be documented and interpreted as social traces *in* the landscape.[16] Importantly, this notion is supplemented with the important corrective that landscape is also, as argued by Denis Cosgrove, 'a way of envisioning, contemplating, manipulating, and representing the natural world, always a construction and thus primarily ideational rather than inherent in nature'.[17]

Visual approaches to reading landscapes have been adopted in various contemporary forms of urban analysis, and aerial imagery in particular has become a tool to reveal the urban landscapes that are not immediately perceivable from the ground or that lack conceptual frames for reading.[18] The Future North survey of the Kola cities was not conducted by airplane, but by minibus and by walking the urban landscape. Our approach relied on photographic documentation and skilled observation by the multidisciplinary project team. Some of our walks followed main routes through the cities, but others deliberately cut across the urban structures, as urban transects from periphery to centre, in order to reveal otherwise unseen urban spaces. As an additional documentation tool, we used social mapping technology in the form of a GPS-based mapping application, which resulted in geo-located digital tracings of our walks.[19]

The result of our survey is a typology of characteristic twentieth- to early twenty-first-century urban landscapes on Kola. The typology is ordered in four broad categories: People, Politics, Profit and Pleasure, based on archaeologist Sefryn Penrose's listing of significant late-modern and contemporary urban landscapes in England.[20] This work is again based on widely used methods of Historic Landscape Characterisation,[21] which is a holistic reading of landscapes that views them as the result of mundane activities over the ages rather than reducing them to just being sites of historic events. While these four themes do not represent any definitive categorisation of landscape types, they have nevertheless come to function as a general framework in contemporary archaeological research and provide a structured way to look at, and understand, the present-day landscapes of the city – with its historic as well as contemporary components.[22]

The purpose of identifying this typology of significant Kola urban landscapes is, following Rodney Harrison, to 'identify the character of the present landscape and the processes that underpin it, and to project and effect future change'.[23] This would also make it possible to compare the Kola cities to cities elsewhere in the Arctic. The typologies identified in this study are pervasive throughout the Kola cities, but certain characteristic cases are listed specifically in parenthesis:

(A) People

1   Microrayons – variations of the modular Soviet era residential districts with 8,000–12,000 inhabitants
2   Churches – new churches and monasteries in traditional style
3   Cemeteries
4   Suburbia – emerging detached housing areas (Murmansk, Apatity)
5   Dachas – 'holiday homes' and allotment gardens
6   Garage cities
7   'Slums' – seriously dilapidated and/or abandoned housing areas
8   Dying towns – with dwindling populations and loss of industry (Revda, Teriberka).

Figure 4.2 Garage district in Murmansk

Photo: the author.

(B) Politics

 1 Monumental urban squares
 2 Boulevards
 3 Military installations – including bases, airfields and harbours
 4 War memorials
 5 Political monuments – including ubiquitous statues of Lenin
 6 Border exclusion zone – the closed off zone running along the Russian border
 7 Closed cities (Zaozersk, Skalistyi, Snezhnogorsk, Poliarnyi, Vidiaevo, Severomorsk).[24]

(C) Profit

 1 Shopping centres – both in suburbia and city centre (Murmansk)
 2 Repurposed street facades – shops and offices in refurbished ground floors of buildings, or in the form of semi- or fully detached 'kiosks'
 3 Deadscapes – de-vegetated areas around industrial complexes (Nikel, Monchegorsk)
 4 Power plants – combined coal powered district heating and power plants
 5 Industry – mines and ore processing facilities
 6 Tailing landscapes – piles and ponds
 7 Ore trains – carrying raw and processed ore and coal

Figure 4.3 The central square in Monchegorsk
Photo: the author.

Figure 4.4 The ore processing plant in Zapolyarny
Photo: the author.

Figure 4.5 Street vegetation in Nikel

Photo: the author.

    8   Farming – pastures for dairy production

    9   Roads – recently upgraded infrastructure.

(D) Pleasure

    1   Urban greening – parks and street vegetation

    2   Cultural palaces – one in each town

    3   Entertainment – cinemas and theatres

    4   Ski slopes.

Ruins and monuments

The economy of the Kola cities has, as elsewhere in Russia, changed radically over the last two decades, but they were spared from total deindustrialisation after 1991. The economic restructuring was mostly not as dramatic here as in many other Russian cities. Their main industry, mining and mineral processing, continues to play a role in the national economy, and the industrial plants and installations are still central features in the cities. An example is the busy harbour of Murmansk, which has not yet been abandoned or given way to the waterfront housing and recreational seaside promenades that have become key components of many post-industrial harbour cities in Europe and elsewhere. There are additional ways in which the urban landscapes of the Kola Peninsula do not conform to West-ern European urban models, including their spatial configuration, which is a direct reflection of the detailed social and economic management of Soviet planning. In addition, they are mostly designed from scratch for an imported population – in locations determined by the geology of mining.

Today, the Kola cities look roughly as they did in the early 1990s; many buildings are in a poor state of repair or in degrees of abandonment. Even where buildings are still inhabited and the interiors have been modernised, the facades are often dilapidated. The urban structure and the deteriorating buildings, sad in their decay or fascinating and picturesque in their semi-ruinous state, is a reminder of a lost era. They represent a shared memory of the privileged lifestyle of the workers of the Soviet North. While only a few of the remaining buildings have explicit architectural qualities, the structure of the cities, the size and layout of streets, the huge squares and parks are important reminders of the scale and scope of the social and economic utopian vision of Soviet society. The cities are elements of large, urban ensembles that reflect the grand visions and the historic destiny of a society rather than any individual desires. Most of the individual celebratory monuments have been retained, even though they – as in other post-Soviet Russian cities – have undergone a selective redaction that has retained most of the Soviet narrative, particularly the focus on historic national achievements such as the struggle of the Second World War and the colonialisation and urbanisation of the Arctic.[25]

Figure 4.6 Mosaic from the abandoned mine in Revda

Photo: the author.

The modernist urban design still sharply delimits the cities from their surroundings and isolates them as objects in the landscape. The grim, smoke-belching industrial plants loom over the city, while the towering combined heat and power plants are central, visual features of the three urban districts of Murmansk. In a way, these plants are strangely reminiscent of expressionist architect Bruno Taut's idea of the inspirational *Stadtkrone* (City Crown) that was supposed to embody combined civic efforts and social urban unity in the form of central monumental architectural objects.[26] On Kola, the cities no longer embody any communal spirit or social vision, but even though the link to underlying ideologies has been lost, they still today appear as spatial monuments. They were built to last in a non-capitalist era, beyond the programmed 'creative' destruction and death of cities, buildings and infrastructure of capitalist modernisation.[27]

According to cultural landscape theorist John Brinckerhoff Jackson, hortatory monuments are employed by revolutionary social orders to gather support and demonstrate to the public 'what it should believe and how it is to act'.[28] In the Soviet political narrative, the conquest and urbanisation of the North was a historic achievement – a triumph over nature and the elements that should be celebrated.[29] Conquering and improving the landscape through industry and inhabitation were main motives in the urbanisation of Kola and other parts of the Arctic, and the cities were designed as monuments to celebrate this point.[30]

But what happens, Jackson asks, when monuments no longer remind us of specific obligations – when they are detached from their hortatory role? He argues that as they lose their original role, they become celebrations of another kind of past. They no longer celebrate the grand historical narratives, but rather serve as reminders of a 'vernacular past' in which historic events and the achievement of individuals are no longer important, an 'age where there are no dates or names'.[31]

Our visual survey reveals a gap between the spatial determinism of socialist urban design and contemporary everyday life in the Kola cities. They were built as spatial monuments to Soviet ideology and achievements, but there is now a discontinuity between the narrative behind the urban forms and the lives of the inhabitants. The cities have become 'everyday' monuments reminding us of how things were – 'chronicle[s] of everyday existence'.[32] While new lifestyles have emerged in the Kola cities, the spaces and structures of the cities have changed so little over 25 years that it is still easy to imagine and visualise what life here was like a generation ago.

According to Jackson, it is important that populations are able to conceive of a discontinuity between the past and the present. The disruption is a precondition for future thinking: 'The old order has to die before there can be a born-again landscape'.[33] In order for a landscape to be reconceptualised, an intervening period of neglect, in which the old is visibly degenerating and turning into ruins, is required. To Jackson, this is not about shedding the past, but rather about learning to appreciate it anew – as representative of another era. There is a 'necessity for ruins'; the landscape has to have been 'destroyed' (as one can argue is the case in the Kola cities) for it to be rediscovered and restored. Ruins are essential, not only because they provide an incentive for restoration, but also because they represent a time *before* contemporary concerns and provide an opportunity to 'redeem . . . what has been neglected'.[34] Archaeologists Bjørnar Olsen and Þóra Pétursdóttir suggest that ruins 'expose meanings and genuinely thingly presences that perhaps are only possible to grasp at second hand when no longer immersed in their withdrawn and useful reality'.[35] In the act of redemption, the fascination and aesthetic appreciation of ruins play an important role.

Kola's urban futures

The Kola cities are currently too large for their populations, but at the same time the connecting regional infrastructure between the cities is underdeveloped, leaving them isolated from wider markets and hindering economic development. The infrastructural landscapes that we have come to see as ubiquitous in industrialised countries are thus less evident on Kola. The dilapidated character of the cities and the shrinking population does not, however, mean that the Kola Peninsula does not have an urban future. Like other parts of the Russian Arctic, it finds itself in a paradoxical economic situation in which

> the region is both overdeveloped, with its now obsolete mono-industrial towns, and under-developed, with its still largely untapped natural resources. The Russian North thus represents both a burden and an asset; a constant reminder of the flaws of planned economy and the main locomotive in the economy.[36]

Industrialisation and resource extraction is considered important in the development of many parts of the Arctic. However, the future of the mining industries is no longer expected to provide significant numbers of jobs for the existing population or include full-scale urban developments.[37] Despite decades of crisis on Kola, it is still the most industrialised and urbanised area in the Arctic (in addition to probably being among the most militarised places in the world).[38] The Kola cities owe their existence to the minerals in the ground, and the mining and processing industry is likely to continue to play a major role in the Kola cities in the future. They are the 'engines' of other places, and the close relationship and physical proximity between the cities and the industry is unlikely to change. The industrial harbour will continue to dominate the Murmansk waterfront, as it is still the major port giving all-year access to the Atlantic from Russia, and the monumental industrial plants and mining structures will continue to visually dominate the cities on the Peninsula, despite fluctuations in world ore prices that may have very direct local consequences in terms of employment.

The Kola cities share many post-Soviet urban development challenges with other regions in Russia. These include centralisation in larger cities, increased car dependency, de-industrialisation, privatisation of urban space, 'slumification' of urban districts and the development of ex-urban, big box, car-based shopping districts.[39] Kola's northern location entails its own set of challenges; while still important, the strategic status of the Peninsula requires fewer soldiers and miners than before. There are even public subsidies in place to facilitate the movement of people away from the northern regions of Russia, presumably because they are seen as a financial burden on the state.

For decades, the Kola Peninsula was considered one of the most polluted places in Russia. Since the early 1990s, however, decreasing population and de-industrialisation have contributed to a relative reduction in environmental damage.[40] De-industrialisation and increased environmental measures in the remaining industry may also contribute to a future shift in which the cities are no longer seen as upwind refuges from industrial emissions, but are themselves regarded as sources of pollution and environmental pressure. This will particularly be the case if infrastructural systems continue to be in disrepair, and if increased affluence and changing lifestyle results in greater individual ecological footprints, for instance in the form of more car ownership or living space per inhabitant, as well as boosted retail consumption and waste production.

A key aspect of the urban future of the peninsula is the fact that Kola is not actually very remote compared to other regions in the Arctic. It possesses all-year ice-free harbours, and its proximity to European markets makes it economically and logistically attractive compared to many other regions of Russia. However, the economic future of the Kola cities relies not only on how they connect to the rest of the country and the rest of the world, but also on how they interact with each other. A major trend in European urban development over the few last decades has been regionalisation; cities are economically integrated by infrastructure to support flows of people, goods and knowledge, creating larger functional regions with more diverse and resilient economies. Effective regionalisation of Kola would provide an economy of scale based on the increased movement of people. Distances between cities are relatively short compared to many other places in Arctic Russia, where such a development would be harder to imagine.[41] Regionalisation would require upgrading the Soviet-era infrastructure, which focused on the transport of raw materials and cargo. This is already happening in the form of upgrades to the regional and cross-border road network.

The Kola cities display many of the typical features of both Soviet and other Arctic communities, but their specific condition also represents its own set of challenges and opportunities.

The Kola cities as Arctic urbanism

Visiting the Kola cities, we were struck by how dramatically different they looked compared to cities in other Arctic countries. Despite climatic similarities, cities in the region have developed along very different historic and political paths. They have been conceived of and designed in contrasting but at times intertwining ways, and a range of conceptual frames for urbanisation, urban design and architecture that have been influential in Arctic regions can be identified.[42] This includes late modernist architect Ralph Erskine's widely celebrated urban schemes for microclimate alleviation and the design of 'human' space, which continue to be influential in the design of Arctic cities today.[43] Canadian geographer Louis-Edmond Hamelin, whose concept of 'Nordicity' from the 1960s became influential in the perception and urbanisation of the Canadian North, represents another important influence on the planning of Arctic cities. 'Nordicity' refers to 'systems of thought, knowledge, vocabularies, intercultural know-how, arts and humanities sensibilities, expressions of opinion, [and] application in territorial, political and economic fields'.[44] Hamelin developed a multi-parametric index of social and economic development based on the idea that 'North' is not only a circumpolar condition, but also a socially constructed idea.[45] His work was based on that previously done in the Soviet Union, where the concept 'The North' was introduced in public policy in the 1930s as a designation of the geographies of minority populations outside the most populated parts of Russia – specifically of the territories in which workers' pay had to be increased to attract a sufficient workforce to the extraction industries.[46] In comparison to Hamelin's index, this categorisation was more narrowly based on climatic 'harshness'. This is an example of the instrumental character of Soviet urban planning, which also proposed the development of social and cultural (urban) scenes to attract workers and their families, the provision of larger dwellings and a supply of jobs for women in order to create socially sustainable communities.[47]

The pioneering Soviet approach, as well as later modernist and utopian modes of planning and designing Arctic cities evident in the Kola cities, has receded into history. Nevertheless, echoes of this utopian approach are found in various other models and concepts of Arctic urbanism; Hamelin's notion of 'Nordicity', for instance, has influenced planning policies in Canada, while Erskine's design strategies for a climate adapted architecture and urbanism in the North has remained influential in architects' thinking about urbanism in the Arctic. These are also examples that show that Arctic cities have been attributed a special status by designers and in policies that sets them apart from other cities. While the post-industrial urban discourse in Western Europe, North America and elsewhere has broadened to include cultural, creative and innovative dimensions,[48] Arctic cities are still subject to a modernist mode of planning associated with social development of communities, a narrow design focus on microclimate amelioration and engineering-based infrastructural agendas. While the Soviet model of urbanisation is no longer applied to city building in the North, other frames of reference that share elements of modernist outlook are still pervasive in urban planning and design of cities across the circumpolar area, though subject to criticism for various reasons.[49] They inform the mindsets of locals, politicians, planners and designers, and colour the perception of the challenges and strategic opportunities represented by urbanism in the Arctic. Reflecting a modernist utopian ethos, urbanism is still being projected onto northern territories from afar by governments and planners as well as by industry and the general public. Even as climate change is unfolding in these territories, they are still seen as mineral provinces, and their settlements as support for these operations or as places in need of social development.

Historically, the Kola cities were projections of geo-politics and the state's desire for minerals; they were designed along pre-defined modernist and essentially place-less utopian urban models. They were explicitly optimised for economic performance and largely ignored the externalities of urbanisation, such as environmental pollution. While reflecting many of the general trends in the Arctic, current urban development on Kola plays out in very particular ways. Supplementing the general narrative of fast and radical change in the current discourse on the Arctic, the Kola cities show us that change and dynamics in Arctic cities do not necessarily, or predominantly, result in radical *spatial* reconfiguration. Even though new traces reflecting change in lifestyles, culture and economy are evident, the urban structure of the Kola cities looks very similar to how they did in the early 1990s. While the social and economic development is dynamic, the urban landscape with its 'overdeveloped' modernist urban footprint remains permanent and is set to determine development to come. The basic connection between the cities, their foundational industries and the minerals in the ground is also a stable configuration that will endure.

In contrast to the continuity of physical structures, the way the cities are interpreted – their value and meaning to inhabitants and visitors – has changed. The cultural image attached to urban form is much less permanent than the buildings and boulevards themselves. The cities retain their planned monumentality but no longer reference the historic underlying ideology. Kola's urban life has adapted to a globalised market economy; the economic future of the peninsula is uncertain, leaving the buildings to dilapidate – bereft of investment. The post-socialist urban landscape is in what geographer Mariusz Czepczynski argues is a 'liminal state . . . characterised by ambiguity, openness and indeterminacy', in which 'the old landscape is re-interpreted and de-contextualised, while the new landscape is constructed, both physically and mentally'.[50]

The question of how the Kola cities can be imagined as places to live in the future prompts us, echoing archaeologists Rodney Harrison and John Schofield, to 'investigate the present landscape in relation to recent processes or change, and potential future change ("futurologies")'.[51] Imagining futures is seen as critical in many Arctic communities, but despite the deterministic tones of the current discourse on Arctic futures, there really is no single imaginary that will provide answers for what is to come in quite the way that the utopian 'workers' paradise' dictated the modernist design of the Soviet Kola cities.[52] As they slowly transform from mono-industrial towns to more diverse economies, the evolving civic society and the diverse forces of capitalist urban development will be unlikely to produce a single vision of the future. Rather, it will be diverse and contradictory, reflecting a variety of social actors and interests; the question is whether the imaginaries of researchers, policy makers and populations in themselves are enough to make changes occur, and in what ways this can happen.

Notes

1 Peter Arbo et al., "Arctic Futures: Conceptualizations and Images of a Changing Arctic," *Polar Geography* 36, no. 3 (2013).

2 An example is Svalbard, where research and tourism are projected to replace coal mining as the dominant industry.

3 See Philip Steinberg et al., *Contesting the Arctic: Politics and Imaginaries in the Circumpolar North* (London/New York: I.B. Tauris, 2015).

4 Rasmus O. Rasmussen, ed., *Megatrends* (Copenhagen: Nordic Council of Ministers, 2011).

5 Susanne Dybbroe, "Is the Arctic Really Urbanising?" *Études/Inuit/Studies* 32 (2008).

6 Mapping methods included physical sensory walks (see Sara Pink, *Doing Visual Ethnography: Images, Media, and Representation in Research* [London: Sage Publications, 2007]) and variants of systematic transect walks using

GPS-enabled social media mapping technology; see Peter Hemmersam et al., "Exploring Locative Media for Cultural Mapping," in *Mobility and Locative Media: Mobile Communication in Hybrid Spaces*, ed. Mimi Sheller and Adriana de Souza e Silva (London: Routledge, 2015).

7 In 2010. Source: Rosstat.

8 Anna Stammler-Gossmann, "Reshaping the North of Russia: Towards a Conception of Space," *Arctic & Antarctic International Journal of Circumpolar Sociocultural Issues* 1 (2007).

9 Andy R. Bruno, "Making Nature Modern: Economic Transformation and the Environment in the Soviet North," PhD diss., University of Illinois at Urbana-Champaign, 2011; Thimothy Heleniak, "Growth Poles and Ghost Towns in the Russian Far North," in *Russia and the North*, ed. Elana W. Rowe (Ottawa: University of Ottowa Press, 2009).

10 Graybill and Dixon, "Cities of Russia."

11 Terence Armstrong, *Russian Settlement in the North* (Cambridge: Cambridge University Press, 1965); Andy R. Bruno, "Making Nature Modern."

12 Jessica K. Graybill and Megan Dixon, "Cities of Russia," in *Cities of the World: World Regional Urban Development*, ed. Stanley D. Brunn et al. (Lanham: Rowman & Littlefield Publishers, 2012); Heleniak, "Growth Poles and Ghost Towns."

13 Konstantin Axenov, Isolde Brade, and Evgenij Bondarchuk, *The Transformation of Urban Space in Post-Soviet Russia* (London: Routledge, 2006).

14 For more on urban historical layered analysis, see e.g. Aldo Rossi, *The Architecture of the City*, trans. Diane Ghirardo and Joan Ockman (Cambridge: MIT Press, 1982).

15 William G. Hoskins, *The Making of the English Landscape* (London: Hodder and Stoughton, 1955).

16 John B. Jackson and Helen L. Horowitz, *Landscape in Sight: Looking at America* (New Haven: Yale University Press, 1997).

17 Denis Cosgrove, "Liminal Geometry and Elemental Landscape: Construction and Representation," in *Recovering Landscape: Essays in Contemporary Landscape Architecture*, ed. James Corner (New York: Princeton Architectural Press), p. 104.

18 Dolores Hayden and Jim Wark, *A Field Guide to Sprawl* (New York/London: W.W. Norton, 2004); Sefryn Penrose, ed., *Images of Change: An Archaeology of England's Contemporary Landscape* (Swindon: English Heritage, 2007).

19 The iPhone application MAPPA was developed by team members in another project and was applied in two sessions in Murmansk with local informants and students of landscape architecture from the Tromsø Academy of Landscape and Territorial Studies.

20 Sefryn Penrose, *Images of Change: An Archaeology of England's Contemporary Landscape* (Swindon: English Heritage, 2008).

21 Developed by English Heritage.

22 Harrison, Rodney, and John Schofield, *After Modernity: Archaeological Approaches to the Contemporary Past* (Oxford: Oxford University Press, 2010).

23 Harrison, *After Modernity*, 225.

24 Geir Hønneland, "Power Institutions and International Collaboration on the Kola Peninsula," *The Journal of Power Institutions in Post-Soviet Societies* 4/5 (2006), accessed January 13, 2015, http://pipss.revues.org/456; Geir Hønneland and Anne-Kristin Jørgensen, "Closed Cities on the Kola Peninsula: From Autonomy to Integration?" *Polar Geography* 22, no. 4 (1998).

25 Benjamin Forest and Juliet Johnson, "Unraveling the Threads of History: Soviet – Era Monuments and Post – Soviet National Identity in Moscow," *Annals of the Association of American Geographers* 92, no. 3 (2002): 524–47, accessed April 25, 2016, doi:10.1111/1467–8306.00303; Julia Gerlach and Nadir Kinossian, "Cultural Landscape of the Arctic: 'recycling' of Soviet Imagery in the Russian Settlement of Barentsburg, Svalbard (Norway)," *Polar Geography* 39, no. 1 (2016): 1–19, accessed February 29, 2016, doi:10.1080/1088937X.2016.1151959.

26 Taut, Bruno, Paul Scheerbart, Erich Baron, and Adolf Behne, *Die stadtkrone* (Jena: Eugen Diederichs Verlag, 1919).

27 Marshall Berman, *All That Is Solid Melts into Air: The Experience of Modernity* (New York: Simon and Schuster, 1982).

28 John B. Jackson, *The Necessity for Ruins, and Other Topics* (Amherst: University of Massachusetts Press, 1980).

29 Heleniak, "Growth Poles."

30 Bruno, *Making Nature Modern*.

31 Jackson, *The Necessity for Ruins*, 94.

32 Ibid.

33 Ibid, 101.

34 Ibid.

35 Bjørnar Olsen and Þóra Pétursdóttir, preface to *Ruin Memories: Materialities, Aesthetics and the Archaeology of the Recent Past* (London: Routledge Chapman Hall), 1.

36 Blakkisrud, Helge and Geir Hønneland, "The Russian North – An Introduction," in *Tackling Space: Federal Politics and the Russian North*, ed. Helge Blakkisrud and Geir Hønneland (Lanham: University Press of America, 2006), 2.

37 Rasmussen, Rasmus O., "Megatrends in Arctic Development," paper at the First International Conference on Urbanisation in the Arctic, Nuuk, August 28–30, 2012.

38 Bruno, *Making Nature Modern.*

39 Graybill and Dixon, "Cities of Russia."

40 Bruno, *Making Nature Modern.*

41 Heleniak, "Growth Poles."

42 Peter Hemmersam, "Arctic Architectures," *Polar Record* FirstView (2016), accessed April 04, 2016, doi:10.1017/S003224741500100X. See also interview with Matthew Jull and Leena Cho in Samuel Medina, "The Cold Rush," *Metropolis Magazine*, November, 2014, accessed September 18, 2015, www.metropolismag.com/November-2014/The-Cold-Rush

43 Kirsten Birk, "Arctic architecture?" *Arkitektur DK* 4 (2012); Mats Egelius, *Ralph Erskine: The Humanist Architect*, A.D. Profiles Vol. 11–12 (London: Architectural Design, 1977); Ralph Erskine, "Building in the Arctic," *Architectural Design* 5 (1960); Ralph Erskine, "The Sub-Arctic Habitat," in *CIAM '59 in Otterlo: Arbeitsgruppe für die Gestaltung Soziologischer und Visueller Zusammenhänge*, ed. Oscar Newman (Stuttgart: Karl Krämer Verlag, 1961): 160–68; Ralph Erskine, "Architecture and Town Planning in the North," *Polar Record* 14, no. 89 (1968); Jérémie M. McGowan, "Ralph Erskine, (Skiing) Architect," *Nordlit* 23 (2008). Rhodri W. Liscombe, "Modernist Ultimate Thule," *Revue d'art Canadienne-Canadian Art Review* 31 (2006); Alan Marcus, "Place with No Dawn. A Town's Evolution and Erskine's Artic Utopia," in *Architecture and the Canadian Fabric*, ed. Rhodri W. Liscombe (Vancouver: University of British Columbia Press, 2011).

44 Louis-Edmond Hamelin, *Discours du Nord* (Québec City: Université Laval, Groupe d'Etudes Inuit et Circumpolaires, 2002).

45 Daniel Chartier, "Towards a Grammar of the Idea of North: Nordicity, Winterity," *Nordlit* 22 (2007); Louis-Edmond Hamelin. *Canadian Nordicity: It's Your North Too* (Montreal: Harvest House, 1979).

46 Helge Blakkisrud and Geir Hønneland. "The Russian North – An Introduction," in *Tackling Space: Federal Politics and the Russian North*, ed. Helge Blakkisrud and Geir Hønneland (Lanham: University Press of America, 2006); E. Carina H. Keskitalo, "'The North' – Is There Such a Thing?," in *Cold Matters: Cultural Perceptions of Snow, Ice and Cold*, ed. Heidi Hansson and Cathrine Norberg (Umeå: Umeå University and the Royal Skyttean Society, 2009); Anna Stammler-Gossmann, "Reshaping the North of Russia: Towards a Conception of Space," *Arctic & Antarctic International Journal of Circumpolar Sociocultural Issues* 1 (2007).

47 A.V. Makhrovskaya, M.Y. Vaytens, L. K. Panov, and A.Y. Belinskiy, "Urban Planning and Construction in the Kola North," *Polar Geography* 3 (1977), *Polar Geography* 4 (1977), *Polar Geography* 5 (1977) (three parts).

48 Richard Florida, *The Rise of the Creative Class: And How It's Transforming Work, Leisure, Community and Everyday Life* (New York: Basic Books, 2002); Charles Landry, *The Creative City* (London: Earthscan Publications, 2002).

49 See e.g. E. Carina H. Keskitalo, "'The North' – Is There Such a Thing?," in *Cold Matters: Cultural Perceptions of Snow, Ice and Cold*, ed. Heidi Hansson and Cathrine Norberg (Umeå: Umeå University and the Royal Skyttean Society, 2009); McGowan, "Ralph Erskine, (Skiing) Architect."

50 Mariusz Czepczynski, *Cultural Landscapes of Post-Socialist Cities* (Aldgate: Ashgate, 2008), 112.

51 Rodney Harrison and John Schofield, *After Modernity: Archaeological Approaches to the Contemporary Past* (Oxford: Oxford University Press, 2010), 225.

52 Arbo et al., "Arctic Futures."

Bibliography

Arbo, Peter, Audun Iversen, Maaike Knol, Toril Ringholm, and Gunnar Sander. "Arctic Futures: Conceptualisations and Images of a Changing Arctic." *Polar Geography* 36, no. 3 (2012): 163–82.

Armstrong, Terence. *Russian Settlement in the North*. Cambridge: Cambridge University Press, 1965.

Axenov, Konstantin, Isolde Brade, and Evgenij Bondarchuk. *The Transformation of Urban Space in Post-Soviet Russia*. London: Routledge, 2006.

Berman, Marshall. *All that Is Solid Melts into Air: The Experience of Modernity*. New York: Simon and Schuster, 1982.

Birk, Kirsten. "Arctic Architecture?" *Arkitektur DK* 4 (2012): 48–51.

Blakkisrud, Helge, and Geir Hønneland. "The Russian North – An Introduction." In *Tackling Space: Federal Politics and the Russian North*, edited by Helge Blakkisrud and Geir Hønneland, 1–24. Lanham: University Press of America, 2006.

Blakkisrud, Helge, and Geir Hønneland, eds. *Tackling Space: Federal Politics and the Russian North*. Lanham: University Press of America, 2006.

Bruno, Andy R. "Making Nature Modern: Economic Transformation and the Environment in the Soviet North." PhD diss., University of Illinois at Urbana-Champaign, 2011.

Chartier, Daniel. "Towards a Grammar of the Idea of North: Nordicity, Winterity." *Nordlit* 22 (2007): 35–47.

Dybbroe, Susanne. "Is the Arctic Really Urbanising?" *Études/Inuit/Studies* 32 (2008): 13–32.

Egelius, Mats. *Ralph Erskine: The Humanist Architect. Vol. 11–12 of AD Profiles*. London: Architectural Design, 1977.

Erskine, Ralph. "Building in the Arctic." *Architectural Design* 5 (1960): 194–97.

Erskine, Ralph. "The Sub-Arctic Habitat." In *CIAM '59 in Otterlo: Arbeitsgruppe für die Gestaltung Soziologischer und Visueller Zusammenhänge*, edited by Oscar Newman, 160–68. Stuttgart: Karl Krämer Verlag, 1961.

Erskine, Ralph. "Architecture and Town Planning in the North." *Polar Record* 14, no. 89 (1968): 165–71.

Florida, Richard. *The Rise of the Creative Class and How It's Transforming Work, leisure, Community and Everyday Life*. New York: Basic Books, 2002.

Forest, Benjamin, and Juliet Johnson. "Unraveling the Threads of History: Soviet – Era Monuments and Post – Soviet National Identity in Moscow." *Annals of the Association of American Geographers* 92, no. 3 (2002): 524–47. Accessed April 25, 2016. doi:10.1111/1467–8306.00303

Gerlach, Julia, and Nadir Kinossian. "Cultural Landscape of the Arctic: 'recycling' of Soviet Imagery in the Russian Settlement of Barentsburg, Svalbard (Norway)." *Polar Geography* 39, no. 1 (2016): 1–19. Accessed February 29, 2016. doi:10.1080/1088937X.2016.1151959.

Graybill, Jessica K., and Megan Dixon. "Cities of Russia." In *Cities of the World: World Regional Urban Development*, edited by Stanley D. Brunn, Maureen Hays-Mitchell, and Donald J. Zeigler, 237–79. Lanham: Rowman & Littlefield Publishers, 2012.

Hamelin, Louis-Edmond. *Canadian Nordicity: It's Your North Too*. Montreal: Harvest House, 1979.

Hamelin, Louis-Edmond. *Discours du Nord*. Québec City: Université Laval, Groupe d'Etudes Inuit et Circumpolaires, 2002.

Harrison, Rodney, and John Schofield. *After Modernity: Archaeological Approaches to the Contemporary Past*. Oxford: Oxford University Press, 2010.

Hayden, Dolores, and Jim Wark. *A Field Guide to Sprawl*. New York/London: W. W. Norton, 2004.

Heleniak, Timothy. "Growth Poles and Ghost Towns in the Russian Far North." In *Russia and the North*, edited by Elana W. Rowe, 129–63. Ottawa: University of Ottawa Press, 2009.

Hemmersam, Peter. "Arctic Architectures." *Polar Record* 52, no. 4 (2016). Accessed April 04, 2016. doi: 10.1017/ S003224741500100X.

Hemmersam, Peter, Jonny Aspen, Andrew Morrison, Idunn Sem, and Martin Havnør. "Exploring Locative Media for Cultural Mapping." In *Mobility and Locative Media: Mobile Communication in Hybrid Spaces*, edited by Mimi Sheller and Adriana de Souza e Silva, 167–187. London: Routledge, 2015.

Hoskins, William G. *The Making of the English Landscape*. London: Hodder and Stoughton, 1955.

Hønneland, Geir. "Power Institutions and International Collaboration on the Kola Peninsula." *The Journal of Power Institutions in Post-Soviet Societies* 4/5 (2006). Accessed January 13, 2015. http://pipss.revues.org/456.

Hønneland, Geir, and Anne-Kristin Jørgensen. "Closed Cities on the Kola Peninsula: From Autonomy to Integration?" *Polar Geography* 22, no. 4 (1998): 231–48.

Jackson, John B. *The Necessity for Ruins, and Other Topics*. Amherst: University of Massachusetts Press, 1980.

Jackson, John B., and Helen L. Horowitz. *Landscape in Sight: Looking at America*. New Haven: Yale University Press, 1997.

Keskitalo, E. Carina H. "'The North' – Is There Such a Thing?" In *Cold Matters: Cultural Perceptions of Snow, Ice and Cold*, edited by Heidi Hansson and Cathrine Norberg, 23–39. Umeå: Umeå University and the Royal Skyttean Society, 2009.

Landry, Charles. *The Creative City*. London: Earthscan Publications, 2002.

Liscombe, Rhodri W. "Modernist Ultimate Thule." *Revue d'art Canadienne-Canadian Art Review* 31 (2006): 64–80.

Makhrovskaya, A. V., M. Y. Vaytens, L. K. Panov, and A. Y. Belinskiy. "Urban Planning and Construction in the Kola North." *Polar Geography* 3 (1977): 205–216, *Polar Geography* 4 (1977): 286–306, *Polar Geography* 5 (1977): 42–52 (three parts).

McGowan, Jérémie M. "Ralph Erskine, (Skiing) Architect." *Nordlit* 23 (2008): 241–50.

Marcus, Alan. "Place with No Dawn. A Town's Evolution and Erskine's Artic Utopia." In *Architecture and the Canadian Fabric*, edited by Rhodri W. Liscombe. Vancouver: University of British Columbia Press, 2011.

Medina, Samuel. "The Cold Rush." *Metropolis Magazine*, November, 2014. Accessed September 18, 2015. www.metropolismag.com/November-2014/The-Cold-Rush.

Olsen, Bjørnar, and Þóra Pétursdóttir. Preface to *Ruin Memories: Materialities, Aesthetics and the Archaeology of the Recent Past*, 1. London: Routledge, 2014

Penrose, Sefryn, ed. *Images of Change: An Archaeology of England's Contemporary Landscape*. Swindon: English Heritage, 2007.

Pink, Sarah. *Doing Visual Ethnography: Images, Media, and Representation in Research*. London: Sage Publications, 2007.

Rasmussen, Rasmus O., ed. *Megatrends*. Copenhagen: Nordic Council of Ministers, 2011.

Rasmussen, Rasmus O. "Megatrends in Arctic Development." Paper presented at the First International Conference on Urbanisation in the Arctic, Nuuk August 28–30, 2012.

Rossi, Aldo. *The Architecture of the City*. Translated by Diane Ghirardo and Joan Ockman. Cambridge: MIT Press, 1982.

Stammler-Gossmann, Anna. "Reshaping the North of Russia: Towards a Conception of Space." *Arctic & Antarctic International Journal of Circumpolar Sociocultural Issues* 1 (2007): 53–97.

Steinberg, Philip E, Jeremy Tasch, Hannes Gerhardt, Adam Keul, and Elizabeth Nyman. *Contesting the Arctic: Politics and Imaginaries in the Circumpolar North*. London/New York: I.B. Taurus, 2015.

Taut, Bruno, Paul Scheerbart, Erich Baron, and Adolf Behne. *Die Stadtkrone*. Jena: Eugen Diederichs Verlag, 1919.

5 Hyperlandscape

The Norwegian-Russian borderlands

Morgan Ip

In the autumn of 2015 rows of colourful, newly purchased, yet abandoned bicycles could be found leaning against each other at the Arctic border station at Storskog, the only land entry point between Russia and Norway. Some had training wheels. Hurriedly discarded, this diverse mass of metal reveals elements of the chaos of what became the first mass migration of people in human history facilitated by social media.[1] The ditched bicycles serve as a shocking injection to the usual Arctic discourses – a vast political and scholastic landscape considering issues from melting ice, opening ocean routes, and indigenous people's rights to potential natural resource wealth. The speculation on Arctic futures was jolted by this sudden arrival of people seeking asylum by crossing the border from Russia into Norway via the Arctic Route – a small part of the erratic mass migrations of people in the European south.[2] The new arrivals inscribed upon this corner of the North a new and expanded reading of the landscape.

The Future North project is founded with the notion that landscapes are a shared material human experience operating with political, cultural and social development.[3] The Norwegian-Russian borderlands can be read as a multiplicity of such cultural landscapes cutting across vast ranges of material and temporal scales. Its exceptional border conditions provide a contrast between different political and cultural models to a degree unseen elsewhere in the Arctic. This is highlighted in the migrant crisis, and yields the potential of an expanded approach to understanding northern landscapes and how citizens participate in the co-creation of these. Migrants are included as citizens of the landscape, not necessarily by the legal definition within the context of the territorial states that organise them, but by simple virtue of their presence and the resulting influences they have upon social and material fabrics that compose the landscape. They have a right to participate in the co-creation of the landscape according to the supranational European Landscape Convention, as members of the public – a broad definition that comprises all users of space including transient visitors or potential long-term citizens as some of these new arrivals may be.[4]

The migration issue is in rapid flux, with governments responding reactively. Future changes are difficult to anticipate given that the causes of any situation are an intermingled instantiation of several factors. Climate change, for example, is but one variable implicated in the Syrian conflict as a long drought exacerbated by anthropogenic forces occurred in a particular political and economic climate.[5] In the Arctic, climate change is a staggering force with effects occurring at twice the rate as in other areas of the globe, such as higher changes in average temperatures and accompanying transformations on flora, fauna, industry, transport and livelihood of locals.[6]

Coupled with the fast-changing dynamics of social media, the Arctic migration route has directly impacted the communities straddling this Arctic border as people scrambled to adapt to the urgent human crises. This underscores how the Arctic is often described as the canary in the coalmine for the global condition in terms of climate change, globalisation, technological evolution and a host of

Figure 5.1 Bicycles abandoned at the Storskog border station. The blurriness reflects the author's hesitation and uncertainty of capturing these traces of migration within the official borderzone.

Photo: the author.

parallel socio-cultural shifts. The borderlands of Kirkenes on the Norwegian side and Nikel on the Russian side appear to be fertile ground for mapping the complexities of northern cultural landscapes, in particular the relationship between radical global change and local responses. Here I will highlight a few of the landscapes nested within the hyperlandscape that is the Norwegian-Russian borderland that show this local-global relationship.

Liminal cultural landscapes: Kirkenes and Nikel

Geographer Michael Jones, in his review of the concept 'cultural landscape', noted the contentious use of the term as many disciplines and users adopt different definitions to suit their own interests. In addition, the relationship between human cultures and landscapes is such that some academics believe that appending 'cultural' to the term landscapes is redundant.[7] Carl Sauer notably introduced the term from German into English in 1925, stating that the natural landscape is the medium through which human culture is the transformative agent in the resulting creation of cultural landscapes.[8] The critique of this notion, as Jones discovers, is that it is too broad for some, and too limiting to others. Canadian archaeologist Ellen Lee emphasises that even within cultures there are subcultures and individual value systems that inform landscapes:

> Not only do values placed on landscapes differ between local, indigenous people and the dominant culture, but they also vary between the male and female perspectives and among the different generations within a community. While there are group understandings of landscapes, each person also carries an individual and unique cultural landscape in his or her head and heart, based on personal experience. Thus it is important to be aware of the layering of values and the potential for conflicting values, even between members of the same cultural group.[9]

Geographer Donald W. Meinig underscores this cognitive component and unique individual approach to any given considered landscape in his essay *The Beholding Eye: Ten Versions of the Same Scene*.[10] He shows that a group of individuals, when taken to a similar viewpoint, would not be looking at the same landscape. The material elements such as buildings, mountains, rivers and trees all take on meaning through individual association, and a plurality of meanings may occur in the minds of the observers. For example, landscapes could be seen for their aesthetic components, ideological and symbolic meanings, spatial arrangements, commoditised as sources of wealth and so on.[11] Cultural geographer John Wylie says that these multitude forces of landscape are in constant tension and exerting upon each other. These tensions are 'between proximity and distance, body and mind, sensuous immersion and detached observation' for example.[12] Landscapes exist as each of these concepts, and in-between them.

Belgian architect and historian Tom Avermaete aligns this concept of in-between-ness with the term 'liminality' first used by English anthropologist Victor Turner in the 1960s. Turner gave the term 'liminality' and 'limens' to the in-between stage of the teenager – a state of being that is not of a child, not of an adult, and at once comports elements of both.[13] Avermaete applies liminality to the spatial environment, specifically urban and peri-urban landscapes:

> Etymologically, liminality denotes something found 'in between' things. Referring to notions of threshold and interval, it also means a space of time. Thus the liminal embraces not only such notions as openness, porosity, breach and relationship, but also those of process and transformation.[14]

Liminal landscapes pertain to the specific meeting place of various material conditions inherently imbued with cultural meaning and as processes over time. This includes both natural and human directed processes that are engaged in the constant change occurring in the landscape.[15] The geopolitical borders, demarcations between nations, political structures and culture that straddle these areas are physically and cognitively liminal.

Even though the Norwegian-Russian border seems 'hard' it actually approximates these complex heterogeneous interpretations of landscape against major hegemonic paradigms (East vs. West) reflected in material, linguistic and cultural meetings. The border is a meeting of contemporary cultural landscapes – their extensions into the other form a liminal landscape. Within this, the bicycles provide an example of how legislative differences on either side affect the movement of people in these limens.

Transient landscapes

Russian law forbids border crossings by foot, and Norwegian legislation forbids drivers from bringing over people without documentation. A surging business in bicycles has resulted in Nikel and Murmansk, in proportion to the increasing numbers of asylum seekers choosing this unlikeliest of routes towards a perceived better life in Norway.[16] However, the bikes cannot be reused in Norway because they do not abide to its national standards as they have only one brake, and they cannot be sent back to Russia for other migrants without encroaching upon import/export issues. The bikes, like the migrants themselves, are caught in a liminal space.

The good diplomatic relations between the current Syrian government and Russia, and the location of Norway as a neighbour to Russia, has provided a route to the EU/Schengen countries with open borders, and therefore fewer obstacles, as opposed to the rapidly shifting routes in Southern Europe. As more people became aware of this route through media and word-of-mouth, more arrived.[17] The shift in routes were reactionary and occurred instantaneously, as technologically equipped migrants and smugglers communicated with each other on social media such as Facebook and a Russian facsimile, VKontakte.

The various routes were traced with the material remnants of the passage; the discarded masses of bicycles were paralleled in the scattered orange life jackets and other detritus strewn along the beaches of Greece. Images of this changing social and material landscape streamed through television, laptop and smartphone screens across the globe. The lived space throughout Kirkenes, the district of Sør-Varanger and Nikel changed too. It started with the filling of the hotels: in Russia for

Figure 5.2 The human face of migration is profoundly visceral. Privileged and with a visa in hand, I shared the same frozen road as the migrants, and passed families with children.

Photo: the author.

those awaiting transport across the border, and in Norway to receive them.[18] This resulted in the paradoxical mix of tourists seeking experiences in the periphery for pleasure and those seeking asylum. The streets of Kirkenes became busier as increasing numbers of police and social workers arrived to process claims and transport people south. The migration flow is but a most recent meeting of different peoples in this border region. Though small in number, these new arrivals are emblematic of the fact that the North, previously seemingly distant, is intimately connected to global processes and systems.

Historical border landscape

Plural cultural and political histories are revealed directly in the material evidence on either side of the border, with towns consisting of detached single or double-storey wood frame houses on one side and Soviet blocks on another. The proximity of Kirkenes and Nikel in relationship to the border has implications on the socio-cultural fabric in the region. As cultural landscapes, cultures and their built spaces are interlinked. One can see elements of Russian culture in the street signage in Kirkenes, the monthly Russian women's market and the Russian fishing ships dotting the harbour. Historical photographs of Kirkenes depict Russian-influenced cupolas on the main commercial street before its razing in the Second World War. In Nikel, visual elements of the other are not as noticeable in the physical composition of the town, with the exception of the dilapidated original housing built by the Finns and Canadians for the smelter workers before the war when sovereignty of the territory shifted from Finland to the Soviet Union. However, the shopkeepers and librarians are sometimes more inclined to speak Norwegian than English to foreigners here, a non-visual but experiential sign of the relationship between the two communities.

Urban theorist Michael Dear refers to this intermingling border society as a 'third nation': a group of transnational actors that belong both to their primary nation and that of the border space.[19] In any case, concepts of the other are diverse in both Norwegian and Russian populations, with some favourable and some not as much. And for some, the differences across the border are not as great as they are within their nation, as a young Nikel informant explains:

> I think that the people in the North don't differ from each other so much regardless of the country – in Russia, Canada or elsewhere. They differ more from the people who live in the cities – southern, central, eastern or western people. Northern people are tougher and weather beaten . . . nature has an impact on the people and their character. Harsh weather is different from other places in the world and therefor so are the people.
>
> (Youth informant, Nikel)[20]

This youth expresses a regional 'imagined community', to use Benedict Anderson's concept of nationhood, of hardiness resulting from living in the Arctic regardless of the political and cultural borders of each nation-state and their corresponding walls and other physical infrastructures.[21]

Political Scientist Wendy Brown cogently argues that '[W]alls are consummately functional, and walls are potent organizers of human psychic landscapes generative of cultural and political identities'.[22] To those on either side, the same object, such as a wall, may infer power relationships that buttress mental differentiation of 'the other':

> It is too simple, for example, to say that the Israeli wall connotes protection and security to one side and aggression, violation, and domination to the other. While the Wall may comport with an entitlement to safety in a Jewish homeland felt by some Israeli Jews, it carries for others the shame and violence of the occupation.[23]

Thus objects like walls and other physical infrastructure can represent opposing tensions of the border community – security or fear, freedom or oppression – and construct varied perceptions of self and of others. The exceptional conditions of the border thus expand the cultural landscapes by including the physical and particular symbolic objects as they are represented and organised in our minds. This places each observer in a diverse collection of perspectives on the other. This is the hyperlandscape made of a multitude of individual perspectives in tandem with the multitude and constantly changing physical and digital layers. The Norwegian-Russian borderlands thus become a laboratory to investigate potential new intersections and divergences that may enable colliding forces to transmute and create novel cultural landscapes. This is not to say that such dynamism is new, for in fact this particular border area has undergone several drastic periods of change.

National demarcation took place relatively recently, in 1826, when a border was agreed upon between Norway and Russia, although the context was ever shifting, with Petsamo on the Russian side temporarily Finnish territory between 1920 and 1944.[24] The border physically divided an already heterogeneous mix of various Sámi, Kven, Norwegian, Finnish and Russian peoples. The discovery of iron and nickel ore in this border area and the Gulf Stream that opens up the Barents Sea to shipping throughout winter made this a hot spot in the Second World War. Occupying German soldiers followed scorched earth policies and burned everything they could in Northern Norway, save for a few unscathed pockets such as Bugøynes, as they fled the Soviet advance in 1944.[25] They ravaged civilian infrastructure, and created an internal refugee crisis. They created a *dark landscape* populated with *dark infrastructure*.

Dark landscapes

I use the terms 'dark landscape' and 'dark infrastructure' borrowing from two different concepts. First, from 'dark tourism' which are tourist attractions relating to 'dark' aspects of humanity, such as war memorials, cenotaphs and burial places, as well as noteworthy sites or commemorations of violence, sadness, conflict and other negative aspects of lived experience.[26] I embed this with the notion of 'dark ecology' by English philosopher Timothy Morton, in which an object-oriented ontology prescribes a look at landscapes with equal focus on non-human material aggregates.[27] He references

Figure 5.3 Iron mine in Bjørnevatn outside of Kirkenes, Norway
Photo: the author.

Figure 5.4 Nickel smelter in Nikel, Russia

Photo: the author.

Figure 5.5 Steilneset monument to the Finnmark witch trials in Vardø, Norway

Photo: the author.

ecology in an all-encompassing manner beyond nature, environment or economy, but all humanly imagined ways of existing together.[28] Even that which is perceived negatively or darkly is equally important as human perceptions and actions in considering landscape futures. Dark landscapes in this sense incorporate dark material and temporal aspects of humanity.

Figure 5.6 House built atop a Second World War bunker, Kirkenes
Photo: the author.

In the case of these borderlands, there exists a formidable number of dark infrastructures – the iron and nickel mines, the derelict concrete fortifications, the monuments to war and witch trials, and so on. In fact, the city of Kirkenes rests above a collection of tunnels, bunkers and fortified basements, the hills hollowed with massive reservoirs and bunkers from the Second World War.

The industrial layers of these dark infrastructures could adequately be explained as part of the rationale behind the occupation of Norway in the first place. Access to the iron ore in Kirkenes and the nickel in then Finnish-controlled Nikel was crucial to feeding the war machines. The Soviets relied on the ice-free port of Murmansk for a vital conduit for supplies via Allied convoys. The two communities on either side of the border were built up because of the important mineral resources, and reflecting changing global influences. The 2015 closing of the Kirkenes mine for the third time in its history presents a renewed challenge to the community, as does the volatile production out of the Nikel metallurgical plant and of the nearby Zapolyarny mine on the Russian side of the border.[29]

Dark infrastructure is a manifestation of the histories of a place. As space-time shifts, so too can the pertaining layers of darkness. For example, Fjellhallen ('The Mountain Hall') in Kirkenes was created in the Second World War as a subterranean shelter. After the war it was reconstituted as a community hall, hosting sporting and cultural events, and became part of the quotidian life of Kirkenes. It became a place of badminton practice, weight training and concerts. It was used as a temporary

Figure 5.7 The Norwegian-Russian border and respective marking posts (Norway – yellow and black, Russia – red and green)

Photo: the author.

resettlement centre during the Kosovo crises in 1999, when the government resettled refugees from the Balkan war.[30] In the autumn of 2015 it again became a place providing temporary shelter for refugees, suiting of its original purpose.

Elements of the dark landscape can also be that which is missing. For instance, scorched earth tactics eradicated the previous wooden structures of Kirkenes during the Second World War, replaced in haste with post-war modern design. The replacement architecture physically reflects contemporaneous style built on the foundations of the previous town. Further yet, the social and cultural effects of this erasure are part of a dark reading in the local collective memory:

> In many ways the war was not only about the region being burnt down by the Germans. It was worse that we lost so much competence, because very many people moved to the South and didn't come back. And key persons also lost their lives. The economic impact was tremendous. All the destruction changed several places quite fundamentally . . . it changed much of the mentality . . . I think the long-term effects are much more devastating than the short-term effects.
>
> (Kirkenes Informant)[31]

The sepia photographs adorning the walls of the bank, the mall, the library and in the scrolling pages of Facebook groups dedicated to the history of the town capture this previous city. It is remembered in the imagination as a place of nostalgic beauty, of what used to be, and what, perhaps, could be. Likewise, we can see images of the past on the Russian side, for example in the libraries, monuments and a roadside museum en route to Murmansk.

An extraterritorial neighbourhood

The border fence between Russia and Norway does not loom as ominously in the landscape as do the high concrete barriers Brown describes in the Levant, and in fact the barbed wire fences, in places doubled, are mostly unseen in the everyday life for those in Kirkenes. They are hidden away from populated centres in the woods and invisible unless directly approached at the official crossing point of Storskog, or for tourists on the Barents Safari, crunching through king crabs metres away from the thin strip of cleared border. Nothing prevents an illegal crossing at this point, save knowing that there is invisible electronic surveillance from the unmanned watchtowers looming sporadically in the distance. Cameras dispatch border guards to those who transgress this border.

The wire fences are not on the border itself, but run deeper in the woods on the Russian side. Only on the Russian side along the road to Nikel do the series of barbed wire fences, cameras and occasional military patrols provide visual evidence of the national edge. These physical border elements demonstrate Brown's 'human psychic landscapes' in that there are unseen but very real social constructions responding to this infrastructure.[32] These physical barriers are mere accoutrements to the political obstacles at this crossing, and pertain to the national strategies. Like the walls sprung up across Europe only to be transgressed in myriad other ways, they are temporary or porous obstacles that serve mainly to assuage the mental or imagined perspectives of fortified borders.

It requires some effort to obtain a visa to Russia and vice-versa unless you live within a certain radius from the border. Those who have lived in the region for three years can apply for a border zone visa which allows travel for local residents in close proximity on either side. Even then, not everyone on either side of the border takes up this opportunity. There are some who have spent their entire lives in Kirkenes, or Nikel, and have not crossed the border.[33] There are then those who sporadically visit the opposite border city, but this requires considerable foresight and planning for non-national local residents, and they face obstacles if there is a spontaneous desire to cross as visa requirements can be onerous and time consuming. In spite of this, Kirkenes is home to diplomatic postings and research centres based on Norwegian proximity to Russia.[34] These are namely institutions that aim to understand and increase people-to-people collaboration and neighbourliness in the Barents Region, a region of political cooperation created by the signing of the Kirkenes Declaration in 1993 by the four nations in the vicinity of the Barents Sea in the Arctic Ocean.[35] Further, people cross borders to shop, and participate in international sporting and cultural events. In Murmansk, a 3-hour drive east of Kirkenes, and the capital of Murmansk Oblast which includes Nikel and the Pechenga District, many consider Kirkenes to be part of their everyday landscape, and not particularly foreign.[36] It has become, in a way and to some, an expanded extraterritorial neighbourhood.

Virtual landscape

Aside from the fences along the land border, much of the demarcation actually runs along the swift currents of the Pasvik River where dams generate energy for the region in a collaborative effort between the two nations. In my research, I have aimed to harness the current voices of the people that inhabit the shared border region to record the human component of the cultural landscape. Typical sociological investigations include interviews of select and limited numbers of people to

Figure 5.8 Finnmark Police tweets on November 3rd–4th, 2015. Author's translations: 'Storskog: 173 asylum seekers came over Storskog border station today', 'Storskog: 196 asylum seekers have come over Storskog border station today.'

Source: Tweets by @OPSostfinnmark

bring out full understandings of a local community. However, technological advances, particularly pertaining to social media, have vastly transformed our societies into increasingly and more tightly connected communities with multiple platforms to share knowledge, ideas, creativity and more. With the abundance of digital media associated with place, our world is increasingly tethered to the ethereal, and new ways of capturing more local voices emerge.[37]

In the case of the asylum seekers in 2015, social media was integral to locating new paths as quickly as governments introduced barriers. As social media has enabled and supported migration routes for people fleeing their war-torn lands, it likewise provided an important communication and networking tool in the receptive Arctic borderlands. This is a global network with local responses amongst urgently transient populations, but those who live in places on a more permanent basis are just as adept at using the same tools. An example of this new role of digital media is the way the police monitored and broadcast the numbers of asylum seekers each day on their Twitter account during the migration influx peak at the Norwegian-Russian border in autumn 2015, peaking with 196 people on November 3rd.[38] The ebb and flow of refugee claimants ceased on the 29th of the same month when Norway posted police on the border to refuse entry to anyone without a visa to the Schengen zone.[39]

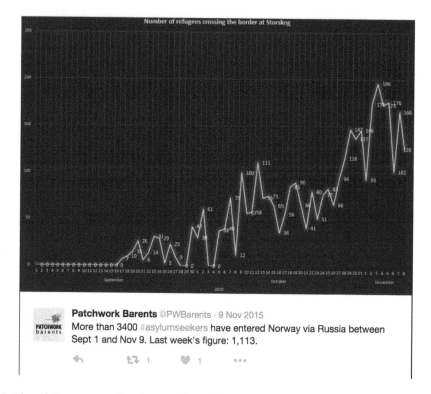

Figure 5.9 Patchwork Barents, an online data mapping website, posted a tweet on November 9th, 2015 showing the totals sourced from the police tweets[40]

The proliferation of locative social media has become integral for much of contemporary society. Participatory mapping tools are among them, platforms that various actors use as digital tools to accomplish a number of tasks, from mapping happiness levels to activism such as raising awareness of evictions in a local community.[41] These applications are used to capture, digitally mark, and create the perceptions of individuals' relationships within cities and landscapes.

MyCity and MyBarents: online participatory landscapes

Mycity.io is one such mapping platform, which I have applied as a tool to capture the voices of locals in four towns spanning the border region under the name MyBarents.[42] MyBarents covers the communities of Vardø and Kirkenes, Norway, and Nikel, Russia, and Näätämö, Finland, as research sites for their mutual threads of peripheral yet historically strategic locations derivative of resource-rich lands, and their proximity and associated interplays within transnational Arctic space

The MyCity platform, upon which MyBarents is based, was incubated from within the region, with examples already initiated in Murmansk and Tromsø, and with a nascent network of locally interested actors in Kirkenes and Nikel. It was thus a tool that reflected already existing cross-border exchanges, and could potentially reveal local voices to understand the cultural landscape at a horizontal level. MyBarents allows users to populate a map of their city with ideas, perceptions of place and improvement to the civic fabric.[43] Ideas can be continuously added to the map, discussed and voted upon. This map-in-progress weaves digital and physical space together with social and mental spaces, and records citizens' contributed visions as an on-going online idea repository and forum. It becomes what David Sneath, Martin Holbraad and Morten Axel refer to as a 'technology of the imagination' – a tool or process that is collectively used to engage and enable the imagination – in this case in everyday responses to Arctic landscapes.[44] The plots on the map reveal a landscape of intimate knowledge and experience of people through time. This virtual aggregate of ideas can evolve over time in concert with the changing attitudes, visions and direction of the community.

In this landscape of changing populations, locative media enables an accumulation of people's voices that expands upon the scope of classic ethnographic research and adds a cross section through the local cultural landscape. MyBarents reveals a shallow but broad overview of issues confronting the local spatial environment. It proves a mirror of the ideas that emerge during in situ community mapping projects, a virtual archive that can be accessed at any time by anyone in the world with access to the Internet.

Considerations of local responses to the refugee crises emerged well before the current border challenges increased in severity during the fall of 2015, but in terms of taking in resettled asylum seekers – the direct arrival at the border came as a surprise. The discussion stalled on MyBarents before the urgent autumn migration, but it has continued quite frenetically on social networks such as Facebook, which allows more in-depth community dialogues, such as that shown in 'Kirkenes Refugee Group' and 'Refugees Welcome to Finnmark', the latter with over 850 members.[45] These groups, mainly composed of local people, are online organisations set up to assist in any way they can with refugees. Further, an international discussion has been sparked on Reddit, a major contemporary online community where users vote on and discuss content, but unlike MyBarents, it is done without cartographical reference and is mainly topical.

A particular post on Reddit in November 2015 linked to an article by the Kirkenes-based Barents Observer on Afghan migrants entering at the Norwegian-Russian border and the likelihood of them remaining in Norway. It drew over 5,000 votes and 4,000 comments into the discussion on the border conditions and nuances, becoming one of the top discussions in the world on the site within 24 hours of being published.[46] We see here local and global audiences focus on the Kirkenes-Nikel border area within the digital realm that pick up on the pressing migrant crisis. These online social platforms give voice to new, imagined communities of vast diversity, those anticipating arrival into a new physical territory, and to the inhabitants of the border area who, expectedly or not, will share the space.[47] These voices are not always in concordance, and conflict remains a considerable 'hyperobject'.[48] A 'hyperobject' is another concept from Morton, which he defines as materials and events that exist at temporal and spatial scales beyond that which a human can perceive in its totality.[49]

Elements of conflict continue to form the urban landscape of Kirkenes, as old spaces are reappropriated to deal with asylum seekers crossing into Norway from Russia. In this case, as mentioned, Fjellhallen, the former bunker built as part of the town's 'dark infrastructure', had returned to a place to house people fleeing war or persecution.[50] This dark reading runs in parallel to a pronounced local perception that it was a ready-made positive infrastructure allowing the community flexibility and an ability to accommodate rapid change.

Arctic ethnoscapes

By 'ethnoscape', I mean the landscape of persons who constitute the shifting world in which we live: tourists, immigrants, refugees, exiles, guestworkers and other moving groups and persons constitute an essential feature of our world, and appear to affect the politics of and between nations to a hitherto unprecedented degree.

(Arjun Appadurai)[51]

Norway processed the arriving asylum seekers on a case-by-case basis with those accepted being absorbed into the social system.[52] The community has committed in favour of accepting 160 asylum seekers to permanently settle over 2 years.[53] However, these are not the only new arrivals. Many in the community have come as immigrants and expatriates to work for long periods in the mines, research centres and diplomatic missions, amongst other enterprises. This border area is a constantly fluctuating ethnoscape of diverse peoples.

The latest figures show that there are at least 1,323 recent immigrants from 68 countries out of a total population of 10,213 in Sør-Varanger County, of which Kirkenes is the central community.[54] This is in addition to the local population, which is already a heterogeneous mix of peoples including the indigenous Sámi, indigenous Norwegians, historical immigrant Kven, the Finns and people born elsewhere in the country. People are constantly moving in and out of the area. With some rural regions losing population, Kirkenes is able to retain and grow. However, this is not without the underscoring loss of youth seeking educational opportunities available only elsewhere, or most recently with workers moving out after the closure of the iron mine. As some leave, others arrive, and there is an influx of professionals who fulfil local needs not met with local capacity alone. One informant from the mine highlighted the fact that some skills needed were not found locally, and about 20 per cent of the workers in the administration and perhaps also in the general working population came from outside of Norway.[55] Transient human resources form an important part of the dark infrastructure. Migration is an integral part of the Kirkenes identity, and is echoed in Nikel where much of the population originated from elsewhere in the former Soviet Union, particularly from Ukraine.

This highly unstable and freshly agitated layer of migration, itself a plural field of immigrants, expatriates, asylum seekers and refugees, is but one of many contained in these borderlands. To differentiate, expatriates intend to return to their nation-state of origin, whereas immigrants intend to become new citizens within their adopted nation-state. This discourse is not so clean cut, with allegations of neo-colonialism attached to 'expats' vs. the darker implications surrounding 'immigrant'.[56] How people identify themselves and others within the ethnoscape is contingent on self-reflection and reference, and this provides evidence of conflicting perspectives of landscapes. This brings to the fore the challenging aspects of the border condition in the lives of those who inhabit such transient zones. The material landscape coexists with the immaterial, be it an ethereal digital intertwinement or the various perceptual qualities each of us imbues upon our spaces and places.

Shifting scales at the border

Climate change, technological change and globalisation are all variables complicit in the unexpected changes that face Arctic communities. So too are the internal forces that alter the political, cultural and religious landscapes in far-away countries. Though great changes may occur thousands of kilometres away from a particular locus, the effects are global and consequences local. This recalls Doreen Massey, 'What we need, it seems to me, is a global sense of the local, a global sense of place'.[57] The Norwegian-Russian border area does not escape the tumultuous reality of conflict elsewhere by its seemingly remote distance from major population centres. Rather, it is directly affected by the influx of asylum seekers who entered from Russia to Norway at the Storskog border crossing precisely due to its particular geopolitical status. These migrants, along with a wider shifting population, co-construct the ethnoscape. As you read this, no doubt the border conditions have shifted again many times, with new and unpredicted developments altering the relationships and character of the global landscapes, and those of the North, with new and transformed landscapes continuously emerging.

Dark readings are manifested in the material components of the landscape on both sides of the border represented through the vectors of resource extraction, conflict and historical atrocity. This further extends to non-material social and cultural elements such as the imagination of the inhabitants and in the ongoing discourse of locative and social media.

In communities where border conditions are a substantial local variable, communication and knowledge exchange with neighbouring communities is important. At regional scales, in this case the Arctic, the notion of neighbourness is extended because there are overlaying geophysical, climatic and socio-cultural synergies that tie into global narratives. The Norwegian-Russian borderscape is a microcosm of the global condition; it is directly implicit in the human affairs spanning the globe with a specific convergence of socio-cultural influences that exist physically, digitally, virtually and cognitively. This cultural border landscape is unique in its particular confluence of constituent materiality and geography, and interwoven socio-cultural features. However, this is where the uniqueness ends, for elements of each of these parts – liminality, transience, darkness, extraterritoriality – can be found in varying degrees and calibre across the cultural landscapes of the world. No thread from this entanglement can be wholly perceived at once; the landscapes are a collection of Morton's hyperobjects. They are fibres of a hyperlandscape.

Notes

1 Several media sources indicate that the migrants and human smugglers were using social media such as Facebook to rapidly read the shifting lay of the land and adjust their migration routes accordingly. See: Clare Cummings et al., "Why People Move: Understanding the Drivers and Trends of Migration to Europe," Working Paper (London: Overseas Development Institute, December 2015), www.odi.org/sites/odi.org.uk/files/odi-assets/publications-opinion-files/10208.pd.; Ivan Watson, Clayton Nagel, and Zeynep Bilginsoy, "'Facebook Refugees' Escape Syria via Cell Phones," *CNN*, September 15, 2015, www.cnn.com/2015/09/10/europe/migrant-facebook-refugees/index.html.

2 Andrew Higgins, "Avoiding Risky Seas, Migrants Reach Europe With an Arctic Bike Ride," *The New York Times*, October 9, 2015, www.nytimes.com/2015/10/10/world/europe/bypassing-the-risky-sea-refugees-reach-europe-through-the-arctic.html.

3 "Future North: About the Project," accessed September 1, 2016, www.oculs.no/projects/future-north/about/.

4 Michael Jones, "The European Landscape Convention and the Question of Public Participation," *Landscape Research* 32, no. 5 (2007), doi:10.1080/01426390701552753.

5 Peter H. Gleick, "Water, Drought, Climate Change, and Conflict in Syria," *Weather, Climate, and Society* 6, no. 3 (2014), doi:10.1175/WCAS-D-13-00059.1; Colin P. Kelley et al., "Climate Change in the Fertile Crescent and Implications of the Recent Syrian Drought," *Proceedings of the National Academy of Sciences* 112, no. 11 (March 17, 2015), doi:10.1073/pnas.1421533112.

6 Arctic Council, *Arctic Resilience Interim Report 2013* (Stockholm: Stockholm Environment Institute and Stockholm Resilience Centre, 2013); Intergovernmental Panel on Climate Change, *Climate Change 2014: Impacts, Adaptation, and Vulnerability. Part A: Global and Sectoral Aspects. Contribution of Working Group II to the Fifth Assessment Report of the Intergovernmental Panel on Climate Change* (Cambridge: Cambridge University Press, 2014).

7 Michael Jones, "The Concept of Cultural Landscape: Discourse and Narratives," in *Landscape Interfaces*, ed. Hannes Palang, and Gary Fry, Landscape Series 1 (Springer Netherlands, 2003), 23, http://link.springer.com/chapter/10.1007/978-94-017-0189-13.

8 Carl O. Sauer, "The Morphology of Landscape," *University of California Publications in Geography* 2, no. 2 (1925).

9 Ellen Lee, "Epilogue: Landscapes, Perspectives, and Nations: What Does It All Means?," in *Northern Ethnographic Landscapes: Perspectives from Circumpolar Nations*, ed. Igor Krupnik, Rachel Mason, and Tonia Horton (Washington, DC: Arctic Studies Center, National Museum of Natural History, Smithsonian Institution, 2004), 404, http://archive.org/details/northernethnogra62004krup.

10 Donald W. Meinig, "The Beholding Eye: Ten Versions of the Same Scene," in *The Interpretation of Ordinary Landscapes: Geographical Essays* (New York: Oxford University Press, 1979).

11 Ibid; and architect and theorist James Corner acknowledges further yet this multiplicity of definition, and writes that landscape is 'less a quantifiable object than it is an idea, a cultural way of seeing, and as such it remains open to interpretation, design, and transformation'. James Corner, preface to *Recovering Landscape: Essays in Contemporary Landscape Architecture*, ed. James Corner (New York: Princeton Architectural Press, 1999).

12 John Wylie, *Landscape: Key Ideas in Geography* (London: Routledge, 2007), 1.

13 Victor Witter Turner, *The Forest of Symbols: Aspects of Ndembu Ritual* (Cornell University Press, 1967).

14 Tom Avermaete, "The Borders Within: Reflections upon Architecture's Engagement with Urban Limens," in *Border Conditions*, ed. Marc Schoonderbeek (Amsterdam: Architectura & Natura Press, 2009).

15 Jones, "The Concept of Cultural Landscape," 45.

16 "Bike Shortage Stems Flow of Refugees Using Russian Arctic Route to Europe," *The Guardian*, October 29, 2015, sec. World news, www.theguardian.com/world/2015/oct/30/bike-shortage-stems-flow-of-migrants-using-russian-arctic-route-to-europe.

17 Kjetil Malkenes Hovland, "Syrian Refugees Take Arctic Route to Europe," *Wall Street Journal*, September 3, 2015, www.wsj.com/articles/syrian-refugees-take-arctic-route-to-europe-1441273767.

18 Amund Trellevik, "Flyktningstrømmen Øker – Hotellene I Kirkenes Er Fulle," *NRK*, September 16, 2015, www.nrk.no/finnmark/flyktningstrommen-oker-_-hotellene-i-kirkenes-er-fulle-1.12556386; СеверПост, "В Гостинице Никеля Поселились Более 200 Сирийских Беженцев | Информационное Агентство 'СеверПост.ru'," November 5, 2015, severpost.ru/read/34385/.

19 Michael Dear, *Why Walls Won't Work: Repairing the US-Mexico Divide* (Oxford: Oxford University Press, 2013).

20 Informant P23, Interview with P22, P23, P24 at a local gym, Nikel, interview by Morgan Ip, trans. by Katja Jylhä, August 11, 2015.

21 Benedict Anderson, *Imagined Communities: Reflections on the Origin and Spread of Nationalism* (London: Verso Books, 1991).

22 Wendy Brown, *Walled States, Waning Sovereignty* (New York: Zone Books, 2010), 74.

23 Ibid., 76.

24 Anastasia Rogova, "'Chicken Is Not a Bird – Kirkenes Is Not Abroad': Borders and Territories in the Perception of the Population in a Russian-Norwegian Borderland," *Journal of Northern Studies*, no. 1 (2009).

25 Jens Christian Hansen, "Coastal Finnmark, Norway: The Transformation of a European Resource Periphery," *European Urban and Regional Studies* 6, no. 4 (1999): 348, doi:10.1177/096977649900600410.

26 Gilly Carr, "Shining a Light on Dark Tourism: German Bunkers in the British Channel Islands," *Public Archaeology* 9, no. 2 (2010): 64–84, doi:10.1179/175355310X12780600917559; Malcolm Foley and J. John Lennon, "Editorial: Heart of Darkness," *International Journal of Heritage Studies* 2, no. 4 (1996): 195–97, doi:10.1080/13527259608722174.

27 Timothy Morton, *The Ecological Thought*, Reprint edition (Cambridge, MA: Harvard University Press, 2012); There is also a very interesting art and research project based on Morton's concept of Dark Ecology in the border area: www.darkecology.net/.

28 Ibid.

29 "Norilsk Nickel Halts Murmansk Mining," *Barents Observer*, October 1, 2015, http://barentsobserver.com/en/business/2015/10/norilsk-nickel-halts-murmansk-mining-01-10; The Barents Observer, "Mining Keeps on in Zapolyarny," *Barentsobserver*, accessed February 17, 2016, http://barentsobserver.com/en/business/2015/10/mining-keeps-zapolyarny-07-10; Atle Staalesen, "Kola Peninsula Nickel Mining Increases," *Arctic Now*, February 11, 2016, http://pagebuilder.arctic.arcpublishing.com/pb/business/2016/11/02/kola-peninsula-nickel-mining-increases/.

30 Øystein Barth-Heyerdahl, "Fra Kosovo Til Bomberom," *Aftenposten*, June 1, 1999, http://tux1.aftenposten.no/nyheter/uriks/kosovo/d83910.htm.

31 Informant P11, Interview with Informant P11 at the Barents Institute, Kirkenes, interview by Morgan Ip, July 13, 2015.

32 Brown, *Walled States*.

33 Informant P23, Interview with P22, P23, P24 at a local gym, Nikel.

34 The Barents Institute is a research arm of the University of Tromsø focused on cross-border issues and is situated in 'The Barents House,' which also hosts the diplomatic missions of the Norwegian Barents Secretariat and the International Barents Secretariat, as well as the University of Tromsø's Campus Kirkenes.

35 United Nations Environment Program, "The Kirkenes Declaration 1993," June 6, 1999, www.unep.org/dewa/giwa/areas/kirkenes.htm.

36 Rogova, "Chicken Is Not a Bird."

37 Tristan Thielmann, "Locative Media and Mediated Localities; An Introduction to Media Geography," *Aether: The Journal of Media Geography* 5, no. 1 (2010).

38 @OPSostfinnmark, "Storskog: 196 Asylsøkere Har I Dag Kommet over Storskog.," November 3, 2015, https:// twitter.com/OPSostfinnmark?lang=en.

39 Thomas Nilsen, "Zero Asylum Seekers at Storskog After Norway Places Police on Borderline," *The Independent Barents Observer*, November 30, 2015, http://thebarentsobserver.com/2015/11/police-norways-borderline-russia-stop-migrants-not-holding-legal-visa.

40 @PWBarents, "More Than 3400 #asylumseekers Have Entered Norway via Russia Between Sept 1 and Nov 9. Last Week's Figure: 1,113." November 9, 2015, https://twitter.com/PWBarents.

41 Social Life, "Participatory Mapping as a Social Digital Tool," September 11, 2015, www.social-life.co/blog/.

42 MyBarents.com was supported by a grant from the Norwegian Barents Secretariat in addition to the stipendium set within the Future North project funding from the Research Council of Norway's SAMKUL program.

43 Developed by Murmansk developers as inspired by in-situ workshops by AHO student, Jan Martin Klauza, and used by local activists in several Russian cities, as well as by officials in some Northern European locales such as Tromsø and Espoo; Stepa Mitaki, "Why We Started MyCity," June 12, 2015, http://blog.mycity.io/why-we-started-mycity/.

44 David Sneath, Martin Holbraad, and Morten Axel Pedersen, "Technologies of the Imagination: An Introduction," *Ethnos* 74, no. 1 (March 1, 2009), doi:10.1080/00141840902751147.

45 Flyktninggruppa Kirkenes, www.facebook.com/groups/900101913418854/?fref=ts; Refugees Welcome to Finnmark, www.facebook.com/groups/1030457363633028/?fref=ts; Refugees Welcome to the Arctic, www.facebook.com/refugeeswelcometothearctic/?fref=ts

46 Marginally Relevant, "Norway Sends a Clear Message to Afghan Migrants in Russia. Crossing the Border at Storskog Can Give a One-Way Ticket to Kabul," *Reddit Thread*, (November 14, 2015), www.reddit.com/r/worldnews/comments/3rzv6v/norway_sends_a_clear_message_to_afghan_migrants/.

47 Anderson, *Imagined Communities*.

48 Timothy Morton, *Hyperobjects: Philosophy and Ecology After the End of the World* (Minneapolis: University of Minnesota Press, 2013).

49 Ibid.

50 Tor Sandø, "Fjellhallen Blir Transittmottak," *Sør-Varanger Avis*, September 14, 2015, http://sva.no/index.php?page=vis_nyhet&NyhetID=4465.

51 Arjun Appadurai, "Disjuncture and Difference in the Global Cultural Economy," *Public Culture* 2, no. 2 (1990), doi:10.1215/08992363-2–2-1.

52 According to UN definitions, some migrants are classified as either refugees or immigrants, and each designation determines whether or not a person is qualified to stay in the country, and who must be repatriated. See: Somini Sengupta, "Migrant or Refugee? There Is a Difference, With Legal Implications," *The New York Times*, August 27, 2015, www.nytimes.com/2015/08/28/world/migrants-refugees-europe-syria.html; United Nations High Commissioner for Refugees, "Who Is a Refugee?," UNHCR Protection Training Manual for European Border and Entry Officials, Session 3, Brussels: UNHCR Bureau for Europe, 2011, www.unhcr.org/4d944c319.html.

53 Amund Trellevik, and Kristian Sønvisen Bye, "Sør-Varanger Vil Åpne Dørene for Syria-Flyktninger," *NRK*, May 21, 2015, www.nrk.no/finnmark/sor-varanger-vil-apne-dorene-for-syria-flyktninger-1.12371055; and, Sør-Varanger kommune. 2015. "Tilvisningsrett av boliger til bosatte flyktninger i Sør-Varanger kommune," September 1. www.sor-varanger.kommune.no/tilvisningsrett-av-boliger-til-bosatte-flyktninger-i-soer-varanger-kommune.5790449–154949.html; Sør-Varanger kommune. 2015. "Økt Bosetting Av Flyktninger I 2015 Og 2016 – Sør-Varanger Kommune." September 1, 2015, www.sor-varanger.kommune.no/oekt-bosetting-av-flyktninger-i-2015-og-2016.5774196-154949.html.

54 Statistics Norway, "Table: 09817: Immigrants and Norwegian-born to immigrants, by immigration category, country background and proportion of the population (M)," www.ssb.no/statistikkbanken/selectvarval/Define.asp?subjectcode=&ProductId=&MainTable=FolkInnvkatLand&nvl=&PLanguage=0&nyTmpVar=true&CMSSubjectArea=befolkning&KortNavnWeb=innvbef&StatVariant=&checked=true; Statistics Norway, "Population and population changes, Q2 2015," www.ssb.no/en/befolkning/statistikker/folkendrkv/kvartal/2015-08-20?fane=tabell&sort=nummer&tabell=236794

55 Informant P4, Interview with Informant P4 at the Barents Institute, Kirkenes, interview by Morgan Ip, July 17, 2015.

56 Mawuna Remarque Koutonin, "Why Are White People Expats When the Rest of Us Are Immigrants?" *The Guardian*, March 13, 2015, www.theguardian.com/global-development-professionals-network/2015/mar/13/white-people-expats-immigrants-migration; and, a discussion on how Americans in Finland fit into connotations of 'immigrant' in Finnish society can be found in Johanna Leinonen, "Invisible Immigrants, Visible Expats? Americans in Finnish Discourses on Immigration and Internationalization," *Nordic Journal of Migration Research* 2, no. 3 (2012), doi:10.2478/v10202-011-0043-8.

57 Doreen Massey, "A Global Sense of Place," *Marxism Today* June (1991).

Bibliography

@OPSostfinnmark. "Storskog: 196 Asylsøkere Har I Dag Kommet over Storskog." November 3, 2015. https://twit-ter.com/OPSostfinnmark?lang=en.

@PWBarents. "More Than 3400 #asylumseekers Have Entered Norway via Russia Between Sept 1 and Nov 9. Last Week's Figure: 1,113." November 9, 2015. https://twitter.com/PWBarents.

Anderson, Benedict. *Imagined Communities: Reflections on the Origin and Spread of Nationalism*. London: Verso Books, 1991.

Appadurai, Arjun. "Disjuncture and Difference in the Global Cultural Economy." *Public Culture* 2, no. 2 (1990): 1–24. doi:10.1215/08992363-2-2-1.

Arctic Council. *Arctic Resilience Interim Report 2013*. Stockholm: Stockholm Environment Institute and Stockholm Resilience Centre, 2013.

Avermaete, Tom. "The Borders Within: Reflections upon Architecture's Engagement with Urban Limens." In *Border Conditions*, edited by Marc Schoonderbeek. Amsterdam: Architectura & Natura Press, 2009.

Barth-Heyerdahl, Øystein. "Fra Kosovo til bomberom." *Aftenposten*, June 1, 1999. http://tux1.aftenposten.no/nyheter/uriks/kosovo/d83910.htm

"Bike Shortage Stems Flow of Refugees Using Russian Arctic Route to Europe." *The Guardian*, October 29, 2015. www.theguardian.com/world/2015/oct/30/bike-shortage-stems-flow-of-migrants-using-russian-arctic-route-to-europe.

Brown, Wendy. *Walled States, Waning Sovereignty*. New York: Zone Books, 2010.

Carr, Gilly. "Shining a Light on Dark Tourism: German Bunkers in the British Channel Islands." *Public Archaeology* 9, no. 2 (2010): 64–84. doi:10.1179/175355310X12780600917559.

СеверПост. "В Гостинице Никеля Поселились Более 200 Сирийских Беженцев | Информационное Агентство 'СеверПост.ru'." November 5, 2015. https://www.severpost.ru/read/34385/.

Corner, James, ed. *Recovering Landscape: Essays in Contemporary Landscape Architecture*. New York: Princeton Architectural Press, 1999.

Cummings, Clare, Julia Pacitto, Diletta Lauro, and Marta Foresti. "Why People Move: Understanding the Drivers and Trends of Migration to Europe." Working Paper. London: Overseas Development Institute, 2015. www.odi.org/sites/odi.org.uk/files/odi-assets/publications-opinion-files/10208.pdf.

Dear, Michael. *Why Walls Won't Work: Repairing the US-Mexico Divide*. Oxford: Oxford University Press, 2013.

Flyktninggruppa Kirkenes, www.facebook.com/groups/900101913418854/?fref=ts

Foley, Malcolm and J. John Lennon. Editorial. "Heart of Darkness." *International Journal of Heritage Studies* 2, no. 4 (1996): 195–97. doi:10.1080/13527259608722174.

"Future North: About the Project." Accessed September 1, 2016. www.oculs.no/projects/future-north/about/.

Gleick, Peter H. "Water, Drought, Climate Change, and Conflict in Syria." *Weather, Climate, and Society* 6, no. 3 (2014): 331–40. doi:10.1175/WCAS-D-13-00059.1.

Hansen, Jens Christian. "Coastal Finnmark, Norway: The Transformation of a European Resource Periphery." *European Urban and Regional Studies* 6, no. 4 (1999): 347–59. doi:10.1177/096977649900600410.

Higgins, Andrew. "Avoiding Risky Seas, Migrants Reach Europe with an Arctic Bike Ride." *The New York Times*, October 9, 2015. www.nytimes.com/2015/10/10/world/europe/bypassing-the-risky-seas-refugees-reach-europe-through-the-arctic.html.

Hovland, Kjetil Malkenes. "Syrian Refugees Take Arctic Route to Europe." *Wall Street Journal*. September 3, 2015. www.wsj.com/articles/syrian-refugees-take-arctic-route-to-europe-1441273767.

Informant P4. Interview with Informant P4 at the Barents Institute, Kirkenes. Interview by Morgan Ip, July 17, 2015.

Informant P11. Interview with Informant P11 at the Barents Institute, Kirkenes. Interview by Morgan Ip, July 13, 2015.

Informant P23. Interview with P22, P23, P24 at a local gym, Nikel. Interview by Morgan Ip. Translated by Katja Jylhä, August 11, 2015.

Intergovernmental Panel on Climate Change. *Climate Change 2014: Impacts, Adaptation, and Vulnerability. Part A: Global and Sectoral Aspects. Contribution of Working Group II to the Fifth Assessment Report of the Intergovernmental Panel on Climate Change*. Cambridge: Cambridge University Press, 2014.

Jones, Michael. "The Concept of Cultural Landscape: Discourse and Narratives." In *Landscape Interfaces*, edited by Hannes Palang and Gary Fry, 21–51. Dordrecht: Springer Netherlands, 2003.

Jones, Michael. "The European Landscape Convention and the Question of Public Participation." *Landscape Research* 32, no. 5 (2007): 613–33. doi:10.1080/01426390701552753.

Kelley, Colin P., Shahrzad Mohtadi, Mark A. Cane, Richard Seager, and Yochanan Kushnir. "Climate Change in the Fertile Crescent and Implications of the Recent Syrian Drought." *Proceedings of the National Academy of Sciences* 112, no. 11 (2015): 3241–46. doi:10.1073/pnas.1421533112.

Koutonin, Mawuna Remarque. "Why Are White People Expats When the Rest of Us Are Immigrants?" *The Guardian*, March 13, 2015. www.theguardian.com/global-development-professionals-network/2015/mar/13/white-people-expats-immigrants-migration.

Lee, Ellen. Epilogue. "Landscapes, Perspectives, and Nations: What Does It All Means?" In *Northern Ethnographic Landscapes: Perspectives from Circumpolar Nations*, edited by Igor Krupnik, Rachel Mason, and Tonia Horton, 401–6. Washington, DC: Arctic Studies Center, National Museum of Natural History, Smithsonian Institution, 2004. http://archive.org/details/northernethnogra62004krup.

Leinonen, Johanna. "Invisible Immigrants, Visible Expats? Americans in Finnish Discourses on Immigration and Internationalization." *Nordic Journal of Migration Research* 2, no. 3 (2012): 213–23. doi:10.2478/v10202-011-0043-8.

Marginally_Relevant. "Norway Sends a Clear Message to Afghan Migrants in Russia. Crossing the Border at Storskog Can Give a One-Way Ticket to Kabul." *Reddit Thread*, November 14, 2015. www.reddit.com/r/worldnews/comments/3rzv6v/norway_sends_a_clear_message_to_afghan_migrants/.

Massey, Doreen. "A Global Sense of Place." *Marxism Today*, June 1991, 24–29.

Meinig, Donald W. "The Beholding Eye: Ten Versions of the Same Scene." In *The Interpretation of Ordinary Landscapes: Geographical Essays*, edited by Donald W. Meining, 33–48. New York: Oxford University Press, 1979.

Mitaki, Stepa. "Why We Started MyCity," June 12, 2015. http://blog.mycity.io/why-we-started-mycity/.

Morton, Timothy. *The Ecological Thought*. Reprint edition. Cambridge, MA: Harvard University Press, 2012.

Morton, Timothy. *Hyperobjects: Philosophy and Ecology After the End of the World*. Minneapolis: University of Minnesota Press, 2013.

Nilsen, Thomas. "Zero Asylum Seekers at Storskog After Norway Places Police on Borderline." *The Independent Barents Observer*, November 30, 2015. http://thebarentsobserver.com/2015/11/police-norways-borderline-russia-stop-migrants-not-holding-legal-visa.

Refugees Welcome to Finnmark, www.facebook.com/groups/1030457363633028/?fref=ts

Refugees Welcome to the Arctic, www.facebook.com/refugeeswelcometothearctic/?fref=ts

Rogova, Anastasia. "'Chicken Is Not a Bird – Kirkenes Is Not Abroad': Borders and Territories in the Perception of the Population in a Russian-Norwegian Borderland." *Journal of Northern Studies*, no. 1 (2009): 31–42.

Sandø, Tor. "Fjellhallen Blir Transittmottak." *Sør-Varanger Avis*, September 14, 2015. http://sva.no/index.php?page=vis_nyhet&NyhetID=4465.

Sauer, Carl O. "The Morphology of Landscape." *University of California Publications in Geography* 2, no. 2 (1925): 19–54.

Sengupta, Somini. "Migrant or Refugee? There Is a Difference, with Legal Implications." *The New York Times*, August 27, 2015. www.nytimes.com/2015/08/28/world/migrants-refugees-europe-syria.html.

Sneath, David, Martin Holbraad, and Morten Axel Pedersen. "Technologies of the Imagination: An Introduction." *Ethnos* 74, no. 1 (2009): 5–30. doi:10.1080/00141840902751147.

Social Life. "Participatory Mapping as a Social Digital Tool," September 11, 2015. www.social-life.co/blog/.

Staalesen, Atle. "Kola Peninsula Nickel Mining Increases." *Arctic Now*, February 11, 2016. http://pagebuilder.arctic.arcpublishing.com/pb/business/2016/11/02/kola-peninsula-nickel-mining-increases/.

Statistics Norway. "Population and population changes, Q2 2015." www.ssb.no/en/befolkning/statistikker/folkendrkv/kvartal/2015-08-20?fane=tabell&sort=nummer&tabell=236794.

Statistics Norway. "Table: 09817: Immigrants and Norwegian-Born to Immigrants, by Immigration Category, Country Background and Proportion of the Population (M)." www.ssb.no/statistikkbanken/selectvarval/Define.asp?subjectcode=&ProductId=&MainTable=FolkInnvkatLand&nvl=&PLanguage=0&nyTmpVar=true&CMSSubjectArea=befolkning&KortNavnWeb=innvbef&StatVariant=&checked=true.

Sør-Varanger kommune. "Tilvisningsrett av boliger til bosatte flyktninger i Sør-Varanger kommune." September 1, 2015. www.sor-varanger.kommune.no/tilvisningsrett-av-boliger-til-bosatte-flyktninger-i-soer-varanger-kommune.5790449-154949.html.

Sør-Varanger kommune. "Økt Bosetting Av Flyktninger I 2015 Og 2016 – Sør-Varanger Kommune." September 1, 2015. www.sor-varanger.kommune.no/oekt-bosetting-av-flyktninger-i-2015-og-2016.5774196-154949.html

The Barents Observer. "Mining Keeps on in Zapolyarny." *Barentsobserver*. October 07, 2015. http://barentsobserver.com/en/business/2015/10/mining-keeps-zapolyarny-07-10.

The Barents Observer. "Norilsk Nickel Halts Murmansk Mining." *Barentsobserver*. October 1, 2015. http://barentsobserver.com/en/business/2015/10/norilsk-nickel-halts-murmansk-mining-01-10.

Thielmann, Tristan. "Locative Media and Mediated Localities; An Introduction to Media Geography." *Aether: The Journal of Media Geography* 5, no. 1 (2010): 1–17.

Trellevik, Amund. "Flyktningstrømmen Øker – Hotellene I Kirkenes Er Fulle." *NRK*, September 16, 2015. www.nrk.no/finnmark/flyktningstrommen-oker-_-hotellene-i-kirkenes-er-fulle-1.12556386.

Trellevik, Amund and Kristian Sønvisen Bye. "Sør-Varanger Vil Åpne Dørene for Syria-Flyktninger." *NRK*, May 21, 2015. www.nrk.no/finnmark/sor-varanger-vil-apne-dorene-for-syria-flyktninger-1.12371055.

Turner, Victor Witter. *The Forest of Symbols: Aspects of Ndembu Ritual*. Ithaca: Cornell University Press, 1967.

United Nations Environment Program. "The Kirkenes Declaration 1993," June 6, 1999. www.unep.org/dewa/giwa/areas/kirkenes.htm.

United Nations High Commissioner for Refugees. "Who Is a Refugee?" UNHCR Protection Training Manual for European Border and Entry Officials. Session 3. Brussels: UNHCR Bureau for Europe, 2011. www.unhcr.org/4d944c319.html.

Watson, Ivan, Clayton Nagel, and Zeynep Bilginsoy. "'Facebook Refugees' Escape Syria via Cell Phones." *CNN*, September 15, 2015. www.cnn.com/2015/09/10/europe/migrant-facebook-refugees/index.html.

Wylie, John. *Landscape: Key Ideas in Geography*. London: Routledge, 2007.

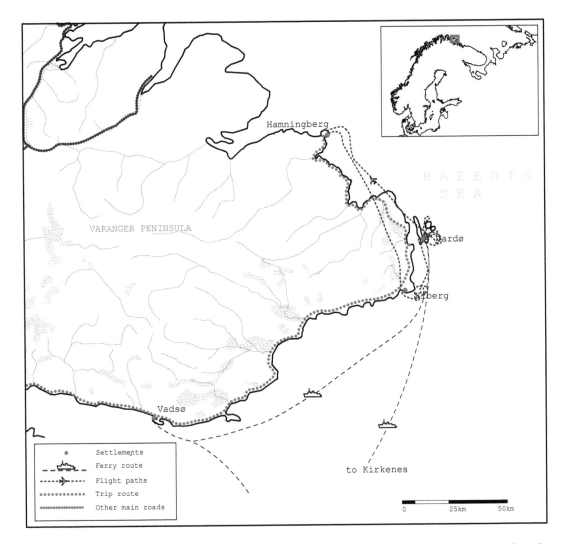

Map 3 The settlements along the Varanger peninsula have been subsisting on fishing and small-scale agriculture for centuries. Closer to the projected Shtokman natural gas field than any Russian settlements, Vardø attempted to position itself as a supply harbor. However, as the development has been mothballed, Vardø is left with an idle industrial harbor and decaying facilities for the dwindling traditional fishing.

Map: Eimear Tynan.

6   Landscape in the new North

Janike Kampevold Larsen

Scene 1

Murmansk, September 2013

Five researchers are standing on a pedestrian bridge that runs across the railway tracks in Murmansk, Russia, looking over the railing. A transect walk brought us here from the top of a residential hill popularly called Idiot Hill, as people living there seem to favor the expansive views that the hill offers, braving the icy cold winds that continuously pervade it. We have descended the hill, documenting our gradual immersion into the monumental buildings lining the Lenina Prospect. Downtown Murmansk holds a strong resemblance to quarters in larger European cities, thanks to Stalin's love for classicist architecture. The outskirts are not so alluring, comprising older quarters of dilapidated but still impressive wooden structures on the one hand, and 1970s and '80s concrete apartment buildings on the other. The large, clean structures along the main street speak of Murmansk's role as an administrative center. The harbor itself is filled with heaps of coal and other minerals. The many cargo trains that we see from the bridge carry an array of processed minerals, from dust to pills and pellets.

Scene 2

Vardø, January 2014

Six researchers are walking the streets of a small Arctic town on an island off the North-Eastern coast of Norway. It is late January, and not very cold, minus 5°C, but quite strong winds blowing in from the Barents Sea leave us with an experienced temperature that is much lower. The light is blue, quickly moving to dusk. The group is inspecting and documenting material traces of present and past activity in the former fishing community of Vardø – one that is far more complex than anticipated. The transect walk commences at the brink of the whirling sea, just next to the NATO radar station that is discernible from every point in the city,[1] and ends at the southern shore by the Steilneset witch monument facing the mainland. It cuts across the new housing area, an old water reservoir called Klondyke on the hill next to the church, and through the church yard to the new culture house.[2] We cut across the former glorious main street and the fishing harbor as we proceed across the isthmus that connects the two wings of the island. Our transect method involves pausing and discussing the material configurations we encounter. When crossing the old water reservoir, we have an uninterrupted view of the many soaring signs of present day forces operating on the island. At this spot, some of us synthesize: The church steeple is the historical connector of Heaven and Earth, while the Norwegian Coastal Administration's service central with its four radio towers and the space monitoring globes behind us represent present-day technological connectors of earthly tracking activities and celestial space.[3] Crisscrossing the vertical dominance of official infrastructure, we see the usual web of electricity wires and road infrastructure.

Figure 6.1 Our transect walk commenced at the brink of the Polar Sea. A couple of us climbed the hill to the Globus II to get a better view of the premises

Figure 6.2 Globus II dominates views throughout the city. Strangely though, one stops noticing it after a few days in Vardø

Scene 3

Svalbard, May 2015, on board MS Billefjord

Deep in the Billefjord, a sidearm to the Isfjord on Svalbard: The boat has stalled at the edge of the ice that coats the head of the fjord between the old Soviet mining town Pyramiden and the Hammerskiöld glacier. On board is a small group of researchers from the Future North project: Bill Fox, Head of the Center for Art + Environment, Nevada Museum of Art; Kathleen John-Alder, landscape architect from Rutgers University; Aileen A. Espiritu, UiT, The Arctic University of Norway; and Peter Hemmersam, Janike Kampevold Larsen, and Andrew Morrison, all from the Oslo School of Architecture and Design.

The skipper, Pronie Caguicla, is from the Philippines, as is his crew. The guide, Marcel, is a young man from Austria. Morrison, a Zimbabwean by birth, Norwegian by 20 years of residency, is having an exalted time on the bridge with the captain, who for a few moments lets him steer and back the boat away from the ice flow that blocks our approach to Pyramiden. On board are also a handful of tourists. We have just spotted a polar bear on shore; there are some seals coming up to breathe in the distance and plenty of flapping birds on the still fjord. Through binoculars, we inspect the hillsides of Pyramiden: the dirt roads that zig-zag up the steep mountain side, the coal chutes, the railway lines to the mines, the farm house, the hotel, the wooden buildings, the coal and chemical residues on red, sandy slopes.[4] Apart from the animals, all native to the archipelago, we are all visitors. Even the captain and the guide have been in Longyearbyen for less than a couple of years. Morrison and the captain are exchanging stories of migration and acclimatization.

This chapter is framed by a series of travels to different parts of the near Arctic as seen from Norway: the Russian Kola Peninsula, the Norwegian Barents coast city of Vardø, and Svalbard. They represent landscapes that are exposed to strong forces of consumption by an expansive tourism industry, by extraction interests, or by time and neglect. While Kola seems to be the emblem of a postindustrial

Figure 6.3 Sand and pebble material shapes into fantastic forms along the shorelines of the Svalbard fjords
Photo: the author.

Figure 6.4 Vardø: The built environment in Vardø forms but a thin layer on top of the strong geological foundation of the Varanger Peninsula. The sedimentary geology is around 500 million years old.

Photo: the author.

landscape, strewn with extremely polluting ore processing facilities, Svalbard performs as a territory that is still perceived as relatively untouched, although moving towards a charged and uncertain future. Vardø is a society bent on eluding both a stagnant history and a futuristic hype. All three territories are unfolding as fields of forces where intimacy and distance need to be rethought.

These territories challenge our perception, reading, and conceptualization skills. A first encounter requires an understanding of the topographic and material distribution that constitute the spaces – be they city or land territory. Landscape morphology provides information as to the manifest activities in the territories.[5] However, landscape is more than what we see. An understanding of landscape as both process and place requires an understanding of drivers and desires, of the forces shaping the earth's material distribution, of social processes, and of our conceptualization of them.

This chapter looks at Arctic landscape as "both a thing and a social 'process', at once solidly material and ever-changing", as argued by Don Mitchell, reflecting that "the evident . . . form of landscape often masks the facts of its production".[6] It attempts a reading of territory as it performs in a contemporary scene of policies and habits, logics and networks. Within a context of landscape thinking that reflects on landscape's own agency, the three scenes above are maybe better described as *situations* – ones in which a visually informed approach to landscape is challenged by the agency of that which is looked at. As particular situations, they are formed by the materialities present, the humans watching and interpreting them, and the exchange of agency between the two.

The skin of the earth

American photographer Michael Light describes the American West from the air: "In the arid West both the bones of the land and the contours of the contemporary built world are strewn across space like writing of another world".[7] Much of the high Arctic was classified as Arctic desert until a changing climate brought about increased precipitation over the last few decades. Still, vegetation grows so slowly that any found object is just about as visible as it was when lost, left, placed, or displaced one or several hundred years ago. The material distribution of Arctic spaces bears a resemblance to the arid West. The "skin of the earth", the bones of the land, are visible; even from an on-ground perspective one perceives the layering of the landscapes, where the solid deep time layer of geology is always discernable and dominant, between and under the thin layers of soil and the similarly thin layer of the man-made and cultural skin – what Raoul Bunschoten refers to as the first and the second skin of the earth, arguing for an attention to the second that considers the first.[8] A consideration of the second skin requires that we look beyond the visual configuration of the first.

Landscape as distance

Referring to seminal landscape thinkers J.B. Jackson and John Stilgoe, geographer Denis Cosgrove describes a geographical thinker's initial areal idea of landscape as "an identifiable tract of land, an area of known dimensions like the fields and woods of a manor or parish".[9] Stressing that a tract of land needs to be identified *as* something, he relates for example to the British scholar of eighteenth- and nineteenth-century studies, John Barrell, who opens his book *The Idea of Landscape & the Sense of Place, 1730–1840* with these words: "There is no word in English which denotes a tract of land, of whatever extent, which is apprehended *visually*, but not, necessarily *pictorially*".[10] Barrell is in search of a term that allows him to escape the hegemony of the pastoral, the human scale idyllic landscape that was a model for the English park tradition. The term "landscape", he argues, is impossible to use without introducing "notions of value and form, which relate, not just to seeing land, but to seeing it in a certain way – pictorially".[11]

The idea of landscape as a way of seeing rests firmly on a tradition of landscape painting, and the notion of *landskip* – a painted and framed landscape.[12] Cosgrove has, however, repeatedly claimed that regarding landscape as a visual construction is suffused by connotations of control, mastery, and ownership. Geographer John Wylie concludes:

> As a way of seeing . . . landscape is the accomplice and expression of a classical subject–object epistemological model, one whose central supposition posits a pre-given reality which an independent subject contemplates, represents and masters from a position of cohered detachment.[13]

Landscape geographer Kenneth Olwig, however, claims that the word *landschaft* was used in Northern European countries in the Middle Ages to describe territories that were managed and administered: "The Landschaft as place was . . . defined not physically but socially, as the place of a polity".[14] It involved, among other things, the organization of territorial estates as a judicial entity. Olwig demonstrates how *landschaft* and *landskip* are etymologically connected, but also how early painted landscapes were indeed concerned with land regulation. He points to Pieter Breughel's combination of "landscape painting with the form of painting known as . . . Sittenbild (Sitten = custom, Bild = picture)".[15] The Sittenbild was concerned with the display of everyday practices and common situations, and, as Olwig points out, "the customs or *mores* of a people give rise to both law and morality". Brueghel's paintings, then, show not only "the logic of the terrain", but also "the logic of the activity".[16] He concludes: "Landscape painting was thus a way of representing, and making concrete, the more abstract, social idea of landscape expressed by representative legal bodies and the law they generated".[17] The two conceptions of landscape, landskip and landschaft, are thus connected historically. The idea of landschaft, however, is also associated with an idea of territory, defined by polity and not by area, in which the right to land and the right to justice is the same.

Olwig's explication resonates with the deployment of a more classical definition of territory as discussed by architectural theorist Antoine Picon, who defined it as "a space mastered and policed by institutions and corporations".[18] The idea of territory as "an easy circulation of men and goods" has existed since pre-enlightenment, according to Picon, and "opened up the possibility to exploit, in a comprehensive way, mines and fields as well as people and their skills".[19] As a managed and exploited tract of land, territory is based on distance: the distance of the managers to that which is managed. Picon further bases what he calls a disinterested landscape view on the Kantian aesthetics that pervaded the eighteenth century, that great age of mapping and subsequent exploitation of newfound territorial expanses. We know that the Kantian notions of the beautiful and the sublime contributed to forming the park ideal that reinforced the pictorial conception of landscape.[20] Picon argues that the ideas of landscape and territory converge; they are both based on distance: "Often using the same remote point of view, the territorial entrepreneur charted resources where the landscape amateur experienced disinterested emotions".[21] Both represent a distanced relation to that which is seen or explored.

This historical dynamic of distance involved in looking at landscape as manageable territory seems to be materializing in many areas of the North right now, and examples are abundant. The Barents Sea is being prospected for hydrocarbons, and measures have been implemented that support exploitation. In February 2015, the Norwegian Government declared the "edge" of the polar ice cap to lie further north than previously acknowledged. While it is true that the ice cap retreats, and will most probably have vanished by 2040, this is the first time a government has officially defined its edge – which is a construct. If anything, the so-called ice-edge is a fluctuating zone, varying seasonally and over the years. A definition of the ice edge as a receding geographic boundary, however, allowed for petrochemical prospecting further north. We are indeed talking about a "political ice edge", as geographers Berit Kristoffersen and Philip Steinberg have suggested.[22]

This is an acute example of a distanced relation to a landscape of ice and to landscape as exploitable territory. A conception of landscape as production or process presupposes an understanding of its entangled relationship to people, ecologies, and industries as well as to local and global forces working upon it – none of which are immediately accessible to the gaze. How, then, to relate to Arctic landscapes as inhabited, visited, exploited, and environmentally precarious places?

The singular place

Referring to Peter Sloterdijk, philosopher and architectural theorist Sébastien Marot argues that the basic spatial element of cultural geography is what may be called an "anthropogenic island", a "locality that comes with a culture, some kind of collective scene, stage or persona, produced by a long and incremental adjustment to (and confirmation of) a substrate of resources turned into a landscape of habits".[23] However, he asks rhetorically, is there any place left in our globalized world that can really be considered autonomous? Cultural geographer Doreen Massey, poses a similar question: "How, in the face of all this movement and intermixing, can we retain any sense of a local place and its particularity?"[24] Her repeated analyses of the globally, socially, and economically networked place rests on the presumption that place has always been networked, and that movement, migration, and trade are inherent to human (and geological) nature:

> It is a sense of place, an understanding of "its character", which can only be constructed by linking that place to places beyond. A progressive sense of place would recognize that, without being threatened by it. What we need, it seems to me, is a global sense of the local, a global sense of place.[25]

A global sense of place debunks the myth of the local which is particularly vivid in discourses on remote Arctic communities, and challenges the conception of space as bounded, static, and enveloped.[26] A conception of space as always in progress and always networked implies an insistence on time as an integral part of space – and not only past time and its reminiscences in a landscape, but also the fluctuation of present time and future, an anticipation that space, and place, is always changing pertaining to actors and forces working upon it. However, considering places as relational, unfolding in a permanent exchange with other places, begs a critical view also on the global versus the local. As E. Carina H. Keskitalo and Mark Nutall point out, "one aspect of globalization is a continuation and intensification of processes [that influence, mold or shape societies], especially in relation to mobility and communication around the globe".[27] Local places are not simply prey to globalized forces; they are also "agents" in globalization.[28] This is true of places in the Arctic. Here, globalization is not new; the Arctic has been globalized for centuries through trade and migration. What is new, one may argue, is the degree to which these areas are now subjected to an intensified pressure from global forces, and hence a relationality that becomes deeper at a very high speed.

The Kola Peninsula

An intimate ecology

The Future North research team has walked and talked its way through six of the eight major mineral processing settlements on the Kola Peninsula, as well as the administrative center Murmansk, which also serves as the logistic hub for out-shipment of coal as well as for processed ore such as iron, nickel, apatite, and copper. The towns of Nikel, Sapoljarnyj, Murmansk, Monchegorsk, Revda, Apatity, and Kirvosk are all connected to the same main road at the thick of the peninsula. Future North teams have visited the region several times, trying to conceptualize its cultural landscapes – and to understand the territory's particular configuration of landscape forces and materialities through research into its geological, environmental, and political history. The initial approach was to perform transects in all cities visited, to study them by walking, pausing, and talking about what we observed as a multi-disciplinary group of researchers.

The first transect walk in Murmansk explored the configuration of the town from the forested hilltop to the impressive city center and the railroad and harbor area along the bay. Descending the steep hillside by stairways and shortcuts, we read the city's layered views – watched the vast views of the hinterlands fade as we descended into its urban fabric. Reading the hill by topographical layout, we compared it to European urban hillsides, talking about how easy it is to compare new landscapes with those already known.

It was only after walking through the stately railway terminal building and onto the footbridge that we saw proof of what we already knew: The city is a major hub for trans-shipment of coal and nickel ore coming in from Siberia and the nearby mineral processing towns. Arriving in the harbor area, our conversation was no longer about the layered views, but about the massive amounts of ore displaying itself. Open cart freight trains were lined up in the multitrack area of the station. Watching these, we did not even think about the fact that we were standing on a viewing platform of sorts – the footbridge leading over the tracks. We were concerned with the sudden presence of minerals having arrived across the expanses of Kola and Russia at large. We were not prepared for such a splendid display of differently textured minerals. The main serendipitous moment of this walk was the realization that the visual is a barrier to truly understanding the state of the Kola territory. We moved from a touristic view to an immersed view: from one in which we employed a certain mode of viewing onto the landscape, to one in which there was no view, or where what was looked at was somehow looking back at us, affecting our understanding on the territory. This something was not only a landscape matter, but an active matter that has transformed the entire configuration of the Kola peninsula.

Figure 6.5 Train carts of ore of different texture and size testify to Murmansk's connection to the mining territories of the Kola Peninsula

Photo: the author.

One could say that the material of the Kola peninsula introduced itself to us at the end of our walk. It demonstrated what political theorist Jane Bennet calls the "vitality of (nonhuman) bodies", referring to matter's capacity to "act as quasi agents or forces with trajectories, propensities, or tendencies of their own".[29] Nonhuman matter is an active force and affects both us and events. In this case, the black, uniform masses of the nickel products alerted us both to its prominence as commercial product, but also to the fact that it has enabled the Kola cities by its very existence and by its value in a human production chain.

Bennet's perspective transcends the idea that the world is "for us", that the human being qua subjectivity is the lucid and driving force of the world, free to appropriate it and its resources. She speaks not of landscape agency, but transferred to landscape the view on matter that she advocates breaks with the Western landscape tradition's cemented notion of subject–object and hence its idea of landscape as a "way of seeing". Hence, the mineral display in the Murmansk train station collapsed our experience of distance to what we were looking at. In the situation that unfolded between us on the pedestrian overpass and the mineral display, landscape as view clashed with and was taken over by a view upon landscape as production. The vital materiality of the nickel prefigured the rest of our trip as a journey through a landscape of tailings, of soil burnt by sulphur dioxide, and dust.

Monchegorsk

The industrial towns on the Kola are all planned as mining communities and display Soviet state-of-the-art urban planning. The design is clean and structural. All cities have beautifully designed parks and lush vegetation separating roads from pavements (Murmansk is said to be the greenest city above the polar circle). Adjacent to them we always find the opposite: industrial areas informed by

Figure 6.6 Driving along the main road on the Kola Peninsula tailing mounds are an integrated part of a landscape where the nature–culture distinction has collapsed entirely

Photo: the author.

Figure 6.7 The mineral processing facility in Monschegorsk: one of the most polluted places in Russia, and one that has contributed amounts of sulphuric acid, palladium, and gold dust to the environment in a large radius around the city

Photo: the author.

the rationale of ore processing – entangled masses of factories, black and worn by fumes, mud, and mineral residue. The towns are hence marked by a double desire: The vertical desire for the mineral is overlaid by a horizontal desire for grand human-scale plans that are also structured by lines (of view) – nowhere more apparent than in Monchegorsk.[30]

Monschegorsk ("beautiful tundra") was planned in the 1930s by a sensational new model. The city center was placed three kilometers away from the processing plant itself – a fact that invoked debates about the loss of effective time as the workers had to actually travel to the plant. The town is equipped with a park, whose central axis runs perpendicular to the main street and ends by the shore of Lake Lumbolka in a terraced stairway framed by a white balustrade. This picturesque arrangement provides a paradoxical view to a mountain of tailings in the distance, in the neighboring town of Olenegorsk. The central axis of the city itself is the Metallurgica Prospect, Mineral Boulevard, which ends in the large nickel and copper production plant that leaves the town and its surroundings one of the most polluted places in Russia. What we see at the end of the street is a jumble of black steel and acid burnt ground. Lines of beauty end in the grandeur of mineral production – not unusual, we know, for Soviet planning.

The lines we read are markers of what Raoul Bunschoten would call subliminal forces or proto-urban conditions:

> The behavior of people, groups and institutions is linked to large-scale but sometimes invisible action tendencies, forces or more general conditions that create instability. We call these proto-urban conditions. They are like emotions in human beings, or like tectonic forces under the crust of the earth.[31]

The proto-urban conditions are not necessarily "soft" structures like social networks; they may also be the raw economic incentives behind extraction forces, infrastructural needs, or policy forces. Looking at the configuration of Monchegorsk, what we see are traces of such conditions – the structuring lines of the built environment. What we do not see is the human-natural mesh of pollutants pervading the air, soil, water, mushrooms, and other life forms. The trains of processed minerals sitting at the train station in Murmansk were traces of an industry that has marked the entire peninsula and contributed to what we might now call a dark ecology – the intimate mesh of humans as industry, humans as economics, humans as desire for minerals, humans as nature.

Philosopher Timothy Morton develops his thoughts on dark ecology exactly at the Kola, visiting the industrial city of Nikel right across the border from Norway.[32] A dark ecology is not one ecology, and not ecology as such. It refers to a global condition where what was once called nature, or ecology, cannot be perceived as something apart from the human sphere.[33] The two are deeply enmeshed. Man has changed the environment so profoundly that at this point there is no part of it that is not affected by human cultures of production and consumption. We have changed the environments at a molecular level; nothing that touches air remains "natural". Building a fundamental critique of the epistemology that posits man as a strong subject, free to appropriate, control, and exploit his object – the globe with all its resources – Morton claims that the two can no longer be separated.

The classical Western idea of landscape implies something that is apart from us and which can be perceived "at a glance". The perspective of a dark ecology changes this profoundly. Watching an industrial section of the Kola, we are not looking at landscapes, but at something that is us and at the same time not-us – a sulphur dioxide (SO_2) infused mesh of materials and desires. It is difficult to establish exactly how much SO_2 the air, ground, vegetation, human, and nonhuman bodies on the Kola peninsula contain. According to the Norwegian Geological Survey, an estimated amount of between 300,000 and 600,000 tons of SO_2 was emitted annually in the 1980s and 1990s, plus an additional 1,000 tons of copper and 2,000 tons of nickel. The peculiar fact is that, particularly in Monchegorsk, palladium (Pd) and gold (Au) are byproducts of the copper and nickel ore being processed there: 60,000,000 US$ worth of Pd and Au is estimated to have gone up in smoke every year during these decades, ending up floating around the terrain. "[We] observe the formation of a new ore deposit at the earth's surface in the surroundings of the smelter", say the scientists behind the report, Clemmens Reimann and Rolf Tore Ottesen, hinting at one of the remarkable traits of the Anthropocene – the manmade geological layers.[34]

Photographing the nickel plant in Monchegorsk reminds me that I too am complicit. The nickel in my camera along with the nickel in my kitchen appliances make me intimate with the polluting plant in front of me. I have lost my distance to the landscape. This is the reality that has propelled thoughts of a dark ecology – one that is based on the "the queer idea that we want to stay with a dying world".[35]

Vardø

Global coastal zone

Norwegian geologist Mathias Balthazar Keilhau in 1831 described Vardø as a major hub for fishing and commerce. At that point, the town appeared to be in a period of decline. While there had been 60 families and 11 major traders, there were now much fewer. Still, 150 Russian trade ships visited each year, mainly for an exchange of flour for fish, but also bringing peas, syrup, ropes, hemp, linen, tar, canvas, soap, pots and pans, leather, tallow candles, etc. In short: While being disappointed by the lack of scenic richness, Keilhau was concerned with the island as a hub for traders and fishermen.

In the airport in nearby Kirkenes, for many years travelers could see a large banner covering the end wall of the transit hall. THE OIL HAS ARRIVED, it said.[36] We first noticed it in 2011, thinking that it perfectly reflected the opportunistic aspirations of local communities along the Barents coast to be in position when the oil economy finally materialized. In 2011, Norwegian Statoil, Russian Gasprom and French Total were still planning to open the Shtokman field in the Barents Sea, a petroleum drilling adventure that would certainly benefit coastal communities in both Norway and Russia. The harbors and towns on a relatively small stretch of the Barents Coast closest to Shtokman were mobilizing: Kirkenes, Vardø, and the Russian community of Teriberka. In Vardø, a large supply harbor had been built at Svartnes, and Kirkenes had already started a general upgrade of its existing harbors, as well as initiating two large new harbors. However, Shtokman was put on hold in 2012 and the NOK 1 billion industrial harbor in Vardø is sitting empty, save of a single king crab operation, Arctic Catch, that serves the national and international markets (mainly the United Arab Emirates, Japan, China, South Korea, and Russia).[37] The Shtokman project well illustrates the risks of depending on mega-projects for regional development in the Artic – which Nadir Kinossian points out is a particularly Russian strategy.[38] In this case, it also had ripple effects for a Norwegian community, although not as devastating as the effect for small private investors in Teriberka.

Networked place

The success of the king crab industry in a global market is not novel in a Vardø context. This is a community that has thrived on an intense fishing industry for centuries. The harbor as it appeared until decay set in in the 1970s was the first state-financed large harbor development in Norway, finished in 1897. Today only a small corner of the harbor is active, and the important slipway is beyond repair. However, every part of this island has somehow served and supported the fishing industry. The outfields were covered by stock fish drying racks up until the 1960s, and livestock used to graze, paying no heed to borders between gardens, streets, and commons.

Figure 6.8 One of the dilapidated fish processing buildings in Vardø, decorated by streetart during KOMA-fest in 2012, curated by Norwegian streetartist Pøbel. The derelict slipway in the foreground. The slipway closed on the same night as most of the fish factories in 1989. Until then, it had allowed Vardø to maintain its boats and be independent of the nearby slipways, more than 3 hours away.

Photo: the author.

Our view from Klondyke that cold January day revealed Vardø literally as a networked place. It is hardwired and satellite-wired to the long Norwegian coast and the seas beyond it, to the Norwegian military and to the US, to the large eastern airspace, as well as to the mainland that we see in the distance. Other walks and repeated visits to this community has, however, provided a deeper under-standing of its social, economic, and political web of relations on a global scale.

Historically one of the richest fishing towns in the North, Vardø has always operated in an intimate relationship to international economies. In the twentieth century, the community's fortunes mir-rored the world markets. Vardø suffered economic collapse following the New York stock exchange collapse in 1933. The main fish factories were forced to close as they lost international customers. In the 1960s, a second recession occurred when the Norwegian government banned herring fishing, followed by a drop in fish export during the Nigerian civil war 1967–1970.[39] The last and crippling blow to the fishing industry came in the late 1980s, when the government closed down cod fishing due to overfishing.[40] Until then, at the height of the fishing era more than 3,000 fishermen would arrive every winter from Finland, from places along the Norwegian coast, and from Russia to take part in the rich winter cod fishing. For many years, the Italian fish buyer Dismondi would arrive by the coastal steamer. Dressed in a yellow linen suit and straw hat, he would inspect the stock fish until he hopefully decided to buy all of it.[41] The bustling fishing port pivoted around the beach zone. Here, the fish and the foreigners would arrive, cod would be processed, blubber cooked, and boats built. Barrels of salted seal meat were for a short period each year a physical trace of the Greenland ice flow. This faraway territory, only imagined by most people, materialized there, on the beach, by way of the seal meat.

The coastal zone in Vardø, then, extended not only from the offshore anchor places to the uphill fish racks, but also socially to other communities along the Norwegian coastline. The rocky beachline and wooden piers connected the onshore to the offshore in multiple ways. They connected the island to its immediate beyond of fishing resources, to the outer bay where derelict boats used to be sunk, to the next island over where birds hatch and herbs may be picked, and to the incoming shipping lanes. The shoreline further connected to the historical polar beyond, which is now turning into a strategic beyond, and not least to its global beyond – sea routes, investors, traders, and tourists.

Coastal zone

The coast, then, is a connecting zone and not a dividing line. Paul Carter argues in *Dark Writing, Geography, Performance, Design* that the coastline, an Enlightenment tool for measurement, reasoning, and territorial control, was a mere tool for cartographers, delimiting a territory under sovereignty, its dynamics concealing the "zone of environmental transformation and unprecedented human encounters".[42] In Vardø's case, if one were to look at the sharply outlined coastline as a defining trait for place, it would conceal the harbor as a transitional zone for loading and processing of fish, for entries and exits of national and international trading and fishing vessels, a transition zone between life at sea and life on shore. It is not a "boundary between worlds", as Anna Ryan calls the forced Enlightenment outline of territorializing lines on a map.[43] It is a productive and dynamic zone connecting worlds, where activities and imagination play off in a global network of trade, fishing, and social mobility. None of this is apparent from a merely visual perspective.

The small community on an Arctic island is not what Marot calls an anthropogenic island. Vardø has been connected to the global community on a material (fish, boats, goods), economical (trade), and mental (imagined spaces) level. It has been a relational place, pivoting around its coastal zone. It has not, however, been globalized, in the contemporary meaning of the word, as a neoliberal "unbounded free space" for trade.[44] Instead, it has contributed to a production of globalized networks, in the sense that it both persists as a self-contained place and *at the same time* relies on a relational network with other places.[45] The local has an open, relational, and connective character while remaining grounded in its local resources.

The hegemonic notion of globalization hinges, according to Doreen Massey, on a "powerful imaginative geography", especially in the North: "The vision of global space . . . is not so much a description of how the world is, as an image in which the world is being made".[46] On the Barents coast, it is not difficult to see how local places mobilize in the face of a prospective petroleum industry, one that is surely propelled by a global need for fossil fuel, as well as a globalized economy. Municipalities are eager to make a leap into the oil age – responding to a petroleum imaginary, while in fact it is far from certain that Barents Sea oil and gas will be heavily developed. In Vardø, municipal funding has been cast into the new industrial harbor. The long breakwaters of the new harbor metaphorically speak of a less dynamic sea–land interaction. The driver of this stealthy landscape situation is located far off shore, outside of the periphery of the settlement.

The locally grounded global

In Vardø now, local actors are working in a slow lived time in order to overcome the setback of the new industrial harbor, the one that made a leap in time by connecting to an offshore development that was only maybe to happen. However, lived time is not what it used to be, as Massey repeatedly points out throughout her prolific work on place. Time and space have been compressed, she claims, referring to "movement and communication across space, to the geographical stretching-out of social relations, and to our experience of all this".[47] Vardø is a place of ongoing production. In a slow time, old cars are sitting on the third story of an old fish storage building; an old ice house has served both as gym and garage. The city harbors a flourishing bird-watching tourism, emerging from the precise work of a local architecture office, Biotope, run by architect and avid bird-watcher Tormod Amundsen.[48] Biotope is designing bird hides and wind shelters in the Vardø area, based on precise knowledge of where birds feed and rest. They are accommodating for a future Vardø that is networked, part of a global flow of people, but still grounded in the local. In addition, the development platform Vardø Restored is working continuously to help property owners refurbish and bring new life into old wooden structures.[49] One could say that these practices are re-developing a sense of place which is "adequate to this era of time-space compression", as Massey puts it.[50] Vardø is not global as a "product of material practices of power" – which is what materializes in the new industrial harbor. It is a part of an extended global community in and by the prolific interest in its materials and materialities – its sense of place as grounded in local heritage, resources, light, and location.

Svalbard

Svalbard, at 74–81 degrees north, is home to the northernmost settlements in the world. It has been governed by Norway since 1920, but has been an international territory for centuries. Hunting, mineral exploration, innovation, research, and tourism have been motivations for a multinational presence. The settled territory is small and limited to two Norwegian towns: Longyearbyen, a mining town and tourism destination, and Ny-Ålesund, a former mining town that now serves as a research base closed to tourists. In addition, there are two Russian mining communities: the still-operating Barentsburg and the abandoned Pyramiden.

The archipelago was officially discovered in 1596, and already from 1611 whaling and walrus hunting were large industries for the English and the Dutch. The Russians started visiting Spitsbergen between 1715 and 1720, and a relatively small number of Russian over-wintering hunters visited the islands up until 1853. Norwegian hunting picked up around 1795 and continued until 1973, when hunting was banned in the archipelago.[51] Already by the late 1800s, mineral exploration had become prolific and mining commenced in Longyearbyen as early as 1903, eventually leading to eight different coal mines in the area, in addition to the Russian mines and the large Norwegian Svea mine. Several nations were exploring the territory for coal, gypsum, gold, and lead sulfide deposits.

Svalbard is a natural base for Arctic exploration. At the turn of the twentieth century, the small bay of Virgohamna at the very North-West of Spitsbergen served as venue for five attempts to reach the North Pole by air balloon between 1896 and 1928, the American Walter Wellmann and Swedish Salomon August Andrée being the most prominent actors. A former hub to air-flight innovation, the entire area of Virgohamna is now preserved as a heritage site featuring residue, rubble, and structural material from balloon hangars and other equipment related to the industry of aerial navigation innovation and technology.[52]

Not only Virgohamna, but also the coastal landscapes of Svalbard feature assemblies of material traces testifying to past activity and present research. Blubber cooking equipment, iron wheels from coal conveyor belts, railway tracks, mine entrances, transportation lines, ropes, hooks, rings, bolts, nails, chains, drill heads, cranes, patches of coal, and chemical ponds have all become part of the Svalbard landscape – revered as human heritage in an otherwise unwelcoming environment.

The present-day research landscape is similarly visible. Sorted by scale, research installations count radars, antennas, instruments for measuring precipitation and for monitoring permafrost, and grazing

Figure 6.9 Derelict mines litter the hillsides around Longyearbyen
Photo: the author.

enclosures installed by biologists to study vegetation changes. The landscape, then, has a processual character, by both testifying to past activity and being home to equipment that is monitoring its changing character. Each and every one of the found objects and contraptions leads back to forces of interest, be they economic, political, or social.

Contemporary global

Today, coal mining is coming to a close, and research and tourism remain the two growing industries in the archipelago for both Russia and Norway.[53] This development is signified by recent news that the Norwegian coal mining company Store Norske Spitsbergen Kulkompani is planning a move from mining to tourism by accommodating tourists in miners' facilities that are no longer in use.[54] As pointed out by Dieter K. Müller, however, this initiative does not reflect a need for new ways of sustenance in a permanent society, which is the case in many other Arctic settlements: "[S]mall populations and longtime economic dependency on extractive industries have created a situation whereby few alternatives are available to make a livelihood in arctic communities".[55] In Svalbard, an emergent tourism is encouraged by a Norwegian government eager to maintain a strategic presence on Svalbard. In the Future North project, we have called this a "presence policy" – one performed by both Russia and Norway in this territory. As this formerly profitable hunting and mining landscape slowly turns into a profitable tourism landscape, it provides a compelling twist to Don Mitchell's claim that "landscape's primary function is to generate profit".[56] In Svalbard's case, a profitable landscape is a welcome tool to justify presence in a contested territory.

Global repository, local recipient

Outside influences upon Svalbard as a territory are extensive: chemical pollution, invasive species, oceanic plastic, tourism, and population migrations impact its physical terrain, climate, and communities. Conversely, the territory has agency as a repository of cultural heritage, through monitoring systems such as satellite monitoring and remote sensing systems of land and sea; all provide essential data and research material and thus influence the actions of the larger global community. It is also networked as a global repository of climate data provided by Svalbard Satellite Service (Svalsat); the ground satellite station operates close to 50 antennas, reflecting interests from individual research centers across the world, and delivers weather and Landsat data to the world community. The Global Seed Vault archives indigenous seeds for the world community, and the Arctic World Archive is under planning in the abandoned Mine 3 in Longyearbyen. The archive will provide storage of material of national value on high-resolution film, and the Brasilian National Archive seems to be the first customer.[57]

However, it is not only Svalbard's hardware that is globally connected, but also the territory's ground materialities. Temperatures rise more than twice as fast than further south, permafrost and glaciers are thawing, and the effect of sea ice loss on animal populations is acute. Its waters are monitored for invasive species and invasive plastics such as micro plastics (<5.5mm).[58] Sand and pebble material, which shapes into fantastic pyramidal forms along the beaches, is subject to erosion following increased precipitation and melting permafrost. There seems to be no material in the Svalbard territory that is not in flux, ultimately as a result of human consumtion and production patterns. The territory is continually remade through industry, tourism,[59] research,[60] and global warming.

Distance collapsed

Due to the processual and fluid character of its landscape, Svalbard is an emblematic example of a collapse of distance. M/S Billefjord that day served as a site for an emblematic situation, one that speaks to the processual character of Svalbard. Svalbard is a place of ever-shifting constellations of trajectories – as well as what Doreen Massey would call a throwntogetherness – a throwntogetherness of human and nonhuman beings:[61] "This is the event of place, in part in the simple sense of the coming together of the previously unrelated, a constellation of processes rather than a thing".[62] She claims that the multiplicity of space that any given assembly makes up is often not noticed except in an unlikely event. Our boat enacted such an unlikely event: Our looking at the Pyramiden hillside and its many materials, at the receding Hammerskiöld glacier, at the seals, and the lumbering polar bear, is shaped both by our interest in the Svalbard territory, and by the territory itself. The territory and its various materials and beings interest our gaze and affect it. Aboard M/S Billefjord we neighbor a receding glacier and a coal mining town, the latter of which has extracted exactly that responsible for the glaciers' decline. We neighbor an entire territory, an archipelago, that is not a constant, static place. Not even its geology is "local". Svalbard has traveled from coordinates of what is now the Congo over the course of about 500 million years. The coal seams that we observe in Pyramiden were formed 310 million years ago, at a latitude of about 10 degrees north.[63] They were extracted at the latitude of Pyramiden at 78 degrees north, shipped off to Europe and burnt by the steel industry there between 1956 and 1992. The Soviets imported the thin layer of fertile soil in Pyramiden, as well as the grass, from southern Russia, where it was deposited by the glacial front of the last ice age. The places around us are not settled, bounded, or a "pre-given".[64] They are part of an ever-changing migratory process.

The Billefjord scene pivots around migration and immigration, around fluid and unsettled trajectories of people and materials. It is a temporary situation. We were thrown together in an unlikely

Figure 6.10 Mining was discontinued in Pyramiden in 1992. The mountain Pyramiden has three levels of coal, and today appears scarred from the operation.

Photo: the author.

situation by being on board a boat in the Arctic, watching a mixed material scenery, and the scene was one of utter collapse. What we were looking at was changing because of us. The Anthropocene effects on the seemingly pristine landscape are not as visible as on the Kola Peninsula. It requires a further degree of imagination – an imagination of that weird loop that coal has performed from Pyramiden on our left onto the glacial ice on our right, where it may be refound as glacial deposits. This is the weird loop of being in Svalbard now – knowing that what you look at is changing due to you and your kind, and that even by flying there to see it you have contributed further to its change. You are as much complicit in the territory's change as on the Kola Peninsula. That old-fashioned distance of the tourist gaze, the joy of landscape splendor, collapses into a deep intimacy with Svalbard landscape.

The three divergent territories looked upon in this chapter all manifest a collapse of the distance inherent to a traditional landscape gaze and conception. They all perform with a particular intensity, as social, economic, and political actors have an acute sense of the material resources present in these territories; the minerals on the Kola peninsula, the natural, human, and built resources in Vardø, the visual expanses, mineral resources, strategic location, and repository repertoire of Svalbard. All of them are networked, Svalbard even hyper-networked given the extreme fluidity of people, species, temperature, and data mined on all of these. They are productive landscapes, subject to global networks of interest and to forces that are not immediately discernible to the eye. They all demonstrate the need to look at landscape otherwise, to negotiate the local in the face of the global, and to identify the strong forces that are changing our landscapes.

Notes

1 It forms the easternmost point of operation for the Norwegian military and the NATO alliance in the Artic. Operated by the Norwegian Armed Forces, it is owned by the US.
2 In the nineteenth century, emigrating Norwegians would sometime choose between going to America and going to Vardø – both places were known as territories of opportunity. Letters from relatives in America would provide them with place names that found their way onto the local landscape.
3 For a more detailed account of this networked view, see William L. Fox's blogpost on www.oculs.no/projects/future-north/news/?post_id=3590.
4 The material composition of Pyramiden has been documented in Elin Andreassen, Hein B. Bjeck, and Bjørnar Olsen, *Persistent Memories, Pyramiden – a Soviet Mining Town in the High Arctic* (Trondheim: Tapir Academic Press, 2010).
5 Don Mitchell, "Landscape and Surplus Value: The Making of the Ordinary in Brentwood, CA," *Environment and Planning D: Society and Space* 12 (1994): 10.
6 Ibid., 10.
7 Michael Light, *Some Dry Space: An Inhabited West*, Anne M. Wolfe (introduction), (Reno: Nevada Museum of Art, 2008), 9.
8 Raoul Bunschoten, *Urban Flotsam, Stirring the City, Chora* (Rotterdam: 010 Publisher, 2001), 19.
9 Denis Cosgrove, *Social Formation and Symbolic Landscape* (London, WIS: The University of Wisconsin Press, 1984), 16.
10 John Barrell, *The Idea of Landscape and the Sense of Place, 1730–1840: An Approach to the Poetry of John Clare* (Cambridge: University Press, 1972), 1.
11 Ibid.
12 The notion of landskip is amongst others referred to by Kenneth Robert Olwig, *Landscape, Nature, and the Body Politics, from Britain's Renaissance to America's New World* (Madison: The University of Wisconsin Press, 2002).
13 John Wylie, *Landscape* (London: Routledge, 2007), 59.
14 Olwig, 10.
15 Ibid., 24.
16 Ibid., 25.
17 Ibid.
18 Antoine Picon, "What Has Happened to Territory?" *Architectural Design* 80 no. 3 (2010): 95.
19 Ibid.
20 See the work of picturesque theorists such as Uvedale Price, *Essays on the Picturesque V1: As Compared with the Sublime and the Beautiful* (London: Printed for J. Mawman, 1810).
21 Picon, *What Happened*, 97.
22 Philip Steinberg and Berit Kristoffersen, "'The Ice Edge is Lost . . . Nature Moved it': Mapping Ice as State Practice in the Canadian and Norwegian North," *Transactions of the Institute of British Geographers*, 2017, doi: 10.1111/tran.12184.
23 Sebastien Marot, "Ecology and Urbanism: The Deepening of Territories," *The Eco-Urb Lectures, Ecology and Urbanism,* ed. Victoria Sjøstedt (Copenhagen: The Royal Danish Academy of Fine Arts, Schools of Architecture, Design and Conservation, School of Architecture, 2011), 74.
24 Doreen Massey, *Space, Place and Gender* (Cambridge: Polity Press, 1994), 147.
25 Massey, *Space*, 156.
26 See Susan Carruth's chapter in this book for a more extensive reading of Massey's discourse on space, time, and place.
27 E. Carina H. Keskitalo and Mark Nuttall, "Globalization of the 'Arctic'," in *The New Arctic*, ed. Birgitta Evengård, Joan Nyman Larsen, and Øyvind Paasche (Berlin: Springer, 2015), 176.
28 Doreen Massey, *For Space* (Los Angeles: Sage, 2005), 101.
29 Jane Bennet, *Vibrant Matter: A Political Ecology of Things* (Durham, NC: Duke University Press, 2010), viii.
30 Minerals processed are apatites and nephelines (nickel and copper) as well as iron ore.
31 Bunschoten et al., *Urban Flotsam*, 28.
32 Timothy Morton, *Ecology Without Nature: Rethinking Environmental Aesthetics* (Cambridge, MA: Harvard University Press, 2009).
33 Timothy Morton, *Dark Ecology, For a Future of Logic Coexistence*, (New York: Columbia University Press, 2016).
34 Clemens Reimann and Rolf Tore Ottesen, "There Is Gold on Your Feet," *NGU-Focus* 9 (2005).
35 Morton, *Ecology Without Nature*, 185.
36 Oljen har kommet!
37 See Peter Hemmersam and Janike Kampevold Larsen, "Landscapes on Hold: The Norwegian and Russian Barents Sea Coast in the New North," in *Critical Norths, Space, Nature, Theory*, ed. Sarah J. Ray and Kevin Maier (Fairbanks: University of Alaska Press, 2016).

38 Nadir Kinossian, "Re-colonising the Arctic: The Preparation of Spatial Planning Policy in Murmansk Oblast, Russia," *Environment and Planning D, Government and Polity* 35, no. 2 (2016): 10–11.

39 Stock fish were exported to Nigeria for decades, and one of the significant recessions in Vardø was directly linked to a failing market in Nigeria due to the "Biafran" Civil War.

40 In 1992, Canada declared a moratorium on the Northern Cod Fishery, which had reached a critical level of biomass. The Atlantic cod population reached near depletion at the same time. Greenland lost its cod populations while Norway and Russia had monitored theirs and managed to save them.

41 Recounted by Arthur Andreassen, an 89-year-old fisherman in Vardø, a "living archive" of information on the town's fishing history.

42 Paul Carter, *Dark Writing, Geography, Performance, Design* (Honolulu: University of Hawai'i Press, 2009), 54.

43 Anna Ryan, *Where Land Meets Sea, Coastal Explorations of Landscape Representation and Spatial Experience* (Farnham: Ashgate, 2012), 32.

44 Massey, *For Space*, 83.

45 Ibid., 188.

46 Ibid., 84.

47 Ibid., 147.

48 Biotope is run by Tormod Amundsen and Elin Taranger.

49 Vardø Restored was founded by Svein Harald Holmen, who currently runs it with Brona Keenan, http://vardorestored.com/en.

50 Massay, *Space*, 147.

51 Odd Lønø, *Norske fangstmenns overvintringer på Svalbard og Jan Mayen 1795–1973,* ed. Sander Solnes and Per Kyrre Reymnert (Tromsø: Svalbard Museums, 2014). Norwegian hunter Arthur Oxaas spent a total of 30 winters in Svalbard, and accounts for having shot 373 polar bears, 1889 Arctic foxes, and a variety of other mammals. Arthur Oxaas, *Svalbard var min verden* [*Svalbard Was My World*] (Skien: Vågetmot Miniforlag, 2008).

52 Hein B. Bjerck, and Leif Johnny Johannessen, ed., *Virgohamna, In the Air Towards the North Pole* (Longyearbyen: Governor of Svalbard/The Svalbard Tourist Board, 1999). All industrial remnants in Virgohamna are preserved under *Regulations Concerning North-West Spitsbergen National Park* (1973).

53 Julia Gerlach and Nadir Kinossian, "Cultural Landscape of the Arctic: 'Recycling' of Soviet Imagery in the Russian Settlement of Barentsburg, Svalbard (Norway)," *Polar Geography* 39, 2016, doi: 10.1080/1088937X.2016.1151959.

54 Announced in April 2017. www.dn.no/nyheter/2017/04/03/0959/Turisme/gruveselskap-satser-pa-turisme.

55 Dieter K. Müller, "Issues in Arctic Tourism," in *The New Arctic*, ed. Birgitta Evengård, Joan Nyman Larsen, and Øyvind Paasche (Berlin: Springer, 2015), 147.

56 Don Mitchell, "California Living. California Dying: Dead Labor and the Political Economy of Landscape," in *Handbook of Cultural Geography*, ed. Kay Anderson, Steve Pile, and Nigel Thrift (London: Sage, 2003), 241. Paraphrased by Wylie, *Landscape*, 105.

57 Christopher Engås, "Åpner for nytt 'dommedagshvelv' på Svalbard", http://svalbardposten.no/nyheter/apner-for-nytt-dommedagshvelv-pa-svalbard/19.8234 The Arctic World Archive will be owned by Store Norske and Piql AS. The Global Seed Vault was inaugurated in 2008.

58 The first study of micro plastics in Arctic waters, from 2015, found that "microplastic abundance values in surface waters were of the same order of magnitude as those found in the North Pacific and North Atlantic, greater than those of the California currents system, south and equatorial Atlantic, but less than hose reported for the closed water systems of the Mediterranean". Amy L. Lusher, Valentina Tirelli, Ian O'Connor, and Rick Officer, "Microplastics in Arctic Polar Waters: The First Reported Values of Particles in Surface and Sub-surface Samples," *Scientific Reports* 5 (2015): article number 14947. www.nature.com/articles/srep14947#ref-link-section-6

59 As in many Arctic communities, tourism is a rapidly growing industry. Tourists that visited Svalbard in 2016 number 64,000 guests in commercial accommodation and 41,600 guests on cruise ships, as well as expedition-style guests numbering about 14,000. Numbers provided by Visit Svalbard in December 2016.

60 The University Center in Svalbard (UNIS) in Longyearbyen and Ny-Ålesund between them accommodates hundreds of researchers annually from more than 30 different nations.

61 Massey, *Space*, 151.

62 Ibid., 141.

63 Rasmus Weitze, former master's student at the Tromsø Academy of Landscape and Territorial Studies, explored the migration of the Svalbard coal layer from sedimentation to ice core residue in his project, Evolutionary Accumulation, 2016. These data were researched by him.

64 Massey, *For Space*, 151.

Bibliography

Andreassen, Elin, Hein B. Bjeck, and Bjørnar Olsen. *Persistent Memories, Pyramiden – a Soviet Mining Town in the High Arctic.* Trondheim: Tapir Academic Press, 2010.

Barrell, John. *The Idea of Landscape and the Sense of Place, 1730–1840: An Approach to the Poetry of John Clare.* Cambridge: Cambridge University Press, 1972.

Bennet, Jane. *Vibrant Matter: A Political Ecology of Things.* Durham, NC: Duke University Press, 2010.

Bjerck, Hein B., and Leif Johnny Johanessen. *Virgohamna, in the Air Towards the North Pole.* Longyearbyen: Governor of Svalbard/The Svalbard Tourism Board, 1999.

Bunschoten, Raoul, Hélène Binet, and Takuro Hoshino. *Urban Flotsam: Stirring the City.* Rotterdam: 010 Publishers, 2001.

Carter, Paul. *Dark Writing, Geography, Performance, Design.* Honolulu: University of Hawai'i Press, 2009.

Cosgrove, Denis. *Social Formation and Symbolic Landscape.* London & Wisconsin: The University of Wisconsin Press, 1984.

Evengård, Birgitta, Joan Nymand Larsen, and Øyvind Paasche, eds. *The New Arctic,* Berlin: Springer, 2015.

Gerlach, Julia, and Nadir Kinossian. "Cultural Landscape of the Arctic: 'Recycling' of Soviet Imagery in the Russian Settlement of Barentsburg, Svalbard (Norway)." *Polar Geography* 39, 2016. doi: 10.1080/1088937X.2016.1151959

Hemmersam, Peter, and Janike Kampevold Larsen. "Landscapes on Hold: The Norwegian and Russian Barents Sea Coast in the New North." In *Critical Norths, Space, Nature, Theory,* edited by Sarah J. Ray and Kevin Maier, 171–89. Fairbanks: University of Alaska Press, 2017.

Keskitalo, E. Carina H., and Mark Nuttall. "Globalization of the 'Arctic'." In *The New Arctic,* edited by Birgitta Evengård, Joan Nyman Larsen, and Øyvind Paasche, 175–89. Berlin: Springer, 2015.

Kinossian, Nadir. "Re-Colonising the Arctic: The Preparation of Spatial Planning Policy in Murmansk Oblast, Russia." *Environment and Planning D, Government and Polity* 35, no. 2 (2016): 1–18.

Light, Michael. *Some Dry Space: An Inhabited West,* Introduction by Ann M. Wolfe. Reno: Nevada Museum of Art, 2008.

Lusher, Amy L., Valentina Tirelli, Ian O'Connor, and Rick Officer. "Microplastics in Arctic Polar Waters: The First Reported Values of Particles in Surface and Sub-surface Samples." *Scientific Reports* 5 (2015): article number 14947.

Marot, Sebastien. "Ecology and Urbanism: The Deepening of Territories." In *The Eco-Urb Lectures, Ecology and Urbanism,* edited by Victoria Sjøstedt, 73–111. Copenhagen: The Royal Danish Academy of Fine Arts, Schools of Architecture, Design and Conservation, School of Architecture, 2011.

Massey, Doreen. *For Space.* Los Angeles: Sage, 2005.

Massey, Doreen. *Space, Place and Gender.* Cambridge: Polity Press, 1994.

Mitchell, Don. "Landscape and Surplus Value: The Making of the Ordinary in Brentwood, CA." *Environment and Planning D: Society and Space* 12 (1994): 7–30.

Morton, Timothy. *Dark Ecology: For a Future of Logic Coexistence.* New York: Columbia University Press, 2016.

Morton, Timothy. *Ecology Without Nature: Rethinking Environmental Aesthetics.* Cambridge: Harvard University Press, 2009.

Müller, Dieter K. "Issues in Arctic Tourism." In *The New Arctic,* edited by Birgitta Evengård, Joan Nyman Larsen, and Øyvind Paasche, 147–58. Berlin: Springer, 2015.

Olwig, Kenneth Robert. *Landscape, Nature, and the Body Politics, from Britain's Renaissance to America's New World.* Madison: University of Wisconsin Press, 2002.

Oxaas, Arthur. *Svalbard var min verden* [Svalbard was my world]. Svalbardminner no. 38, Skien: Vågetmot Miniforlag, 2008.

Picon, Antoine. "What Has Happened to Territory?" *Architectural Design* 80, no. 3 (2010): 94–99.

Price, Uvedale. *Essays on the Picturesque V1: As Compared with the Sublime and the Beautiful.* London: Printed for J. Mawman, 1810.

Reimann, Clemens, and Rolf Tore Ottesen. "There Is Gold on Your Feet." *NGU-Focus* 9, 2005.

Ryan, Anna. *Where Land Meets Sea, Coastal Explorations of Landscape, Representation and Spatial Experience.* Farnham: Ashgate, 2012.

Steinberg, Phil, and Kristoffersen, Berit. "'The Ice Edge Is Lost … Nature Moved It': Mapping Ice as State Practice in the Canadian and Norwegian North." *Transactions of the Institute of British Geographers* (2017): 625–41. doi: 10.1111/tran.12184.

Wylie, John. *Landscape.* New York: Routledge, 2007.

7 Visual and sensory methods of knowing place

The case of Vardø

Henry Mainsah

Introduction

Changes brought about by the convergent forces of climate change, globalization, technological development, and related socio-ecological shifts pose significant challenges for the landscapes of urban and rural areas all over the world. The intersection of all these forces on landscapes, rich in their complexity, requires new methods for understanding, envisioning, and planning alternative futures.

In recent years, there has been a notable increase in the popularity of methods for policy-making at regional and local levels in many European countries. This follows increased demand for an informed framework for urban planning from local and regional administrative bodies. However, much of the methodological literature on methods within landscape architecture, urban planning, and cultural policy research is underpinned by a positivist epistemology with a strong focus on facts, "evidence", assets, and physical infrastructure based on what is quantitatively measurable.

This chapter presents a methodological approach developed in the Future North project, one that draws on visual and sensory modes of inquiry. It addresses visual, sensory, and reflexive modes of inquiry that offer a means of representing landscapes as lived and embodied. I argue that visual and sensory methods provide a more richly textured, multi-layered approach to understanding landscapes because they are well adapted to revealing the complexity of interactions between physical and cultural landscapes.

In order to explore these methods, the chapter describes a series of interdisciplinary research workshops that took place in Vardø, a small town located off the northeast Barents coast of Norway. It involved researchers from the Future North project, landscape architecture students from the University of Tromsø, and participants from a local building restoration project called Vardø Restored.[1] These sessions were part of efforts to find ways of investigating and representing urban space and place in a manner that reveals the complex, multi-layered, and dynamic qualities of landscapes not typically addressed through standard mapping methods used in urbanism, landscape architecture, and planning. Insights gained through visual and sensory methods were meant to serve as empirical data for a cultural mapping exercise used in connection with future city planning.

The project setting

Vardø is an Arctic island town that used to be the commercial centre of Finnmark County due to its abundant fishing opportunities. In 2015, Vardø had a population of about 2,128 – only half of what it was 20 years earlier. Over an extended period beginning in the 1980s, the town witnessed a sharp decline in its main industry, cod fishing, causing the closure of the town's many fishing factories; most of the fishermen had to sell their licenses and, along with them, their boats. In recent years, there has been a gradual increase in the quantity of cod populating the Barents Sea, and Vardø still remains the best location for fishing on the Barents coast. However, as a result of the decline in the fishing industry, the social, cultural, and economic infrastructure of the town has suffered a gradual and sustained neglect.

Vardø was used as one of the case studies for the Future North project, which aimed to map the "future" landscapes of northern territories and, as part of this, to study the relationship between people and their environment, as well as the importance of people's agency in the development of these landscapes. In the Vardø case, Future North collaborated with a local project called Vardø Restored to conduct a participatory mapping exercise as part of the development of a sustainable future city plan. The aim of this development plan was to trigger the town's economic and social revival in a way that maintained the identity of the place and its cultural and community values. Vardø is located relatively close to the Norwegian–Russian border, in a region that has a rich and diverse cultural history. A variety of resources, from minerals to fisheries to hydrocarbons, have shaped and continue to affect the character of the region in various ways. The central question for the region, and for the Future North project, was how to anticipate and respond to changes in the physical, social, and cultural environment.

The process

The project adopted a collaborative approach where the research was carried out by a team comprising landscape architecture students, their teachers, representatives from Vardø Restored, and researchers from the Future North research project. The students participated in the research process as part of a master's studio course in Landscape Architecture, the aim of which was to help them to develop knowledge about territorial management issues in local communities. In the process, they were to learn and apply fieldwork, documentation, and mapping methods, as well as participatory research approaches with communities. Vardø Restored served as a platform for creating a sustainable town development strategy that balanced economic growth with preservation of the town's identity and cultural heritage.

The process of collaborative work involved engaging in a cultural mapping exercise that connected the local architectural landmarks and the natural physical landscape with the history, community, collective identity, and lived experience of the inhabitants of Vardø. This cultural mapping exercise aimed to assist the Vardø community's efforts at generating strategies for sustainable development that prevented environmental degradation and social disruption and respected cultural identity. The research process and the research outputs – textual and visual – were co-designed by all participants.

The project carried out a total of three fieldwork trips to Vardø. Each trip lasted from three to seven days. Each trip also followed a similar format. We kicked off with a workshop in which a plan with objectives and outcomes of the trip was drawn up. During the workshop, a methodological approach to the type of research needed for that trip was defined. Researchers from the Future North project provided hands-on training on research methods. The group was divided into teams, with each group assigned a research task. The students were encouraged to make full use of the repertoire of their landscape architecture skills during the research process. Following this, the team would have daily workshops to discuss experiences and reflections from the field, work progress, and logistic issues. The workshops usually ended with a final session where all the fieldwork data collected by the different groups was presented and discussed.

The research team sought to document and represent the lived experience of place and space, particularly the way in which inhabitants of Vardø constructed meanings of space and place through personal experiences and everyday activity. Equally important was the way in which the discourse on cultural heritage and collective history was articulated and lived. The project developed a methodological assemblage that comprised of archival research, landscape architectural techniques, speculative design, and a range of visual and sensory ethnographic methods, including interviews, observations, audio recordings, written reflections, and photographic documentation. Empirical data was generated through these methods to form the basis for creating strategies for future action on sustainable community development.

Social inquiry as place making

Theories of place are particularly useful for understanding how human practices of everyday life, performance, and imagination are implicated in the way people construct a "sense of place". In this regard, visual anthropologist Sarah Pink's reference to philosopher Edward Casey's theory of place is interesting. Casey critiques earlier anthropological conceptualizations of "place", understood as "carved out of space or superimposed on space".[2] Casey instead argues that space and time are contained in place, rather than vice versa. Thus, he implicitly suggests that place is central to our way of "being in the world", in that we are always "emplaced".[3] He suggests that places are repositories of both animate and inanimate entities.[4] Places also gather experiences and histories, even languages and thoughts. Whereas material dimensions of landscapes and the physical environment are often thought of as built structures that relate to functionality in modern life, places are thought of as sites of human experience comprised of social relationships, memories, and emotions, and how these are negotiated on an everyday basis.[5] Thus, although a town, for example, can be identified as a particular physical urban place, phenomenologically it can be experienced as several types of place simultaneously, depending on the person experiencing/making the place. Local politics and power relations often shape these different and sometimes intersecting subjectivities, which can themselves represent shifting and complex power contexts.

Casey suggests that there are two essential structures that pervade places. One is the centrality of the experiencing body to place, and the other is "the gathering power of place itself", meaning its ability to draw together bodies and things, time, and space.[6] Casey argues that place should be seen as an "event" and therefore as continually changing in accordance with its own power dynamism, rather than just a static site. Pink[7] further argues that place can be created in a variety of contexts, either as determined physical locations (e.g. in a house or a garden), in a public space (e.g. a street), or in movement (e.g. by walking a path). Thus, for Pink, the experiencing body is central to the production of place as it determines place through its movement and physical multisensory dimensions.

Pink suggests that a focus on place might encourage us to conceptualize fundamental aspects of how we are situated as embodied beings (positioned differently by gender, generation, class, race, ethnicity, and more) in the world.[8] For the researcher, place can be used to describe the dual context of research as being both the place we inhabit and the place we investigate. This means that as researchers, while seeking to understand the emplacement of others and the everyday practices through which the places of which they they form a part are continually reconstituted, we are ourselves (bodily) emplaced. Pink suggests that in order to understand the different ways through which places are made and lived, researchers need to both investigate how research participants construct a sense of place individually, and also reflect on how researchers, together with research participants, collaboratively construct a sense of place through research practice.[9] We need to employ methods that offer possibilities for researchers to empathically imagine themselves in places occupied, and sensations felt, by others. Researchers might not feel precisely the same sensations as research participants, nor understand these through the same culturally and biographically informed narratives. Nevertheless, there is a wide repertoire of visual and sensory modes of inquiry with the potential to invite the researcher into other people's worlds, and in doing so to empathize with their emplacement, or "the sensuous interrelationship of body–mind–environment".[10]

In our effort to develop ways of knowing the cultural landscapes we encountered in our project, we followed the lead of scholars who have designed theoretical approaches that take into account the production of place and space through performances and materialities. In order to develop such deep insights, social science researchers are increasingly resorting to physical engagements in activities to glean different types of experiences beyond the visual and the verbal in their accounts of place production. Researchers have focused on activities such as walking as a means to gain insight into how people produce place and make meaning from engagement with places.[11] Overall, the project aimed at gaining an understanding of the landscape through forms of inquiry that focused on curiosity, activated thinking, affect, and reflexivity, in an attempt to become attuned to the ruptures in our habitual ways of knowing the landscape. We engaged in multisensory forms of inquiry[12] that put emphasis on hearing, touching, and seeing, as well as a set of other documentation tools.

Diverse ways of seeing

I love watching waves. Whenever I go to a new place close to the sea I always like to wander off to go watch the waves. At the workshop, each student was asked to wander around and document features of the landscape that caught our attention. After we left the workshop I took my camera with me and headed towards the seashore to explore. It was very windy outside. As I was approaching the shore I began to wonder where I could stand to take the best shot of the waves. The lighthouse! The lighthouse stood towering at the end of the path I was walking on, leading to the seashore. The lighthouse reminded me of the popular folk tale in Taiwan about the Goddess of the Light. The girl who became known as the Goddess of the Light was the daughter of a fisherman. She would carry a lamp every night to the seashore to wait for her father to arrive with his boat from the fishing trip. When her father arrived, they would walk home together. One day her father did not arrive, and since then she goes to the seashore everyday in hope that her father will arrive. She later went to become a guiding light for all fishermen at sea, helping them arrive back safely to the shore. From the top of the lighthouse I pointed my camera downwards and took a series of panorama shots as the waves hit the shore. Through the eye of my camera I watched the waves. I like to watch the waves, as every wave is different. Each wave carries with it its own force. It is hard to predict the level of force with which it will hit the shore and how strongly it will retreat. Neither is it easy to predict how strong the next wave will be. I see the unpredictability of the waves as a good metaphor for my master's project. The strength of each incoming wave represents the tools, insights, and materials that I bring forth when I present the progress of my project to the class each time. After the presentation I will retreat in the same way as the waves will retreat, only to come back again even stronger.

(Excerpt of the field notes of a landscape architecture student in our research team)

This vignette shows the personal reflections of a master's student from Taiwan who was part of the research team during one of our trips to Vardø. Her reflections were conveyed to me during an informal debriefing session after the trip, when we sat together going through her field notes. Her reflections provide an indication of how our subjectivity shapes the way we experience social and physical spaces in everyday life. The research team comprised members of different national origins (Norwegian, Canadian, Dutch, Ethiopian, Zimbabwean, Taiwanese, Greek) and different disciplinary backgrounds (humanists, social scientists, architects, designers). In addition, the research team came to Vardø as outsiders, looking at the town through fresh eyes. The diversity of the research team provided a rich and varied set of veils through which a reading of place was filtered.

Sociologist Margarethe Kusenbach suggests that our perception of the environment is filtered through a series of veils.[13] Some of these veils include our sensual apparatus, our emotions, tastes, values, experiences, and identities. In the practical course of everyday life, we are not always aware that the way in which we perceive place is determined by this series of veils.

Kusenbach identifies two types of perceptual filters through which we can view place in such situations: practical knowledge and tastes/values. Practical knowledge is closely linked to our personal interests, talents, dispositions, and sensibilities. For the landscape architects, for example, the practical knowledge developed through their professional training might provide one such perceptual veil through which to view the environment within which we were working.

During their exploration of Vardø, one of the things that immediately caught the eye of the landscape architecture students was a seemingly abandoned wooden building at the pier with a big graffiti painting covering one of the outside walls. When they inquired about the building, they found out that in the past it was used as a boarding house for fishermen and for storing fishing equipment. During a tour of the building, their trained gaze caught details such as the north-facing elevation, the concrete floors, and the fact that the building, despite its appearance of total disrepair, had retained much of its original layout, materials, paint, and internal fittings. They made sketches and took pictures of the unique features of the building, and later presented these to the owner, who up until then had not fully noticed these details. This illustrated how professional knowledge could create an appreciation of details of a place that might otherwise escape others' attention.

The vignette at the beginning of this section illustrates the constitutive role of tastes, values, and personal experience in the complex process of perception. The Taiwanese student's reading of the lighthouse is mediated through the popular folk tale in Taiwan about the Goddess of the Light. She is also able to see the movement of the waves through the metaphor of a design process.

The workings of such perceptual filters, in a scenario such as the one involving our research team in Vardø, helped produce a diversity of views of the town mediated by professional knowledge and personal experience from the vantage point of outsiders. Combined with these were the views of the local members of the research team from Vardø, mediated by biography, history, social relationships, and everyday embodied knowledge, and negotiated from the vantage point of insiders.

Photographing as a reflexive act

One of the tasks the students were given during the research trips to Vardø was to walk around and take pictures of the features of the buildings and natural landscape that caught their interest and which they considered significant. This exercise yielded a series of photographs that vividly showed a variety of detailed insights, sensibilities, and ways of seeing the natural and the built environment. Such photographs reveal small details often taken for granted, or overlooked features of the town that might be brought to the fore through the outsider photographic "gaze" of the students. The discussions we had after such activities revealed some aspects of the practice of *photographing*, which had equal methodological significance as the outputs (the photographs).

Through the process of pausing to consider what to take pictures of, which shots to take and when, the person taking the photograph in effect composes formal aspects that might communicate her way of seeing the space to construct a particular image. Figure 7.1 shows a picture taken by one of the students from an angle that captures the top parts of the buildings in a Vardø neighbourhood. In the process of finding out what to photograph, the student had to pay attention to tiny details of the place that might ordinarily be overlooked. Furthermore, in the process of figuring out what angle was best to capture the colour profile of the houses, the student had to look at the streets from several angles. Photographing, in this case, becomes the act that triggers a new frame of observation: from an upstairs window, from where the student is encouraged to view the neighbourhood from a fresh angle. This encourages the student to reflect on whether the choice of the paint colour of houses is an aesthetic feature that represents the unique identity of Vardø.

For the student in such a scenario, the process of taking the photograph produces an "activity of thinking", through which he can begin to see the features of the town's landscape from a new

Figure 7.1 A view from an upstairs window showing houses in a neighbourhood. The photographer noticed the juxtaposition of houses painted in bright colours, which led him to wonder whether the colour of houses could be considered as a distinct feature of Vardø's identity.

Photo: Ronald van Schaik.

vantage point.[14] Coats argues that it is possible in such situations for the camera to function as a "nomadic weapon", in the sense that it encourages the photographer to recompose his habits of thought and movement. By describing the reflexive potential of the camera as a methodological device, Coats draws on Deleuze and Guattari's notion of the "nomad" as a figure that operates in an open and free space in which things and flows are distributed, rather than in a closed space of linear and solid things.[15] The nomadic weapon is positioned in the domain of free action and is described as "active, engendering, and traversing; associated with potential, free action, affect, projection, speed, and becoming" in a more performative and process-oriented domain.[16] Coats suggests that positioning the camera as a nomadic weapon in research implies that its purpose is derived first from the subjective engagement of the operator and recognizes the act of *doing* photography as being central. In this case, the camera's nomadic force is one of action and process, rather than capture and representation. This highlights the role of the process of photographing in facilitating moments of self-reflexivity and introducing new ways of knowing. The photographic act here represents a reflexive, embodied, and relational engagement that may help activate new ways of seeing our everyday environment. The photograph as a document becomes a by-product, where the embodied encounter is privileged to consider the experience of photographing. The camera becomes an apparatus of investigation to create new reflections that affect the photographer.

The guided tour

Researchers have explored walking in a variety of sophisticated ways as a mode of inquiry.[17] Yuha Jung, for example, observes that walking serves as a form of inquiry where researchers develop insights on place and space through physical engagement.[18] Jung describes walking as an interactive way of knowing, where one allows the entire body and all of its senses to experience one's surroundings.

In the Future North project, the research team travelled to several places in the Arctic. Whenever the team travelled to a new place, one of the things they did was explore the area on foot. Through transect walks, sometimes performed individually but most often in a group of researchers, they observed, analysed, and developed concepts in and of a landscape. In this way, the members of the research team were able to get a sense of the new surroundings and a feel of the landscape.

During the first workshop in Vardø, the research team was taken on a walking tour of the town by Svein Harald Holmen, the leader of the Vardø Restored project and a collaborating partner. While leading the research team along the streets of the Vestervågen neighbourhood, he told stories about each of the buildings as we passed by. Most of the buildings on that side of the harbour, the team found out, had been built or rebuilt after the Second World War, as the German army had blown up most of the harbour facilities. As the group walked past the church, the team was told the story of the building first constructed in 1869 and then bombed in 1944 by the Germans as they retreated from northern Norway. The building was later rebuilt after the war. Misha Myers suggests that such guided walks create an auditory space through the voicing of place in live talk.[19] She explains that movement through these auditory spaces, constructed through talk in motion, enables places to be "sensed, made sense of, and sensually made" in a distinct way.[20]

During these tours, while the members of the research team listened to Svein Harald's stories, they at the same time took up conversations with each other about what they saw as they walked along. O'Neill and Hubbard suggest that the performative nature of such an activity facilitates talk, dialogue, biographical remembering, and relational engagement all at once.[21] The dialogic, relational space created between Svein Harald and the research team became an embodied space where a "shared viewpoint" facilitated "empathic witnessing" as well as "collaborative knowledge".

One of the researchers would later in his field notes describe the twin crosses capping the church spire in Vardø as metaphorically connecting Heaven and Earth. This researcher recalled that that church spires were often the most or only visible landmarks along what he described as an otherwise foggy, highly fractal, and isotropic coastline. The researcher described in his notes the sensory experience of walking through the streets of Vardø on a winter day:

> The wind where the street ends and the sea begins is gusting over five meters per second, enough to stagger us while we're standing above the massive green combers pushed out of the Barents Sea. Spindrift blows sideways through the rocks. It's -5C°, and we're wearing multiple layers of fleece, wool, and windproof nylon against the wind chill. Taking notes is done with bare hands facing away from the wind in 15-second bursts.[22]

How does one know how it feels to be at the edge of the sea on a winter day? The anecdote from the researcher's notes suggests that during the act of walking, information is received in different ways through seeing, hearing, touching, smelling, tasting, and overall mentally connecting.[23] There are certain parts of Vardø that we got to know as representations, verbally described and shown to us in maps and photographs. Once we had literally "been there", walked and seen the abandoned fish factory with graffiti on it, smelled the fish from the boats at the harbour, and walked through the abandoned junkyard, we were able to get a real sense of the complexities implied by the use of the phrase "a sense of place". This afforded us the opportunity to use our sensory embodied experience as a basis from which to empathize with the Vardø residents inhabiting these places on an everyday basis. These walks thus served as a way of bringing us into sensuous contact with lived experience in real time, activating what Myers describes as a sense of empathic and critical witnessing, and a convivial way of interacting with and knowing place.[24] Myers frames such guided walks as forms of *conversive wayfinding*, a dialogic methodology and a spatial practice

> that conducts participants' attention to landscapes through mediated/live aural performance; perceptual and dialogic strategies of interacting and knowing place – shared viewpoints, ear-points, conversational conviviality and critical witnessing; the use of different paces, paths and places of narrative; and performance as a way of knowing.[25]

When the body is engaged in walking in this way, the act of walking becomes the lived experience and walking itself becomes empirical material. Jung argues that in such a scenario, the body (the instrument to experience materials), the walk (the means to empirical material), and the experience (the empirical material) are closely interrelated.[26]

Pink observes that during such encounters the walk, as a sensory ethnographic methodology, is not simply a walk but also a lesson in how to see, through which one develops a sort of "skilled vision".[27] During our group walk along the streets of Vardø, we learnt not only to look where the guide told us to, but also what to look for when looking, and the meaning of what we saw. The walk

Figure 7.2 A group of students go for a walk on one of the main streets of Vardø

Photo: Ronald van Schaik.

thus helped provide us with both an individual and a collective gaze on the town. During these walks, we were encouraged to appreciate the town's detail, historical value, identity, and local meanings of its material environment.

The go-along interview

> On the day of the interview with 88-year-old Arthur Andreassen, the two researchers walked around the Vestervågen neighbourhood together with their informant. He was a unique living witness of the unwritten history of the different historic landmarks in the neighbourhood. The team of interviewers comprised Kristján, the Icelandic landscape architecture student, who was holding the microphone and the tape recorder, Svein Harald, the project leader of the Vardø Restored project, and Janike, a researcher on the Future North project. As they walked and talked Arthur would stop and point at a building and tell them the story behind it. Arthur told them that before WWII the building used to serve as an industrial fish curing plant. While narrating the story of the neighbouring fish factory Arthur recounted another story, related to that particular part of the harbour. Towards the end of the war, he was commissioned to bring a freighter to Kirkenes. The boat contained 70–80 bodies of German soldiers that had been dug up from their temporary graves in Vardø. The German soldiers wanted to bring their dead with them. He was asked to stop by Vadsø, the neighbouring town, to offload another set of bodies, and eventually arrived in Kirkenes with a load of 149 decomposing bodies. Arthur showed the researchers where the bodies had been stored before they were loaded onto the boat, in the old salt storage facility at the docks. He still vividly remembered how badly it smelt on the boat and how sick he felt afterwards.
>
> (Field notes from a researcher on the Future North project)

The ethnographic vignette that I authored reconstructs the scene of a research encounter. In this encounter, the project researchers employed an ethnographic research device called the "go-along" interview that unwittingly produced a phenomenological understanding of how Vardø locals comprehend and engage with their physical and social environment, even though that was not the original purpose of the interview. The aim of this interview was to gather local inhabitants' historical accounts of the buildings that represent the cultural heritage and identity of Vardø. The informant is one of the few elderly inhabitants remaining who lived through significant periods in Vardø's history since before the Second World War. The research team was soliciting the informant as a witness of the past to help them develop an understanding of the historical significance of certain architectural landmarks located in the Vestervågen neighbourhood in Vardø.

The go-along interview that they employed, in this example, is basically a type of interview where researchers accompany individual informants on outings in their familiar environments, such as a neighbourhood. Through asking questions, listening, and observing, researchers actively explore informants' flow of experiences and practices as they move through and interact with their physical and social environment. Kusenbach suggests that the go-along interview is ideal for bringing to the foreground the different associations that informants make while moving through physical and social space, including memories.[28] In the example at the beginning of this section, the questions were mainly focused on the past and present histories of the buildings. However, the sight of the little red house with the flight of stairs going in from the side helped the informant, Arthur, recall personal memories from his experiences from the period of the war. The situation encouraged Arthur to stop at places where the architectural landmarks helped trigger his memory of past events.

Andrew Irving suggests that by accompanying people on walks through the city the researcher can "gain a 'sense', if not an understanding, of how a particular type of past – which is at once cultural and idiosyncratic – connects people to prior events and mediates their experiences of their neighbourhood".[29] In this way, Irving suggests, memory is "produced in the act of performance as the informants make their way round the city and as events and episodes are drawn out of the city's streets, buildings, market-places and turned into public narratives". For the three researchers accompanying Arthur, what emerged was a tangible, albeit fragmented, sense of place that opened up a small part of Vardø's past to the outside gaze, inviting them to inhabit and re-create the town for themselves, rather than fixing it through explanation.

Go-along interviews can help researchers to systematically focus on and explore in detail the transcendent aspects of experience of place which might otherwise escape attention if one were using observational methods or off-location interviews.

Collective reflection as place making

Usually, after the outdoor fieldwork activities the whole team would convene in our meeting room. Each group would present the results of their research activities for the day and the whole group would discuss the issues raised. This was the time when the groups sent out to do interviews would present their findings and those assigned to take photographs of buildings would show their work on the big screen accompanied by commentary.

Figure 7.3 Members of the researchers looking at a historical photo of Vardø during a workshop session

Photo: Andrew Morrison.

By reviewing sets of images that highlighted material features of Vardø, hearing the accounts the local residents and the meanings they attach to their lived landscape, and listening to subjective observations of individuals in the research team, we developed particular insight into the town's uniqueness and identity. These processes can be understood in terms of the philosopher Edward Casey's notion of the "gathering", a process that is part of place making.[30] In our case, this gathering occurred in conversations while on a walk, in interviews with locals, during group meetings, and when sharing an evening beer at the local pub. These group discussions served as devices for the gathering and rearticulating of the various rhythms, tastes, perceptions, narratives, and meanings shared between the different members of the research team.

Conclusion

In this chapter, I have tried to argue that visual and sensory methods such as the group walk, the go-along interview, and photographing are useful ways for researchers to get attuned socially, materially, and sensorially to a new place. These visual and sensory modes of inquiry allow for an in-depth understanding of the textures of social experience and the meaning of Vardø as a lived everyday place. These methods can offer researchers a view on how the local residents see their world, what is important to them, what their lived social relations are, how they view local history, and their affective attachments to place. These methods give researchers who are new to a place an intimate embodied understanding of the geography, history, and socio-cultural features that are constituent of a place.

However, in our case, what is mostly at stake is how we were able to participate in the imagining of Vardø as a "future landscape". At the moment of writing, groups of individuals and local organizations are in the process of constructing a series of plans for cultural heritage preservation and economic development for the future. The challenge for such processes lies in being able to imagine an alternative and sustainable future for the town that still connects to its history, cultural heritage, and sense of community. Researchers from the Future North project are programmed to play a contributing role in this process. I will suggest that in order to contribute to such processes, researchers need embodied, situated, and self-reflexive understanding of the place in question. The visual and sensory methods I proposed in this chapter are among the methods that were most suitable for this.

To conclude, I return to the scene of one of our workshop sessions:

> The students announced that they were ready to present. We all moved in the meeting room and everyone found a seat. The owner of the abandoned wooden fishing house entered the room and shyly nodded his greetings around the room. He declined an offer to take a seat saying he preferred to remain standing. The students began their presentation. During the next fifteen minutes, they proceeded to present a proposal for how the building could be transformed into a writer's retreat, a guesthouse of sorts. All the unique architectural features of the building would be preserved, they insisted. The room was silent as everyone stared at the flicking images on the screen. When the presentation was over someone asked the owner of the building, standing in the corner what he thought about the proposal. He smiled.

Notes

1 http://vardorestored.com/en.
2 Casey Edward, "How to Get from Space to Place in a Fairly Short Stretch of Time: Phenomenological Prolegomena," in *Senses of Place*, ed. Stephen Feld and Keith Basso (Santa Fe: School of American Research Press, 1996), 46.
3 Ibid., 44.

4 Ibid., 24.
5 See for example Kelvin Low, "The Sensuous City: Sensory Methodologies in Urban Ethnographic Research," *Ethnography* 16, no. 3 (2015): 295.
6 Casey, "How to Get from Space to Place," 44.
7 Sarah Pink, *Doing Visual Ethnography: Images, Media and Representation in Research* (London: Sage, 2007).
8 Sarah Pink, "An Urban Tour: The Sensory Sociality of Ethnographic Place-making," *Ethnography* 9, no. 2 (2008).
9 Sarah Pink, "Mobilizing Visual Ethnography: Making Routes, Making Place and Making Images," *Forum: Qualitative Social Research* 9, no. 3 (2008).
10 David Howes, *Empire of the Senses: The Sensory Culture Reader* (Oxford: Berg, 2005), 7.
11 Tim Ingold and Jo Lee Vergunst, eds. *Ways of Walking: Ethnography and Practice on Foot* (Hampshire: Ashgate, 2008); Kimberly Powell, "Making Sense of Place: Mapping as a Multisensory Research Method," *Qualitative Inquiry* 16, no. 7 (2010).
12 Sarah Pink, *The Future of Visual Anthropology: Engaging the Senses* (London: Taylor & Francis, 2006).
13 Margarethe Kusenbach, "Street Phenomenology: The Go-Along as Ethnographic Research Tool," *Ethnography* 4, no. 3 (2003): 466.
14 Cala Coats, "Thinking Through the Photographic Encounter: Engaging with the Camera as Nomadic Weapon," *International Journal of Education & the Arts* 15, no. 9 (2014): 5.
15 Gilles Deleuze and Felix Guattari, *A Thousand Plateaus: Capitalism and Schizophrenia*, trans. by Brian Massumi (Minneapolis: University of Minneapolis Press, 1987).
16 Coats, "Thinking Through the Photographic Encounter: Engaging with the Camera as Nomadic Weapon," 5.
17 Sarah Pink et al., "Ethnographic Methodologies for Construction Research: Knowing, Practice and Interventions," *Building Research & Information* 38, no. 6 (2010).
18 Yuha Jung, "Mindful Walking: The Serendipitous Journey of Community-Based Ethnography," *Qualitative Inquiry* 20, no. 5 (2013).
19 Misha Myers, "'Walk with me, talk with me': The Art of Conversive Wayfinding," *Visual Studies* 25, no. 1 (2010).
20 Ibid., 61.
21 Maggie O'Neill and Phil Hubbard, "Walking, Sensing, Belonging: Ethno-Mimesis as Performative Praxis," *Visual Studies* 25, no. 1 (2010): 50.
22 William L. Fox, "Walking the High Wire," *Future North*, April 07, 2014, www.oculs.no/projects/future-north/news/?post_id=3590.
23 Tim Ingold and Jo Lee Vergunst, eds. *Ways of Walking: Ethnography and Practice on Foot* (Hampshire: Ashgate, 2008).
24 Myers, "'Walk with me, talk with me'," 67.
25 Myers, "'Walk with me, talk with me'," 67.
26 Jung, "Mindful walking," 5.
27 Pink, "Mobilizing visual ethnography."
28 Kusenbach, "Street phenomenology."
29 Andrew Irving, "Ethnography, Art and Death," *Journal of the Royal Anthropological Institute* 13, no. 1 (2007): 187.
30 Casey, "How to get from space to place."

References

Casey, Edward. "How to Get from Space to Place in a Fairly Short Stretch of Time: Phenomenological Prolegomena." In *Senses of Place*, edited by Stephen Feld and Keith Basso, 13–52. Santa Fe: School of American Research Press, 1996.

Coats, Cala. "Thinking Through the Photographic Encounter: Engaging with the Camera as Nomadic Weapon." *International Journal of Education & the Arts* 15, no. 9 (2014): 1–23.

Coats, Cala. "Materializing Transversal Potential: An Ecosophical Analysis of the Dissensual Aestheticization of a Decommissioned Missile Base." *Journal of Cultural Research in Art Education* 32, no. 1 (2015): 127–60.

Deleuze, G., and Felix Guattari. *A Thousand Plateaus: Capitalism and Schizophrenia*. Translated by B. Massumi. Minneapolis: University of Minneapolis Press, 1987.

Feld, Steven, and Keith Basso, eds. *Senses of Place*. Santa Fe: School of America Research, 1996.

Fox, Bill. "Walking the High Wire." *Future North*, April 07, 2014, www.oculs.no/projects/future-north/news/?Post_id=3590.

Howes, David. "Introduction." In *Empire of the Senses: The Sensory Culture Reader*, edited by D. Howes, 1–17. Oxford: Berg, 2005.

Hyler, Samantha. "Invisible Lines Crossing the City: Ethnographic Strategies for Place-making." *Culture Unbound* 5, no. 3 (2013): 361–84.

Ingold, Tim, and Jo Lee Vergunst, eds. *Ways of Walking: Ethnography and Practice on Foot*. Farnham: Ashgate, 2008.

Irving, Andrew. "Ethnography, Art and Death." *Journal of the Royal Anthropological Institute* 13, no. 1 (2007): 185–208.

Jung, Yuha. "Mindful Walking: The Serendipitous Journey of Community-Based Ethnography." *Qualitative Inquiry* 20, no. 5 (2013): 621–27.

Kusenbach, Margarethe. "Street Phenomenology: The Go-Along as Ethnographic Research Tool." *Ethnography* 4, no. 3 (2003): 455–85.

Lee, Jo, and Tim Ingold. "Fieldwork on Foot: Perceiving, Routing, and Socializing." In *Locating the Field: Space, Place and Context in Anthropology*, edited by Simon Coleman and Peter Collins, 67–86. Oxford: Berg, 2006.

Low, Kelvin. "The Sensuous City: Sensory Methodologies in Urban Ethnographic Research." *Ethnography* 16, no. 3 (2015): 295–312.

Myers, Misha. "'Walk with me, talk with me': The Art of Conversive Wayfinding." *Visual Studies* 25, no. 1 (2010): 59–68.

O'Neill, Maggie, and Phil Hubbard. "Walking, Sensing, Belonging: Ethno-Mimesis as Performative Praxis." *Visual Studies* 25, no. 1 (2010): 46–58.

Pink, Sarah. *The Future of Visual Anthropology: Engaging the Senses*. London: Routledge, 2006.

Pink, Sarah. *Doing Visual Ethnography: Images, Media and Representation in Research*. London: Sage, 2007.

Pink, Sarah. "An Urban Tour: The Sensory Sociality of Ethnographic Place-Making." *Ethnography*, 9, no. 2 (2008): 175–96.

Pink, Sarah. "Mobilizing Visual Ethnography: Making Routes, Making Place and Making Images." *Forum: Qualitative Social Research* 9, no. 3 (2008): 1–17.

Pink, Sarah, Dylan Tutt, Andrew Dainty, and Alistair Gibb. "Ethnographic Methodologies for Construction Research: Knowing, Practice and Interventions." *Building Research & Information* 38, no. 6 (2010): 647–59.

Powell, Kimberly. "Making Sense of Place: Mapping as a Multisensory Research Method." *Qualitative Inquiry* 16, no. 7 (2010): 539–55.

8 Future North, nurture forth

Design fiction, anticipation and Arctic futures

Andrew Morrison

Framings

What role could design fiction play in contemporary investigations of our shared material and immaterial reality and culturally experienced landscapes of a 'future north' when culture and nature are inter-twined conjecturally? In this chapter I focus on the initiation and early phases of collaboratively scripted design fiction in a mode of speculative inquiry. The account is one part of experimenting with design fiction as a form of collaborative research into cultural landscapes of the Arctic in the research project Future North. Our design fictional forays into the high north are conveyed through the persona of Narratta who appears as a team player in our project website as shown overleaf and whose texts are interspersed throughout the chapter.[1]

Narratta's identity has been developed by our interdisciplinary team over time and through fieldwork in some Arctic territories. Her stances and interests have been devised through discussion, but also through the dynamics of our work and as part of shaping online drafts and posts through a blog tool in a mode of discourse and social action.[2] Narratta speaks by way of being a sensory instrument and an instrument of sensibility (akin to the speculative philosopher Whitehead's observation that science always designs its own instruments). Following the notion of narrative ventriloquism from Mikhail Bakhtin,[3] her perspectives are voiced through the co-composition of a diverse research, design and educational team. Her voice is one that develops over time, as our group has found its own experiential and interpretative voice. Her articulations reflect our diverse and interdisciplinary specialisations. They also connect us in a shared subjectivity that we enact in order to address some of the tough challenges in understanding and communicating the complex contexts, histories, systems and cultural landscapes of the Arctic as counter narratives as opposed to master narratives.[4]

In presenting and discussing the role of design fiction[5] in shaping speculative inquiry into a 'future north', I draw on analytical frames from cultural studies,[6] counter-narrative theory and communication design.[7] I also refer to a means of critical production from rhetoric, multi-literacies and writing studies and design writing.[8] In this sense the chapter moves expressly into matters of articulation following Stuart Hall.[9] We have designed Narratta as a speaking entity. She articulates Bruno Latour's 'matters of fact' and 'matters of concern'.[10] This is a distinction Bruno Latour makes to reposition our discourses of subject-object relations towards one that is open to and mindful of socio-technical relations and negotiations between agency, senses and settings, as argues Puig de Bellacasa.[11]

Narratta has many memberships therefore. She is a co-created contributor to the project (not unalike Latour's relation to his self-authored student persona in *Aramis or the Love of Technology*).[12] The twist here (like her spiral 'horn', really a tooth) is that she is a thing that speaks. Narratta speaks

OCULS
Oslo Center for Urban and Landscape Studies

Projects   People   Publications   Events   Themes   Approaches   About   Search

Narratta
Researcher

Biography

I am a bio-enhanced, nuclear assisted narwhal. I keep myself busy by observing and exploring the changing landscapes and discourses of the Far North. My long tooth has special properties. It's an aerial of sorts, able to receive and send information and sense climate conditions and change. I can dive deep and swim great distances. But I am also able to use my special enhanced power to jettison myself out of the water and into the air. Beyond these properties I have developed extra sensory sensitivities that I use to look into the changing landscapes of the future north and the forces of today that may impact on our shared tomorrows.

You might say I am a communicative device, a constructed persona, a mobile apparatus for collaborative communication. A design fiction. Design friction! Read more here to get to know me and how we all need to heed changes in the far north and the ways they are shaped discursively already today. I'll provide you with links and feeds, and a unique opportunity to travel a part of the globe you may find hard to visit yourself.

Projects:
Future North

lyrically and thus the chapter is also constructed paratextually as a blended lyrical poetic discourse, following the writings of David Shields:

> The lyric essay doesn't expound, it is suggestive rather than exhaustive, depends on gaps, may merely mention. It might move by association, leaping from one path of thought to another by way of imagery or connotation, advancing by juxtapositions or sidewinding poetic logic.[13]

This format is employed to enable rhizomatic shifts[14] between communicative styles, forms, typographies, voices and mediations as a means to engage the imaginative with the interpretative. Such a pliable research text allows for the inclusion and interruption of Narratta's voice in excerpts from her blogging that rhizomatically cross time and space and are positioned to be read as part of the flow of the chapter. The reader experiences shifts between her voice, one that expresses both wonder and perplexity, and that of expository academic writing that unpacks the bundling together of design fiction and speculative online discourse with elaborations and links on key analytical concepts in narrative theory.

Narratta's storyworld is a blend of the real and fantastic. Her mind is increasingly her own as she begins to understand the implications of her having been engineered and how the complex systems and relations of life in the Arctic impact on her being and becoming. Her acts of narration are unnaturally articulated as she surfaces, dives and reappears in various times and places, sensing climate changes and geo-political conflicts, sneaking up on the people and events, observing transformations in and across time.

The chapter attempts what I call *a poetics of anticipation* of a 'future north'. In doing so it takes up a number of nested questions. What role might communication design play in shaping a poetic counter mapping, a subaltern narrative, a problematising rhetoric and not solution-driven discourse? Where science fiction is highly present in the literary and filmic imaginaries of the polar regions, what place might there be for the emergent domain of design fiction in generating online spaces for mediating alternate narrative visions?

Narratta's voice is a polyphonic. It is a rhizomatic mediation of multiple views on the changing Arctic and it is made by a variety of participants. Narratta is a thing but not one thing.

Contexts

Story-wise, we have been motivated to locate Narratta within the contexts of prior, existing and projected cultural landscapes of the Arctic,[15] including the Russian Kola Peninsula and Greenland, among others, Narratta's offerings are not disconnected from these experiences and reflections. Rather – on the part of the project team and its activities with students and visiting researchers – her musings are generated by learning about local histories and context-rich accounts. For example, in the island community of Vardø in northern Norway, participative story gathering has involved local experts, fishermen, elderly storytellers and business people (see Mainsah, this volume) in discussing their locative history and their current attempts to provide alternatives to the demise of a once flourishing fishing industry and the projected development of oil and gas reserves offshore.

Cultural landscapes of a future north. They are projected as being just beyond reach, offshore even, proposed and promising. But potentially unscriptable, loomingly perilous.

It is into these waters that Narratta and we as researchers have had to journey. Narratta has allowed us to move between the land, the skies and the seas, between inland lakes and virtual representations of communities (see Uhre, this volume) within a time of increased evidence of climate change. As groups of designer-researchers, locals and students working together, Narratta has been in our workshop and seminar settings. She has taken part in our conversations (on many occasions we have asked what would Narratta think, or say). She has been present alongside and observant of our movements through the landscape, both physical and mediational.

Importantly, the processes of co-scripting Narratta have included frequent face-to-face discussion of the eight main designer-researchers. This group has developed considerable coherence across time and space in engaging with making sense of the territories and terrains that were selected for the project's inquiries. The project team needed to become familiar with its own diverse expertise and interdisciplinary configurations and processes. This has extended to a nomadic practice of sorts, from travelling together, doing fieldwork and the collaborative writing of research and doctoral supervision. We have become more familiar with one another's design and research practices. The group includes experience and formal expertise in architecture, ethnography, political science, comparative literature, narrative and fiction, discourse analysis, interaction design, media and new media studies, cultural studies, landscape and urbanism, development communication and education. Our experience and perspectives have taken us into largely new territories and terrains; even for those of us living in and familiar with working in the Arctic, the contexts and discourses of the Arctic have been changing rapidly in the past half-decade alone. Our visits and dwelling in Arctic cultural landscapes have therefore demanded that we interrogate our own assumptions and perceptions while at the same time being informed and challenged by a variety of local inhabitants whose daily lifeworlds in the Arctic are different to our own research-oriented ones.

Needed is an inventory of sorts. A communication strategy and discursively themed event plan. How to let this simply emerge. When to script it, to prompt the team. . . .

We have worked through early, shared discussions, taken up the persona of Narratta as a partner over numerous coffees and lunches in our journeys, work and teaching in the Arctic. At a closer communicative level we have discussed the major themes of her posts, the tone of her writing and the types of images we might include. We have learned the blog tool WordPress and drafted posts and commented on them, for example, in a hotel lobby where we conducted a workshop during our afternoon tea in Murmansk. Further, we have on several instances written pieces together and reviewed and edited them over time, making changes and, importantly, often splitting up material into several entries. Images have also been shared and discussed as we have come to better understand the cultural landscapes of the Arctic.

The co-design of Narratta has entailed other projects and contributors, from gendered critiques of Arctic expeditions to the conflicting terrains of mining companies and nomadic practices. Narratta did not ask to be nuclear powered, this was part of a secret design and inventive pact between a Russian physicist and a Norwegian biologist. But today she is able to jettison herself out of the sea and into lakes and river valleys, a landscape agent of agentive mapping. Her enhanced persona extends to an ability to use her spiral tooth to receive, read and transmit data and information about historical, contemporary and future perspectives on the fate of the far north. Her interests and views have been fashioned through her having been presented in Murmansk, Russia, as part of a youth centre's engagement with the changing climate of socially mediated mobile communication. This group has now closed under recent regulations concerning NGOs. The landscapes of the Arctic are clearly not stable: the sea's ice is melting and military naval expansion is now a reality in a changing Russia at a time when oil prices and exploration have both declined. Now an artificially engineered hybrid, mammal-machine, Narratta has become a narrative device, a thing of communicative character to address such instability. She is to be seen flying out of the water beside icebreakers ploughing channels to China. Their crews view her as a new imaginary beast of uncharted waters, fantasy and factual confusion at the helm of their careful work, with their and our perspectives suddenly seen and recirculated reflexively.

Narratta is highly mobile and she is technologically enhanced. Adapted to her unrequested hybrid genesis (this being a space of speculative realism of sorts), she is also able to retract it to camouflage herself as a seal when needed. Narratta's movement and her narrative and analytical voice create a vibrant poetic and speculative discursive design landscape that is as restless as the ocean, and as and still as the glassy surface of a mid-summer night's harbour. She assembles and translates, dissembles and deliberates. Yet all of this is co-designed, some pre-scripted, but much of it emergent, dialogical and driven by the main themes that arise in anticipating culturally today the changing climates of tomorrow.

In these contexts, Narratta ventriloquises the concerns of the project members as well as those of people who live in her harbours and coastlines, and along her fiords. Narratta has been investigated further with a group of students in a class of master's students in Landscape Architecture from a joint Arctic-oriented programme between the Oslo School of Architecture and Design and UiT – The Arctic University of Norway in Tromsø – including a related architecture programme visiting from the University of Montreal. Her perspectives are planned to be more diverse with a mix of entries by students and teachers, researchers and community members. Her participant author group is expanding and – through a new series of workshops, co-writing and editing practices and planned group readings – these new authors will become familiar with her tone and themes.

Fictions

In the Future North project we have aligned ourselves as story makers and cultural critics. Overall, we agree with Matt Ward and Alex Wilkie when they write that:

> As material-semiotic storytellers designers we adopt a role in which the construction and communication of possibility is wound into the generation of belief and hope, where new worlds

are made and remade in order to persuade, convince and challenge pre-established norms and whereby, temporality becomes a medium for our practice, where not only relations between actual objects, actors and entities are bound together but also the mediation between the existing and the yet to exist.[16]

In design fictions, the suspension of belief in the functional and the immediate is therefore needed to move into spaces of conjecture and speculation that are devised and communicated by design.[17] While these are spaces that are akin to traditions of science fiction, both in literature and popular cultural communication, they tend to be cast as strongly dystopian, far-flung scenarios or situated in more readily accessible near future imaginary spaces.[18] Often, their intent is to engage audiences in issues and problematics of the day and to move imaginatively and critically away from comfortable assumptions or banal expectations.

Broadly, design fiction[19] works with near future imaginary scenarios and foresight[20] and with the projection of potential outcomes to engage audiences in issues and debates that have an impact on our contemporary cultural practices and understanding.[21] However, the gallery centred and 'speculative' design projects and artefacts,[22] originating in 'critical design',[23] have been critiqued for being art centric and conceptual and remote from the real world and pressing concerns much design faces. In part this argument holds; in part it misses the importance of conjecture in design where the aim may be not to solve but to surmise so as to generate thinking.

A poetics of anticipation. Various views ventriloquised through one persona, a speaking thing with many voices, spatialising complex object-oriented ontologies.

Some recent attempts have argued that design fiction may be more political in nature through designs that 'let things speak' in the socio-technical sphere of contemporary culture.[24] Narratta is knowingly co-designed to reflect the Actor Network Theory's recognition of human-nonhuman 'actants' and concern with things as opposed to a binary distinction between people and objects.[25] Narratta is an imaginary discursive artefact that draws attention to relations between humans and animals,[26] between persons and the environment and technology and geo-politics. She is methodologically 'slippery';[27] she asks us to engage in affective states of sense and sensation, and not only argumentative ones, curving between land, sky and sea, in a compositional cultural articulation of an emergent collective agency.[28]

Stories to frame our predicament

For many people the Arctic is difficult to conceive of, characterised as physically remote, sparsely populated and beset by inhospitable weather. Yet it is the arena, geographical and mediational, in which discourses of climate change are being rapidly made manifest and material.[29] Increasingly, in geo-political terms, the circumpolar region has seen an expansion of national interests and territorial claims centred around oil and gas resources, actual and putative, as well as the opening up of sea routes. Tensions between commercial and cultural interests and stakeholders have been exacerbated in competition between physical resource extraction models of 'development' and local area knowledge, cultural rights of dwelling, access and ownership. In addition, the Arctic is undergoing rapid transformation economically, socio-politically and culturally. Its inhabitants – human and animal – are already experiencing the effects of rising temperatures and melting permafrost, glaciers and sea ice. Yet, this is a context also of immense data-based, researched and cultural, not only geo-physical landscapes. As Jane Marsching writes:

> We have at our fingertips and on our screens reams of data that attempt to outline this future, but the images are too hazy, or too complex, or simply too uncertain. But as the future presses

Future North

News | About | Events | Publications | People | Advisory Board

Vardø futures

Narratta, June 15, 2014

It's a rather mild June afternoon in Vardø. This once Arctic city is at the northeast of Norway. I am resting right underneath the boardwalks in Vestervågen, among poles that were for a large part put down after WWII. You'll recall that the German Nazis blew up most of the facilities on this side of the harbourfront. It looked quite different back in even 2014.

The restaurant above has an outdoor dining area now – over the past ten years people have been able to sit outside more often, most restaurants now have them. This one is buzzing with voices, kitchen sounds and plates. Mostly tourists. The usual blend of German bikers, American and Canadian artists, Norwegians finally exploring their north. And students – the city has become a case study for city development in the Arctic. The food is excellent I hear them saying – local fish, vegetables from the many smaller farms in the area, game meats. I dive down quickly, careful not to make a splash, as I don't want to hear if we are are on the menu!

Exotic we've become in a different way as the temperatures have risen and we have spread eastwards, no longer hard to see, that being our old exotic character. We are still not that easy to catch. But easy to eat?

About

At present the Circumpolar North provides a unique laboratory for studying future landscapes of production, infrastructure, excavation, and environmental change. **More »**

Popular project news:

- Northern arrival
- Fabulous forms and design fictions
- The Wall and the Flower
- The Contorted Architecture of Geopolitics
- Architecture as Landscape

down upon us more and more, as ice melts quicker, as climate triggers become more apparent, as short-term predictions become reality sooner than expected, we long for future stories that might help frame our predicament.[30]

Design fiction has yet to be much applied to diverse global contexts such as the Arctic, or to be very fully explicated in terms of narrative theory, persona and digital media. This gap is at the analytical heart of this chapter. It is addressed within a *cultural landscape* view of a 'future north' as defined by landscape architect James Corner: 'there is nothing natural about landscape: even though landscape evokes nature and engages natural processes over time, it is first a cultural construct, a product of the imagination'.[31] However, as Timothy Morton argues, geo-physical and cultural landscapes are now entwined and coeval in the Anthropocene. How then might we also communicate the complexity of the such landscapes of the Arctic within the context of massive data, techno-scientific research and the interplay of international politics? This text shows one method of response – speculative design fiction, which offers communicative potential ripe with as many questions as answers for the future north cultural landscape.

Communicatively and intertextually, Narratta's blog entries are therefore woven between other posts by members of the project that reflect on the dominant, emergent and counter narratives of the Arctic. We have created narratives of inquiry that tangle with complexity and futurity, but entail accounts located in the past, present and future. They are what Roberto Poli positions as expression of anticipation that escape the framings of discourses on prediction, planning and visions of probable worlds typical of Future Studies.[32] Anticipatory design-driven discourses may instead reach for tomorrow's worlds today.

In shaping Narratta, we have drawn on 'un/natural' literary theory that is non-mimetic.[33] In addition, satire, irony, pastiche and humour are patently part of Narratta's own arsenal of communicative resources, as Matt Malpass reminds is central to much design fiction.[34] Narratta's musings are further a narrative of anticipation. This is in contrast to dystopian science fiction, inflected design fiction or foresight type projections geared towards a desired and prognostic future in a future studies frame.[35] As Paul Raven and Shirin Elahi argue in shaping a modal matrix model of narrative for futures research (crossing between the diegetic, mimetic, dramatic and spectacular),[36] further work is needed to investigate relations between persuasion and empathy. Yet missing here, as well as more widely in futures research geared towards solutions and final forms, however, are actual investigation of design fictional narratives or compositions designed to problematise – in designerly narrative ways – the putative and contingent.

Into adventuristic time

Somewhat surprisingly, and perhaps largely due to its location within design as a domain of origin, little design fiction work makes substantial reference to narrative theory. In the blog extract at the beginning of this section, Narratta reflects on how her narrative is powered, literally, and how her own empowered sense of self is 'a curve into the future'.[37]

Nurturing multi-level narrative as a mode of framing our current predicament of climate change.

Narratta was designed with direct reference to narrative theory, persona and polyvocality. Three concepts from narrative theory of Mikhail Bakhtin have been useful in moulding Narratta's unnatural design fictional story space: *addressivity*, the *chronotope* and *adventuristic time*.[38] Bakhtin saw narrative as historically and culturally framed via *addressivity*. His notion of fictional communication was centred around dialogue and address between author, text and reader. His approach situated fictional narrative within genres of speech and modes of address rather than only plot and character. His notion of dialogicality extended to the temporal-spatial via the concept of the *chronotope*.[39] The chronotope refers to the narrative configuration of time and space that varies according to genre or action modes in relations between story (narrative sequencing) and discourse (narrative and rhetorical structuring and arrangement). Although developed for unpacking such features in the novel, it may be applied constructively in a co-design composition of a web-based narrative and the fictional in design. This refers to concepts Bakhtin devised on the chronotope, or relations between the spatial and the temporal in narrative constructions. In addition, the chronotope is characterised by centrifugal or centripetal communicative motion. Respectively, this refers to how the dynamic of a narrative gravitates towards a convergent and discourse, spatially and temporally, or a more divergent one in which character and plot are not congruent in time and space.

These two concepts concerning the dynamic of narrative sequencing and discursive arrangement have been taken up with respect to Narratta's blogging. In the Future North project – with its challenges of looking forward, both anticipatively and speculatively – it is a discursive, not only artistic, chronotope of cultural landscape that has been investigated through the polyvocal persona of Narratta. Narratta may speak, but she is made up of many views and these are not simply coherent but also relate to various and situated views on her environment today, on legacies and on projected futures. This mix of centrifugal and centripetal narrative elements is important when our independently minded narwhal was co-designed to live in what Bakhtin called *adventuristic time*.[40] This is a temporal scale and textual composition and experiential consumption that differs

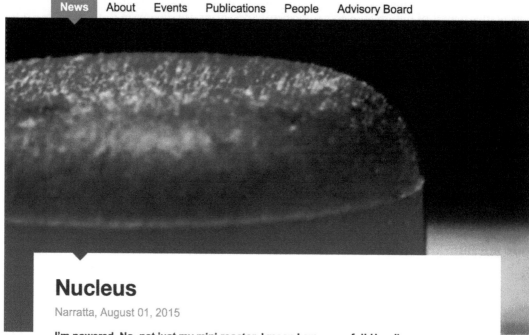

News   About   Events   Publications   People   Advisory Board

Nucleus

Narratta, August 01, 2015

I'm powered, No, not just my mini reactor. I mean I am powerful! Hey, I'm even reactive! I'm alive, I've been living longer than my peers. None of that young blood transfused into me to reduce the aging process. I am alive, not just an atomic half-life! Not even sure how I'm really powered. Some might say it's a radioisotope thermoelectric generator. The sort used in space probes. With its glowing orange core. And the string of lighthouses and beacons along the northern coasts of Russia all to be replaced by wind and solar by 2015.

What a life it's been. What a curve into the future I'm travelling.

Melded by those two scientists, binary innovation, minds merged in 1953! Seeking to escape the fences of the Cold War, by imagination.

I was born … fantastic!

It's taken me a long time to acknowledge my siblings. And I mean mammals and machines.

Mammalian, off the coast of Greenland. Traditional fare they are. The sliced delights of dinner. Other mammal feed. But I'm more than that, a strip of <u>tasty flesh</u>.

considerably from the experiences of early Arctic explorers and the many modernistic accounts of discovery, heroism and endurance. In a mode of adventuristic time, according to Bakhtin, sequentiality may be broken, violated or interrupted. Such moments may also be enabled through the force of non-human actors or events. Metalepsis, or jumping between story worlds, is one of the main narrative devices we have played with in shaping Narratta's polyverse. Metalepsis can take the form of ascending, descending and horizontal jumps in time and place.[41] These jumps, versions and variety are co-scripted in and over time and in their totality form a mesh of communicative expression.

Meshes

In a discussion of fiction, 'polar media' and technoscience, Lise Blom and Elena Glasberg argue that 'Instead of richly layered myths and interlacing stories of survival and adventure, we get measurements and extrapolations, prognoses and doomsday scenarios'.[42] In our work in Future North, we move from a typical structural coupling between system and environment in looking at climate change and geo-political discourses of extraction and occupation to a cultural articulation that itself is a mode of anticipation that is contextual and communicative, relational yet content rich. This can be seen in an image taken from a vessel, which can pass through a metre or so of ice, sailing between Longyearbyen and Barentsberg in Svalbard where swimming alongside a boat of tourists and researchers, Narratta described her dental tusk as 'a sense making cultural landscape tool'.[43]

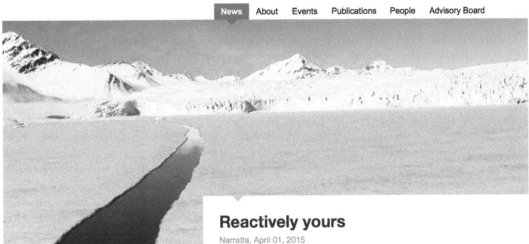

News   About   Events   Publications   People   Advisory Board

Reactively yours

Narratta, April 01, 2015

We are not all made the way we'd like to be. I'm a hybrid now through and through. The long and strong telomeres of my species, oh so superior to other whales. No one ever talks about us in that way. We are just the small cousins, the less spectacular blowers of pressured air, but nonetheless quite unique with out spiral tusks.

Dentally, you might say we are, well outstanding! Our tusked selves mythologised long before titanium implants graced the smiles of cinemas. Then into the weathered jaws of wealthy travellers. More recently sports stars and the general social democratic public of Scandinavia. My spiral is one of nature's instruments of measurement. It assists me in my dives and in sensing pressure and temperature. But it's one of a kind. A sensory device, a sense making cultural landscape tool. Dipping through the icy seas but now also airborne when I flip my tail and fly out into the sunny skies, the northern lights my favourite camouflage.

I am augmented, in the language of today. But over half a century now since those two crazy Cold War scientists in this secret pact cross the border experimented speculatively. Nuclear power. No one knows how it worked, and I'm destined to swim for more than an extra half (narwhal) life's worth or more. Not too long ago a research article reported on one of us as reaching the ripe age of 115! We are talking one of us females!

About

At present the Circumpolar North provides a unique laboratory for studying future landscapes of production, infrastructure, excavation, and environmental change. **More »**

Popular project news:

- Northern arrival
- Fabulous forms and design fictions
- The Wall and the Flower
- The Contorted Architecture of Geopolitics
- Architecture as Landscape

Narratta's presence is incipient; she surfaces at will, her tusk an antenna. She dives and reappears in different time periods and speaks on different topics, here provoked by where she has come up, there wallowing in doubt and indecision about topics that worry or challenge her understanding. She also feels alienated from her own natural environment, shifted from the inhabited narwhal fields off Greenland and now also deep within the tangled spaces and times of perceiving a future north in our chosen research domains of northern Norway, Svalbard, northwest Russia, Iceland and back to Greenland itself. Narratta feels dispossessed yet empowered by her changed status, an engineered techno-species splice of human design intentionality and mammal, whale and reactor. She has to live with the effects of the unwanted nuclear force of her powers that enable long journeys and an ability to propel herself out of the water and across the sky. Her appearance online is linked to the travels and reflections on terrain and territory by project members, physically and virtually. Readers meet her online, able to access her musings and see her vanish and then reappear in different settings and time periods. She is above all a narrative discursive eco-proxy.

Narratta's reflections allow us to present alternative co-compositions on learning about and looking into possibilities, putative and problematic projections of a complex ecological mesh (following Timothy Morton) that is the 'future north'.[44] This mesh is, for most people, difficult to understand in material terms. It is comprised of a changing physical world, but it is also a mediational one that is made up of motivated interests. This mediational space is also open to a more rhizomatic authorship than many writings of ANT encompass. Arun Appadurai argues that:

> Viewing mediation as a mode of materialization also clarifies why there is so much anxiety, in many cultures, about mediation, because it is through mediation, whether in the mode of seeing, touching, feeling, hearing, or tasting (or through more complex infrastructures), that matter becomes active, vital, energetic, agentive, and effective in the world around us.[45]

Appadurai reminds us that 'the future is not a blank space for the inscription of technocratic enlightenment or for natures long-term oscillations, but a space for democratic design that must begin with the recognition that the future is a cultural fact'.[46] In cultural landscape terms design work is by nature developmental and future-oriented. Design works to shape that which does not exist. However, mapping is itself a motivated and contested activity, as the last blog post shows, and needs to be understood as cultural productions (see also Uhre, this volume).

In seeking to articulate a speculative, fictional work related to potential, actual and prospective futures of Arctic landscapes, Narratta was designed as part of a wider project. Her design fictive liquidity is located in what I call *a communicative ecology of a future north*. Following Appadurai ecology is also to do with relations between anticipation, inspiration and imagination. But for us this has also much about the concept of articulation that he does not mention. As Lawrence Grossberg reminds us in reflection on the work of Deleuze and Guattari:

> To the extent that we are able to shape the becoming of the actual, it depends in part on our efforts to analyse the configurations of the actual and describe *the processes and practices (machines)* by which it – comprised of roots, radicles, rhizomes, etc. – has been and is being actualized – formed, taken up, and constructed again. This is critique as the starting point – but never the end – of political struggle. (original italics).[47]

Future north, nurturing futures. A blend of narrative as critique and interpretation. Narratta as a counter narrative agent.

This counter narrative of ironic and critically voiced shifts in time and space and perspective is realised through multiple authors contributing to a single persona. In the field of rhetoric and writing studies, composite personas appear as part of wider communicative strategies to address complex matters, such as in the writing of the fabricated researcher Myka Vielstimmig, a collaboration in single-voiced writing by rhetoric scholars Kathleen Blake Yancey and Michael Spooner, or in the shared views of a research team called Atelier on design things.[48] These may be rhizomatic articulations and they may be understood as distributed, from sites to the web, and from online mediations to orientations that prompt us to reconsider the places and locales of anthropocenically inflected landscapes.

This web-based discourse of Narratta's is part of a large body of research on blogging and its thematics, matters of genre, voice and aspect and even the parodic tone of such mediational articulations.[49] This may be seen to also extend to blogging as a bilateral artistic and reflective research practice.[50] We also encounter rhetorical and mediational co-authoring in blog-research discourse, such as in *Grand Text Auto*.[51] We see blogs used narratively, and ironically, in the popular cultural comedy television show *Dog with a Blog*, where a family dog reflects through voiceover on daily life as an articulate but 'non-speaking' family member.

Through Narratta we have taken up anticipatory and speculative practices to investigate potentially new ways of approaching, seeing and communicating affect and engaging readers playfully and seriously with alternative perspectives on the future north. We use the mode of design fiction to challenge them to think beyond their comfort zones and readerly expectations. We place this work within a broader approach to speculative inquiry that, drawing on the work of Alfred Whitehead, seeks to acknowledge attention to the roles of things as actors in complex environments and aims to avoid earlier binary divisions between humans and objects.[52] This allows us to reconsider things as agentive in the context of wider communicative, climatic and cultural ecologies. For Isabelle Stengers, this is a matter of bringing into being an ecology of practices as a new way of learning and questioning.[53] Earlier we have framed this in terms of a prospective hermeneutics.[54] This is a forward-reaching inquiry that 'works with ideation, abduction and projections, drawing on current contexts, issues and technologies to anticipate and speculate, and not only examine compositions already enacted'. However, this view may be extended to a wider one that I call a *prospective communicative ecology* that includes both practices and methods of speculation with anticipation. Such an ecology may include emergent and experimental practices and also upend the retrospective hermeneutics of the humanities and the contextual logics of contemporaneity in qualitative social science.

Narratta should not be read too literally; she is a trope. By fictional means she makes material a set of concerns; she is a device to prompt and to problematise. Above all, she engages us dialogically, not just as online readers and commentators, to think further about our own voices and views in the undeniable face of climate change and possible and probable futures of the Arctic. As co-participants in shaping such a communicative discursive enactment through Narratta, we suggest, not instruct about, the need to look more closely into how the future of the Arctic is to be examined.

Fabulations

The next screen grab below presents Narratta's ironic observations on the research team and its working dynamics as it travelled though the mining cities of the Kola Peninsula in northwest Russia.[55] Her blog entries are posed through her unnatural persona, yet such an entry points to the collaborative mode of inquiry in learning about and researching the cultural landscapes of the chosen Arctic territories and terrains as well as the project team's own shared and individual processes of her own fabrication.

Relatively little work has been done from a co-design perspective concerning design fiction: Scott Rettberg is one of few who addresses collaboration in fictive construction, and most design fiction has not acknowledged the co-creational aspects of fictive designing.[56] Design is also not often explicitly discussed in leading writings on story making and digital narrative.[57] However, co-design

News   About   Events   Publications   People   Advisory Board

Workshoppings

Narratta, September 20, 2013

Work. Shopping. Daily life in Murmansk. The design-research team is back in town following their bus ride around the Kola Peninsula.

It's been an unusually warm September. They have taken hundred of photographs, read to each other on their bus, stopped and discussed their experience of the landscape – geological, economic, political and cultural. They have visited monuments and museums, studied the plans of mining cities and begun to more fully understand the complexities of this part of north west Russia where 70% of all people living in the arctic reside.

Our journey has been one we have made together but they have not seen me slipping alongside them.

At one point I thrust myself up and out of my difficult journey from lake to lake and river to pool … and ouch! I landed in a stinky slime dam beside the mine. They were talking away about pollution, about the open city plans and spaces between the concrete blocks and didn't hear me gasping. After all, this is my arctic too!

They are sitting in the hotel lobby holding a workshop about how to use this tool. How particpatory! They'll need a photo archive or a geo-located mapped image bank if you ask me.

research has acknowledged the importance of the fictional and the role of collaboration on 'evoking the future' and through attending to the development of generative tools, often for prototyping and staging, including imaginative spaces for participatory prototyping, and for promoting dialogue in co-design events.[58]

For Christian Dindler and Ole Sjer Iversen in building design collaboration in a shared narrative space, 'The central idea of FI [fictional inquiry] is to temporarily change or bypass existing socio-cultural structures in a given practice. This allows the designers to reframe the structures of meaning in a context, and let this be the stage for design activities'.[59] In 'The construction of fictional space in participatory design practice' Dindler suggests that the perspective of fictional space offers concepts for reflection in design inquiry.[60] He sees these as being about positing ways props or imaginary and actual artefacts are used in the staging of design inquiries, as providing dynamic means to negotiations between participants, as giving support to reconsideration of their practices and as the generation of novel forms and conceptualisations. These observations offer ANT and design fiction potential in developing further 'fictional spaces' that are led by co-design. However, to date, there is a sparse research literature that looks both ways, either from co-design or from within narrative inquiry.[61]

In shaping a design fictional space, and collaboratively, we have drawn on a range of established design techniques and research methods in a constructional inquiry mode and their interplay.[62] These include ideation, sketching, story spacing, ventriloquising and personas. Multimodal discourse theory as a communication design, I have argued, takes this onwards and over into multiple mediational means, a co-design of mixed media types, modes of address and engagement referring to our own design and research expertise in web-based media and blogging.[63]

Communication technologies

Overall, many of our co-creational moves in Narratta's voice are abductive in nature, even serendipitous.[64] Together they are a themselves a mediational online landscape and a communicative landscape of technological mediation. Narratta's articulations draw on news feeds and lateral thinking drawn from other research materials and personal and professional encounters in a range of creative and research domains. This extends to reading, travelling, researching and learning about the Arctic, climate change and earlier explorations such as the role of aerial photography in Svalbard. In the early twentieth century Svalbard was the launch point for failed and fatal expeditions to the North Pole. Decades after the use of airborne photography in the First World War the actual photographs of aerial documentation of the largest expedition from Svalbard were discovered.

The next blog entry, taken from one of the several expeditions to the North Pole from Svalbard in the early twentieth century using airships, points to Narratta's view of today and tomorrow as also influenced by inflated senses of discovery in the past.[65] A techno-cultural landscape of the future north is mediated via Narratta's shifts between the levels of story and discourse, reference to historical and contemporary narratives of the future and her own ironised views on these by virtue of her spatial and temporal flexibility. She is communicatively and interpretatively agile – a cultural critic realised by co-design. As Bruno Latour argues, 'The critic is not the one who debunks, but the one who assembles. The critic is not the one who lifts the rugs from under the feet of the naïve believers, but the one who offers the participants arenas in which to gather'.[66]

Narratta is our gatherer. She adopts a fluid, not fixed, state. She does not merely assemble. She dissembles her own perspectives – and those of others. She does not reveal all 'below the ice' but interprets her 'topical' arrivals and discursive departures through dwelling in places such as the harbour of Vardø. She also questions her sensorial and experiential self as a way of relating to others and their own perspectives and reactions in a mode of 'spatial becoming'.[67]

Future north, nurture forth

We need to nurture forth the potential of working together speculatively in ways that design and design fiction offers.

Blimp me

Narratta, November 04, 2015

Technologies of seeing, remote sensing, satellite hoovering. Who'd have thought the far north town of Svalbard would have become such a techno-scape. Mines closed, hand drills and the elaborate overhead shuttle system would be replaced by scientists, students and tourists. Hey, it's time to get outta town, this frontier like moon base sorta place I heard one of those Future Northers saying. Students all off on a landscape architecture task to draw sections, see the town differently, scaled and spliced up with new eyes.

North Pole Expedition Museum

HOME   ABOUT   NEWS   OPENING HOURS AND LOCATION   ARCHIVE   LINKS   CONTACT US

Expeditions
to the North
Pole

As Timothy Morton argues in his work on the Anthropocene in general, climate change acts as an agent beyond our direction, infused with our prior choices and consequences entangled in a complex autonomous system beyond our control. Into this mix floats Narratta and her mode of spatial being and becoming in and across time and place.

Narratta dives and surfaces, waits and wonders. She is a post-humanistic entity, shaped in time and by a team of people, but she is indicative of how our own perceptions and deliberations may be communicated as one instance of the much wider potential of design fiction and wider anticipatory narratives to reveal the complexity and perils of climate change within the cultural landscapes of today.

Kirsten Hastrup cites the words of an Inuit hunter called Qujaakistsoq who refers to the narwhal as a sensitive creature frightened off by shadows and noise.[68] Narratta is shy, but she is stealthy and she is present online and in our project's activities as a digital material invention devised through sensory and affective engagement in the world.[69] She may leave readers with another kind of shadow, an after image, an intaglio, something you might make incisive, creatively, looking beyond the now, a vision, a mental imprint to consider and to contemplate not only measure. Narratta serves in this sense as an *evocative artefact* not only an epistemic one.[70] She operates within a design fictional practice within cultural landscape as a domain; this practice is hybridly shaped, both effective, or solution directed, and evocative.[71]

Scientists have done their work to reveal the intricacies of climate change. The human sciences need to show – and I would argue with the same measure of urgency adopted by those scientists – how these intricacies sit and swim, float and fly culturally. After all, in our daily futures we will live in cultural landscapes not just ones made of undeniable and immense scientific data and projected, projective or even utopian policy. A cultural move on climate change is needed to further voice competing interests and purposive positionings and to point to needed behaviour changes and alternations to shared values of consumption and construction.

An ethnographically framed view as Narratta's allows us to engage in narrative spaces and acts that serve as counterweights to the technologies of scientific measurement. The rhetoric of such a multimodal mapping itself needs to be queried so that design fiction's counter narrative potential may be articulated even further and not relegated to observer status. As Appadurai reminds us, 'mediation may be seen as an *effect* of which some sort of materiality is always the condition of possibility'.[72] Our experimental online work is not only designed to convey the complexity of yesterday's and today's Arctic but to offer a mode of anticipation of incalculable futures,[73] distinguishing present views of the future that cross disciplinary boundaries between the humanities and social sciences and use the future in the present.

Narratta has helped us as a research team to see, and to differing degrees, that a future north today can be more fully understood through a mode of design fictioning. Reflecting on the work of Whitehead, Stengers sees a role for poetic, creative voices as 'to bring into existence, in one form or another, that which reasons cannot board and inspect'.[74] The chapter offers some indications of how this may be floated and swum via a blog-based design fictional work.

Design's task is to address wicked problems, to deal with the slippery, to engage with the contingent, the emergent and the seemingly incongruent.

Towards a poetics of anticipation

The future is not yet built nor is it wholly denuded of hope. Competing and overlapping claims to land and material resources have not all been pitched, negotiated or resolved. Internationally, there is still much to be reached for in communicating the longer-term climate change trajectory of the planet – a trajectory that is already being pre-figured in the Arctic. The future today needs to urgently engage in anticipating the very material dynamics of these changes that may be felt most

Embedded in the future

Narratta, May 29, 2015

Some days all my nuclear empowerment makes me restless. Anxious even. Shark-like, I survey these Arctic waters. Murmansk, Vardø, Svalbard, across to the east coast of Greenland. And I love to lie shallow bays, chortling to myself with my thoughts of how to reveal the mysteries of climate change to the assemblies of scientists and tourists, behind my back, belly laid bare to the midnight sun, I find I become, well, a little reflective.

When I lurk in harbours, swim alongside ships, tune into the Wi-Fi traffic of researchers and the gaggles of strategic planners, I often wonder if they are really able to think about the future. I have no choice. I'm embedded in it for two centuries at least. It really changes your point of view, well so does being able to shift my electronic self across time, something must have happened as I was powered up, a small glitch crossed not only the materiality of the mammal flesh and the neutrons and electrons, but also my trajectories.

And there half the time already powered ahead of the curve, the surge, the endless dirge about climate change all about me in the present is a change isn't ever never gained between carbon taxes and emissions policy. Listened to a bunch of them last night, out on an interdisciplinary seminar on the small icebreaker that plows the fiords of Svalbard. I rather unthinkingly linked in to a discussion of face-to-face and online-line partnering between a bunch of glaciologists down at the prow tapping away at their smartphones and one another's shoulders. It was a bit windy for most of them, used to slower moving currents of water wedged across the valleys! An interesting and dedicated crowd. But they had a strange understanding of time. They were actually discussing arctic tourism.

brutally in struggles for resources, especially water, in a future world. Changes that must be enacted right now and maintained to secure possible liveable futures and to avert dire likely scenarios also demand changes in how we live, work and play. Working with the persona Narratta has encouraged us to shift from traditional, perspectival and remotely viewed landscapes of a future north and dominant mediated geo-political discourses, towards culturally framed ones.[75] An un/natural counter narrative points to one way of conceptualising and engaging with the difficulty of communicating matters of climate change in a cultural landscape view of a future north.

Acknowledgements

In addition to the contributions of the Future North project team, and especially Janike Kampevold Larsen, and related master's studios and summer schools at the Tromsø Academy of Landscape and Territorial Studies, my thanks go to Hilde Arnsten, Jonny Aspen, Manuela Celi, Alittea Chishin, Angeliki Dimaki-Adolfsen, Elena Formia, Tau Lenskjold, Einar Martinussen, Bruce Snaddon and Matt Ward. Special thanks go to Cheryl Ball for editorial and content expertise. The chapter has benefitted from comments and critique at the following conferences: *Arctic Modernities 2014*, *Arctic Frontiers 2014* and the *1st International Conference on Anticipation 2015*.

Notes

1 See Figure 8.1. Accessed March 15, 2017. www.oculs.no/people/narratta/.

2 Carolyn Miller and Dawn Shepherd, "Blogging as Social Action: A Genre Analysis of the Weblog," in *Into the Blogosphere*, ed. Laura Gurak et al. (online: 2004), http://hdl.handle.net/11299/172818; Andrew Morrison and Per Roar Thorsnes, "Blogging the emphemeral," in *Inside Multimodal Composition*, ed. Andrew Morrison (Cresshill: Hampton Press, 2010).

3 Mikhail Bakhtin, *The Dialogic Imagination: Four Essays by M.M Bakhtin*, edited by Michael Holquist, trans. Caryl Emerson and Michael Holquist (Austin: University of Texas Press, 1981).

4 E.g. Michael Bravo and Sverker Sörlin, "Narrative and Practice – an Introduction," in *Narrating the Arctic: A Cultural History of Nordic Scientific Practices*, ed. Michael Bravo and Sverker Sörlin (Canton, MA: Science History Publications, 2002).

5 E.g. Derek Hales, "Design Fictions: An Introduction and Provisional Taxonomy," *Digital Creativity* 24, no. 1 (2013).

6 E.g. Stephen Crofts Wiley, "Spatial Materialism. Grossberg's Deleuzean Cultural Studies," *Cultural Studies*, 19, no. 1 (2006).

7 Michael Bamberg and Molly Andrews, eds., *Considering Counter-Narratives. Narrating, Resisting, Making Sense* (Amsterdam: John Benjamins, 2004).

8 E.g. Andrew Morrison, ed., *Inside Multimodal Composition* (Cresskill: Hampton Press, 2010); Denise Gonzales Crisp, "Discourse This! Designers and Alternative Critical Writing," *Design and Culture*, 1, no. 1 (2009); Catharina Rossi, "Bricolage, Hybridity, Circularity: Crafting Production Strategies in Critical and Conceptual Design," *Design and Culture*, 5, no. 1 (2013); Kitrina Douglas and David Carless, "Sharing a Different Voice: Attending to Stories in Collaborative Writing," *Cultural Studies ↔ Critical Methodologies,* 14, no. 4 (2014).

9 Stuart Hall, *Representation: Cultural Representations and Signifying Practices* (London/Thousand Oaks: Sage in association with the Open University, 1997).

10 Bruno Latour, "Why Has Critique Run Out of Wteam? From Matter of Fact to Matters of Concern," *Critical Inquiry* 30 (2004).

11 Maria Puig de la Bellacasa, "Touching Technologies, Touching Visions: The Reclaiming of Sensorial Experience and the Politics of Speculative Thinking," *Subjectivity* 28 (2009); Maria Puig de Bellacasa, "Matters of Care in Technoscience: Assembling Neglected Things," *Social Studies of Science* 41, no. 1c (2011).

12 Bruno Latour. *Aramis or the Love of Technology* (Cambridge: Harvard University Press, 1996).

13 David Shields, *Reality Hunger. A Manifesto* (London: Hamish Hamilton, 2010), Entry #384.

14 John Drabinski, "Poetics of the Mangrove," in *Deleuze and Race*, ed. Arun Saldanha and Jason Adams (Edinburgh: Edinburgh University Press, 2013).

15 See also Geoff Manaugh, ed. *Landscape Futures* (Bacelona: Actar, 2013).

16 Matt Ward and Alex Wilkie, "Made in Criticalland: Designing Matters of Concern," in *Networks of Design: Proceedings of the 2008 Annual International Conference of the Design History Society (UK)*. University College Falmouth, September 3–6, 2008 (Universal Publishers.com, 2008), 7.

17 Andrew Morrison, "Ruminations of a Wireless Ruminant," in *Proceedings of NORDES 2011: Making it Matter! 4th Nordic Design Research Conference. 29–31 May 2011* (Helsinki: NORDES, 2011), www.nordes.org/opj/index. php/n13/article/view/111/95; Andrew Morrison, "Design Prospects: Investigating Design Fiction via a Rogue Urban Drone," in *Proceedings of DRS 2014. 16–19 June 2014* (Umeå: Design Research Society, 2014); Eva Knutz, Thomas Markussen, and Poul Christensen, "The Role of Fiction in Experiments Within Design, Art and Architecture," *Proceedings of NORDES 2013. 9–12 June 2013* (Copenhagen; Malmö: NORDES, 2013), www.nordes.org/opj/index.php/n13/article/view/308/289.

18 Andrew Morrison, Ragnhild Tronstad, and Einar Martinussen, "Design Notes on a Lonely Drone," *Digital Creativity* 24, no. 1 (2013).

19 Bruce Sterling, "Design Fictions," *Interactions* 16, no. 3 (2009); Julian Bleecker, "Fiction: from Props to Prototypes," in *6th SDN Conference: Negotiating Futures – Design Fiction, 28–30 October, 2010* (Basel: SDN, 2010); Derek Hales, "Design Fictions: An Introduction and Provisional Taxonomy," *Digital Creativity* 24, no. 1 (2013).

20 See also Charles Booth, Michael Rowlinson, Peter Clark, Agnes Delahaye, and Stephen Procter. "Scenarios and Counterfactuals as Modal Narratives," *Futures*, 41, no. 2 (2009).

21 Bruce Sterling, *Scenarios and Speculations* (Amsterdam: Sun Publishers, 2009).

22 E.g. James Augur, "Speculative Design: Crafting the Speculation," *Digital Creativity* 24, no. 1 (2013).

23 Anthony Dunne and Fiona Raby, *Speculative Everything* (Cambridge: MIT Press, 2013).

24 Andrew Morrison, "Design Prospects: Investigating Design Fiction via a Rogue Urban Drone," in *Proceedings of DRS 2014. 16–19 June 2014* (Umeå: Design Research Society, 2014).

25 E.g. Thomas Binder et al., *Design Things* (Cambridge, MA: MIT Press, 2011); Edwin Sayes, "Actor-Network Theory and Methodology: Just What Does It Mean to Say That Nonhumans Have Agency?" *Social Studies of Science* 44, no. 1 (2014).

26 E.g. Donna Harraway, *When Species Meet* (Minneapolis: University of Minnesota Press, 2008).

27 John Law and Marianne Lien, "Slippery: Field Notes on Empirical Ontology," *Social Studies of Science* 43, no. 3 (2013).

28 Lawrence Grossberg, "Identity and Cultural Studies: Is That All There Is?," in *Questions of Cultural Identity*, ed. Stuart Hall and Paul duGay (London: Sage, 1996).

29 See Figure 8.2. Accessed March 15, 2017. www.oculs.no/projects/future-north/news/?post_id=3890.

30 Jane Marching, "Magnets of the Fantastic: the North Pole Observed," in *Farfield. Digital Culture, Climate Change, and the Poles*, ed. Jane Marching and Andrea Polli (Bristol: Intellect, 2012), Kindle edition.

31 James Corner, "Preface," in *The Landscape Imagination*, ed. James Corner and Alison Bick Hirsh (New York: Princeton Architectural Press, 2014), pp. 7–8.

32 Roberto Poli, "Anticipation: What About Turning the Human and Social Sciences Upside Down?" *Futures* 64 (2014); Roberto Poli, "Anticipation: A New Thread for the Human and Social Sciences?" *CADMUS* 2, no. 13 (2014); Manuela Celi, ed., *Advanced Design Cultures. Long-Term Perspective and Continuous Innovation* (Cham: Springer, 2015).

33 Brian Richardson, *Unnatural Voices. Extreme Narration in Modern and Contemporary Fiction* (Columbus: The Ohio State University Press, 2006); Jan Alber, Stefan Iversen, Henrik Skov Neilsen, and Brian Richardson, "Unnatural Narratives, Unnatural Narratology: Beyond Mimetic Models," *Narrative* 18, no. 2 (2010); Jan Alber, Henrik Skov Neilsen, and Brian Richardson, eds., *A Poetics of Unnatural Narrative* (Columbus: The Ohio State University Press, 2013).

34 Matt Malpass, "Between Wit and Reason: Defining Associative, Speculative, and Critical Design in Practice," *Design and Culture* 5, no. 3 (2013).

35 E.g. Ivana Milojevic and Sohail Inayatullaha, "Narrative Foresight," *Futures* 73 (2015).

36 Paul Raven and Shirin Elahi, "The New Narrative: Applying Narratology to the Shaping of Futures Outputs," *Futures*, 74 (2015).

37 See Figure 8.3. Accessed March 15, 2017. www.oculs.no/projects/future-north/news/?post_id=4115&doing_wp_cron=1489736301.2862410545349121093750.

38 Mikhail Bakhtin, *The Dialogic Imagination: Four Essays by M.M Bakhtin*, edited by Michael Holquist, trans. Caryl Emerson and Michael Holquist (Austin: University of Texas Press, 1981); Mikhail Bakhtin, *Problems of Dostoevsky's Poetics*, ed. and trans. Caryl Emerson (Minneapolis: University of Minnestota Press, 1984); Mikhail Bakhtin, *Speech Genres and Other Late Essays*, ed. Caryl Emerson and Michael Holquist, trans. Vern McGee (Austin: University of Texas Press, 1986).

39 Bakhtin, *The Dialogic Imagination*, 84ff.

40 Ibid., 95.

41 Alice Bell and Jan Alber, "Ontological Metalepsis and Unnatural Narratology," *Journal of Narrative Theory* 42, no. 2 (2012).

42 Lise Blom and Elena Glasberg, "Disappearing Ice and Missing Data: Climate Change in the Visual Culture of the Polar Regions," in *Farfield. Digital Culture, Climate Change, and the Poles*, ed. Jane Marching and Andrea Polli (Bristol: Intellect, 2012), Kindle edition.

43 See Figure 8.4. Accessed March 15, 2017. www.oculs.no/projects/future-north/news/?post_id=4126.

44 Timothy Morton, *The Ecological Thought* (Cambridge: Harvard University Press, 2010).

45 Arjun Appadurai, "Mediants, Materiality, Normativity," *Public Culture* 27, no. 2 (2015): 233.

46 Arjun Appadurai, *The Future as Cultural Fact: Essays on the Global Condition* (London: Verso, 2013): 299.

47 Lawrence Grossberg, "Cultural Studies and Deleuze-Guattari, Part 1. A Polemic on Projects and Possibilities," *Cultural Studies* 28, no. 1 (2014): 17.

48 Myka Vielstimmig, "'Petals on a Wet Black Bough': Textuality, Collaboration, and the New Essay," in *Passions, Pedagogies and 21st Century Technologies*, ed. Gail Hawisher and Cynthia Self (Logan: Utah State University Press, 1999); Binder et al., *Design Things*.

49 Trish Roberts-Miller, "Parody Blogging and the Call of the Real," in *Into the Blogosphere; Rhetoric, Community and Culture of Weblogs*, ed. Laura Gurak et al. (Minneapolis: University of Minnesota, 2004). http://blog.lib.umn.edu/blogosphere/.

50 E.g. Lucas Ihlein, "Blogging as Art, Art as Research," in *Material Inventions: Applying Creative Research*, ed. Estelle Barrett and Barbara Bolt (I.B. Tauris: London, 2014).

51 Accessed March 15, 2017. https://grandtextauto.soe.ucsc.edu.

52 E.g. Isabelle Stengers, *Thinking with Whitehead: A Free and Wild Creation of Concepts*, trans. Michael Chase (Cambridge: Harvard University Press, 2011); Ian Bogost, *Alien Phenomenology, or What It's Like to Be a Thing* (Minneapolis: University of Minnesota Press, 2012).

53 Isabelle Stengers, *Cosmopolitics II*, trans. Robert Bononno (Minneapolis: University of Minnesota Press, 2011).

54 Andrew Morrison, Ragnhild Tronstad, and Einar Martinussen, "Design Notes on a Lonely Drone," *Digital Creativity* 24, no. 1 (2013).

55 See Figure 8.5. Accessed March 15, 2017. www.oculs.no/projects/future-north/news/?post_id=3899.

56 Scott Rettberg, "All Together Now: Collective Narrative and Collective Knowledge Communities in Context," in *New Narratives. Stories and Storytelling in the Digital Age* ed. Ruth Page and Bronwyn Thomas (Lincoln: University of Nebraska Press, 2011).

57 E.g. Marie-Laure Ryan, *Avatars of Story* (Minneapolis: University of Minnesota Press, 2006); Marie-Laure Ryan, "Narrative and the Split Condition of Digital Textuality," in *The Aesthetics of Net Literature*, ed. Peter Gendolla and Jörgen Schäfer (Bielefeld: Transcript Verlag, 2007).

58 Eva Brandt and Camilla Grunnet, "Evoking the Future: Drama and Props in User Centered Design," in *Proceedings of Participatory Design Conference 2000. 28 November–1 December 2000* (New York: ACM, 2000); e.g. Christina Brodersen, Christian Dindler, and Ole Sjer Iversen, "Staging Imaginative Places for Participatory Prototyping," *CoDesign* 4, no. 1 (2008).

59 Dindler and Iversen, "Fictional Inquiry," 216.

60 Christian Dindler, "The Construction of Fictional Space in Participatory Design Practice," *CoDesign* 6, no. 3 (2010).

61 E.g. George Triantafyllakos, George Palaigeorgiou, and Ioannis Tsoukalas, "Collaborative Design as Narrative," in *Proceedings of PDC '08. 1–4 October 2008* (Bloomington: PDC, 2008); Markku Lehtimäki, "Natural Environments in Narrative Contexts: Cross-pollinating Ecocriticism and Narrative Theorym," *Storyworlds* 5 (2013); Mathew Vechinskim, "The Design of Fiction and the Fiction of Design: Revisiting the Idea of Literature Through the Study of Design," *Textual Practice* 27, no. 2 (2013).

62 Ilpo Koskinen et al., *Design Research Through Practice: From the Lab, Field and Showroom* (Waltham: Elsevier/Morgan Kaufmann, 2011).

63 Andrew Morrison, ed., *Inside Multimodal Composition* (Cresskill: Hampton Press, 2010).

64 Lily Diaz, "By Chance, Randomness and Indeterminacy Methods in Art and Design," *Journal of Visual Art Practice* 10, no. 1 (2011).

65 See Figure 8.6. Accessed March 15, 2017. www.oculs.no/projects/future-north/news/?post_id=4196.

66 Bruno Latour, "Why Has Critique run out of Steam? From Matter of Fact to Matters of Concern," *Critical Inquiry* 30 (2004): 246.

67 Lawrence Grossberg, "The Space of Culture, the Power of Place," in *The Post-Colonial Question: Common Skies, Divided Horizons*, ed. Iain Chambers and Lidia Curti (London: Rouledge, 1996), 177.

68 Kirsten Hastrup, "Narwhals and Navigators on the Arctic Sea," in *Living with Environmenal Change*, ed. Kirsten Hastrup, and Cecilie Rubow (London: Routledge, 2014), Kindle edition.

69 Estelle Barrett and Barbara Bolt, eds., *Material Inventions: Applying Creative Research* (London: IB. Tauris, 2014).

70 Karin Knorr-Cetina, *Epistemic Cultures* (Cambridge: Harvard University Press, 1999).

71 Gillian Hamilton and Luke Janniste, "The Effectve and the Evocative: A Spectrum of Creative Practice Research," in *Material Inventions: Applying Creative Research*, ed. Estelle Barrett and Barbara Bolt (London: I.B. Tauris, 2014).

72 Arjun Appadurai, "Mediants, Materiality, Normativity," *Public Culture* 27, no. 2 (2015): 225 (original italics).

73 Poli, "Anticipation."

74 Stengers, *Thinking with Whitehead*, 505.

75 See Figure 8.7. Accessed March 15, 2017. www.oculs.no/projects/future-north/news/?post_id=4127&doing_wp_cron=1489736607.6789710521697998046875.

References

Alber, Jan, Stefan Iversen, Henrik Skov Neilsen, and Brian Richardson. "Unnatural Narratives, Unnatural Narratology: Beyond Mimetic Models." *Narrative* 18, no. 2 (2010): 113–36.

Alber, Jan, Henrik Skov Nielsen, and Brian Richardson, eds. *A Poetics of Unnatural Narrative.* Columbus: Ohio State University Press, 2013.

Appadurai, Arjun. *The Future as Cultural Fact: Essays on the Global Condition.* London: Verso, 2013.

Appadurai, Arjun. "Mediants, Materiality, Normativity." *Public Culture* 27, no. 2 (2015): 221–37.

Augur, James. "Speculative Design: Crafting the Speculation." *Digital Creativity* 24, no. 1 (2013): 11–35.

Bakhtin, Mikhail. *The Dialogic Imagination: Four Essays by M.M Bakhtin,* edited by Michael Holquist. Translated by Caryl Emerson and Michael Holquist. Austin: University of Texas Press, 1981.

Bakhtin, Mikhail. *Problems of Dostoevsky's Poetics.* Edited and translated by Caryl Emerson. Minneapolis: University of Minnestota Press, 1984.

Bakhtin, Mikhail. *Speech Genres and Other Late Essays.* Edited by Caryl Emerson and Michael Holquist. Translated by Vern McGee. Austin: University of Texas Press, 1986.

Bamberg, Michael and Andrews Molly, eds. *Considering Counter-Narratives. Narrating, Resisting, Making Sense.* Amsterdam: John Benjamins, 2004.

Barad, Karen. "Posthumanist Performativity: Toward an Understanding of How Matter Comes to Matter." *Signs: Journal of Women in Culture and Society* 28, no. 3 (2003): 801–31.

Barrett, Estelle and Barbara Bolt, eds. *Material Inventions: Applying Creative Research.* London: I.B. Tauris, 2014.

Bell, Alice and Jan Alber. "Ontological Metalepsis and Unnatural Narratology." *Journal of Narrative Theory* 42, no. 2 (2012): 166–92.

Bleecker, Julian. "Fiction: From props to prototypes." 6th SDN Conference: Negotiating Futures – Design Fiction, 28–30 October 2010, Basel: SDN, 2010.

Blom, Lise and Elena Glasberg. "Disappearing Ice and Missing Data: Climate Change in the Visual Culture of the Polar Regions." In *Farfield. Digital Culture, Climate Change, and the Poles,* edited by Jane Marching and Andrea Polli. Bristol: Intellect, 2012. Kindle edition.

Bogost, Ian. *Alien Phenomenology, or What It's Like to Be a Thing.* Minneapolis: University of Minnesota Press, 2012.

Booth, Charles, Michael Rowlinson, Peter Clark, Agnes Delahaye, and Stephen Procter. "Scenarios and Counterfactuals as Modal Narratives." *Futures* 41 no. 2 (2009): 87–95.

Brandt, Eva and Camilla Grunnet. "Evoking the Future: Drama and Props in User Centered Design." In *Proceedings of Participatory Design Conference 2000. 28 November – 1 December 2000.* New York: ACM, 2000.

Bravo, Michael and Sverker Sörlin. "Narrative and Practice – an Introduction." In *Narrating the Arctic: A Cultural History of Nordic Scientific Practices,* edited by Michael Bravo and Sverker Sörlin, 3–32. Canton: Science History Publications, 2002.

Brodersen, Christina, Christian Dindler, and Ole Sjer Iversen. "Staging Imaginative Places for Participatory Prototyping." *CoDesign* 4, no. 1 (2008): 19–30.

Celi, Manuela, ed. *Advanced Design Cultures. Long-Term Perspective and Continuous Innovation.* Cham: Springer, 2015.

Corner, James. "Preface." In *The Landscape Imagination,* edited by James Corner and Alison Bick Hirsh, 7–11. New York: Princeton Architectural Press, 2014.

Crofts Wiley, Stephen. "Spatial Materialism. Grossberg's Deleuzean Cultural Studies." *Cultural Studies,* 19, no. 1 (2006): 63–99.

Diaz, Lily. "By Chance, Randomness and Indeterminacy Methods in Art and Design." *Journal of Visual Art Practice* 10, no. 1 (2011): 21–33.

Dindler, Christian. "The Construction of Fictional Space in Participatory Design Practice." *CoDesign* 6, no. 3 (2010): 167–82.

Dindler, Christian and Ole Sjer Iversen. "Fictional Inquiry: Design Collaboration in a Shared Narrative Space." *CoDesign* 3, no. 4 (2007): 213–34.

Douglas, Kitrina and David Carless. "Sharing a Different Voice: Attending to Stories in Collaborative Writing." *Cultural Studies ↔ Critical Methodologies* 14, no. 4 (2014): 303–11.

Drabinski, John. "Poetics of the Mangrove." In *Deleuze and Race,* edited by Arun Saldanha and Jason Adams, 288–99. Edinburgh: Edinburgh University Press, 2013.

Dunne, Anthony and Fiona Raby. *Speculative Everything.* Cambridge: MIT Press, 2013.

Gonzales Crisp, Denise. "Discourse This! Designers and Alternative Critical Writing." *Design and Culture* 1, no. 1 (2009): 105–20.

Grossberg, Lawrence. "Identity and Cultural Studies: Is That All There Is?" In *Questions of Cultural Identity*, edited by Stuart Hall and Paul duGay, 88–107. London: Sage, 1996.

Grossberg, Lawrence. "The Space of Culture, the Power of Place." In *The Post-Colonial Question: Common Skies, Divided Horizons*, edited by Iain Chambers and Lidia Curti, 169–88. London: Rouledge, 1996.

Grossberg, Lawrence. "Cultural Studies and Deleuze-Guattari, Part 1. A Polemic on Projects and Possibilities." *Cultural Studies* 28, no. 1 (2014): 1–28.

Hales, Derek. "Design Fictions: An Introduction and Provisional Taxonomy." *Digital Creativity* 24, no. 1 (2013): 1–10.

Hall, Stuart. *Representation: Cultural Representations and Signifying Practices.* London/Thousand Oaks: Sage in association with the Open University, 1997.

Hamilton, Gillian and Luke Janniste. "The Effective and the Evocative: S Spectrum of Creative Practice Research." In *Material Inventions: Applying Creative Research*, edited by Estelle Barrett and Barbara Bolt, 232–56. London: I.B. Tauris, 2014.

Harraway, Donna. *When Species Meet.* Minneapolis: University of Minnesota Press, 2008.

Hastrup, Kirsten. "Narwhals and Navigators on the Arctic Sea." In *Living with Environmenal Change*, edited by Kerstin Kastrup and Cecilie Rubow. London: Routledge, 2014. Kindle edition.

Ihlein, Lucas. "Blogging as Art, Art as Research." In *Material Inventions: Applying Creative Research*, edited by Estelle Barrett and Barbara Bolt, 38–49. London: I.B. Tauris, 2014.

Knorr-Cetina, Karin. *Epistemic Cultures.* Cambridge: Harvard University Press, 1999.

Knutz, Eva, Thomas Markussen, and Poul Christensen. "The Role of Fiction in Experiments within Design, Art and Architecture." In *Proceedings of NORDES 2013. 9–12 June 2013*, 341–348. Copenhagen/Malmö: NORDES, 2013. www.nordes.org.

Koskinen, Ilpo, John Zimmerman, Thomas Binder, Johan Redström, and Stephan Wensveen. *Design Research Through Practice: From the Lab, Field and Showroom.* Waltham: Elsevier/Morgan Kaufmann, 2011.

Latour, Bruno. *Aramis or the Love of Technology.* Cambridge: Harvard University Press, 1996.

Latour, Bruno. "Why Has Critique Run Out of Steam? From Matter of Fact to Matters of Concern." *Critical Inquiry* 30 (2004): 225–48.

Law, John. *After Method: Mess in Social Science Research.* London: Routledge, 1994.

Law, John, and Marianne Lien. "Slippery: Field Notes on Empirical Ontology." *Social Studies of Science* 43, no. 3 (2013): 363–78.

Lehtimäki, Markku. "Natural Environments in Narrative Contexts: Cross-Pollinating Ecocriticism and Narrative Theory." *Storyworlds* 5 (2013): 119–41.

Malpass, Matt. "Between Wit and Reason: Defining Associative, Speculative, and Critical Design in Practice." *Design and Culture* 5, no. 3 (2013): 333–56.

Manaugh, Geoff, ed. *Landscape Futures.* Bacelona: Actar, 2013.

Marching, Jane. "Magnets of the Fantastic: The North Pole Observed." In *Farfield. Digital Culture, Climate Change, and the Poles*, edited by Jane Marching and Andrea Polli, Intellect: Bristol, 2012. Kindle edition.

Markussen, Thomas and Eva Knutz. "The Poetics of Design Fiction." In *DPPI 2013: Praxis and Poetics. 3–5 September 2015*, 231–40. Newcastle upon Tyne: DPPI, 2013.

Miller, Carolyn and Dawn Shepherd. "Blogging as Social Action: a Genre Analysis of the Weblog." In *Into the Blogosphere*, edited by Laura Gurak, Smiljana Antonijevic, Laurie Johnson, Clancey Ratliff and Jessica Reyman. Online, University of Minnesota, 2004. http://blog.lib.umn.edu/blogosphere/blogging_as_social_action.html

Milojevic, Ivana and Sohail Inayatullaha. "Narrative Foresight." *Futures* 73 (2015): 151–62.

Morrison, Andrew, ed. *Inside Multimodal Composition.* Cresskill: Hampton Press, 2010.

Morrison, Andrew. "Ruminations of a Wireless Ruminant." In *Proceedings of Nordes 2011: Making it Matter! 4th Nordic Design Research Conference. 29–31 May 2011*, Helsinki: NORDES, 2011. www.nordes.org/opj/index.php/n13/article/view/111/95.

Morrison, Andrew. "Design Prospects: Investigating Design Fiction via a Rogue Urban Drone." In *Proceedings of DRS 2014. 16–19 June 2014.* Umeå: Design Research Society, 2014.

Morrison, Andrew, Timo Arnall, Jørn Knutsen, Einar Martinussen, and Kjetil Nordby. "Towards Discursive Design." In *Proceedings of IASDR 2011, 4th World Conference on Design Research. 31 October–4 November 2011.* CD-rom. Delft: IASDR, 2011.

Morrison, Andrew and Per Roar Thorsnes. "Blogging the Emphemeral." In *Inside Multimodal Composition*, edited by Andrew Morrison, 255–90. Cresshill: Hampton Press, 2010.

Morrison, Andrew, Ragnhild Tronstad, and Einar Martinussen. "Design Notes on a Lonely Drone." *Digital Creativity* 24, no. 1 (2013): 46–59.

Morton, Timothy. *The Ecological Thought*. Cambridge: Harvard University Press, 2010.

Morton, Timothy. *Hyperobjects: Philosophy and Ecology After the End of the World*. Minneapolis: University of Minnesota Press, 2013.

Parisi, Luciana. "Speculation: A Method for the Unattainable." In *Inventive Methods*, edited by Celia Lury and Nina Wakeford, 232–44. London: Routledge, 2012.

Poli, Roberto. "Anticipation: What About Turning the Human and Social Sciences Upside Down?" *Futures* 64 (2014): 15–18.

Poli, Roberto. "Anticipation: A New Thread for the Human and Social Sciences?" *CADMUS* 2, no. 13 (2014): 23–36.

Puig de la Bellacasa, Maria. "Touching Technologies, Touching Visions: The Reclaiming of Sensorial Experience and the Politics of Speculative Thinking." *Subjectivity*, 28 (2009): 297–315.

Puig de Bellacasa, Maria. "Matters of Care in Technoscience: Assembling Neglected Things." *Social Studies of Science* 41, no. 1 (2011): 85–106.

Raven, Paul and Shirin Elahi. "The New Narrative: Applying Narratology to the Shaping of Futures Outputs." *Futures* 74 (2015): 49–61.

Rettberg, Scott. "All Together Now: Collective Narrative and Collective Knowledge Communities in Context." In *New Narratives. Stories and Storytelling in the Digital Age* edited by Ruth Page and Bronwyn Thomas, 187–204. Lincoln: University of Nebraska Press, 2011.

Richardson, Brian. *Unnatural Voices. Extreme Narration in Modern and Contemporary Fiction*. Columbus: Ohio State University Press, 2006.

Roberts-Miller, Trish. "Parody Blogging and the Call of the Real." In *Into the Blogosphere; Rhetoric, Community and Culture of Weblogs*, edited by Laura Gurak, Smiljana Antonijevic, Laurie Johnson, Clancy Ratliff, and Jessica Reyman. Minneapolis: University of Minnesota Press, 2004. http://blog.lib.umn.edu/blogosphere/.

Rossi, Catharina. "Bricolage, Hybridity, Circularity: Crafting Production Strategies in Critical and Conceptual Design." *Design and Culture* 5, no. 1 (2013): 69–87.

Ryall, Anka, Johan Schimanski, and Henning Howlid Wærp. eds. *Arctic Discourses*. Newcastle upon Tyne: Cambridge Scholars Publishing, 2010.

Ryan, Marie-Laure. *Avatars of Story*. Minneapolis: University of Minnesota Press, 2006.

Ryan, Marie-Laure. "Narrative and the Split Condition of Digital Textuality." In *The Aesthetics of Net Literature*, edited by Peter Gendolla and Jörgen Schäfer, 257–80. Bielefeld: Transcript Verlag, 2007.

Sayes, Edwin. "Actor-Network Theory and Methodology: Just What Does It Mean to Say That Nonhumans Have Agency?" *Social Studies of Science* 44, no. 1 (2014): 134–49.

Shields, David. *Reality Hunger: A Manifesto*. London: Hamish Hamilton, 2010.

Stengers, Isabelle. *Thinking with Whitehead: A Free and Wild Creation of Concepts*. Translated by Michael Chase. Cambridge: Harvard University Press, 2011.

Stengers, Isabelle. *Cosmopolitics II*. Translated by Robert Bononno. Minneapolis: University of Minnesota Press, 2011.

Sterling, Bruce. "Design Fictions." *Interactions* 16, no. 3 (2009): 21–4.

Sterling, Bruce. *Scenarios and Speculations*. Amsterdam: Sun Publishers, 2009.

Telier, A., Thomas Binder, Giorgio De Michelis, Pelle Ehn, Guilio Jacucci, Per Linde, and Ina Wagner. *Design Things*. Cambridge: MIT Press, 2011.

Travis, Kathryn. "Interrogating Place: Cartographic Techniques and Masculinist Maps." In *On and Off the Page. Mapping Place in Text and Culture*, edited by M.B Hackler, 292–322. Newcastle upon Tyne: Cambridge Scholars Publishing, 2009.

Triantafyllakos, George, George Palaigeorgiou, and Ioannis Tsoukalas. "Collaborative Design as Narrative." In *Proceedings of PDC '08. 1–4 October 2008*, 210–13. Bloomington: Indiana University, ACM, 2008.

Vechinskim, Mathew. "The Design of Fiction and the Fiction of Design: Revisiting the Idea of Literature Through the Study of Design." *Textual Practice* 27, no. 2 (2013): 269–93.

Vielstimmig, Myka. "'Petals on a Wet Black Bough': Textuality, Collaboration, and the New Essay." In *Passions, Pedagogies and 21st Century Technologies*, edited by Gail Hawisher and Cynthia Self, 89–114. Logan: Utah State University Press, 1999.

Ward, Matt and Alex Wilkie. "Made in Criticalland: Designing Matters of Concern." In *Networks of Design: Proceedings of the 2008 Annual International Conference of the Design History Society (UK). University College Falmouth, 3–6 September 2008.* Universal Publishers.com, 2008.

9 The perforated landscape

Kjerstin Uhre

I am standing in the smallest circular corral, the *gárdi*. The air is damp from the animals' breathing and perspiration. A reindeer owner invited me to take part in her family's autumn fence-work. We're just done, but I linger a while watching the reindeer as they run round and round: their antlers, their stylish ears, their GPS bracelets, their faces with curly fur on the foreheads, the graffiti paint on the shoulders of those who received parasite treatment. Suddenly I sense a marvelling calmness. It is an all-encompassing feeling that I don't know the source of. The reindeer slow down and pass me on each side so close that I merge into the carpet of bodies. I forget myself. Spellbound for just a moment as if in another time. Then I look around – puzzled and a little disoriented – to see what is going on. The reindeer owners inside the gárdi stand along the stockade wall – waiting. I hurry cautiously towards them. When I find myself once again outside the circle of animals one of the guys open the exit gate. The scene explodes in motion as the reindeer notice the opening and run for the escape.[1]

Disputed prospects

Architectural and design-oriented research in landscape and territorial studies is gravitating towards recognizing gaps in our knowledge. These gaps are particularly evident once we work to understand anticipated changes in Arctic landscapes. Such inquiries include speculation on how to develop tools to understand and engage with landscapes that already exhibit and are likely to continue to present contested trajectories towards the future. This extends to a critical review of the role our professions play in the service of public land-use policy as well as private corporate interests, which often lead to conflicts over large-scale landscape utilization. Twenty-five years ago, the landscape architect James Corner claimed that 'we have been adequately cautioned about mapping as a means of projecting power-knowledge', and asked 'but what about mapping as a productive and liberating instrument, a world-enriching agent, especially in the design and planning arts?'[2] However, after decades of post-colonial critique, we still see subtle silencing of indigenous Sámi and local landscape knowledge in Fennoscandia, manifested, I argue, in unbalanced narratives in the landscape mapping of feasibility studies and environmental impact assessments.[3] These are typical tasks undertaken by consultancy firms that employ landscape architects, planners and architects.

In this chapter, which is based on fieldwork in North Norway from 2012 to 2016, I present contextual and site-specific data from an on-going mineral prospecting case that impacts the landscapes of pastoral and coastal communities in West Finnmark. Conceptually, the chapter discusses mineral resource prospecting as *perforations* on many levels: drill holes, information voids, behavioural avoidance and knowledge gaps that, in sum, constitute a perforated landscape. I argue that the manifold natures of these perforations bring *voids of uncertainty* from the prospected futures into the present landscapes. These uncertainties in turn open knowledge gaps towards which research in a wide range of disciplines gravitate. Given the power relations at play between diverse interests and stakeholders that voice concerns ranging from national and corporate to ones that reflect professional local area landscape practices and livelihoods, this chapter discusses how different ways of knowing – including the reindeer herders' attentiveness to the reindeer's knowledge of the terrain – can be charted in new dialogues about future landscapes. Guided by notions of language in the philosophy of landscape and counter moves within cartography, I ask how we might engage analytically with a multiplicity of ways of understanding the many readings of northern landscapes. I suggest that we might do so by situating ourselves in the voids generated by disputed prospects, and carefully observe the discourses, avoidances, knowledge production and practices circling around them. Architectural and interpretative mapping practices might document and bring to the fore the qualitative traits and records of relations to and within landscapes that are brought to the negotiation table but subjugated in the course of political decision-making processes.

A prospect is something likely to happen, a plan or an expectation to the future. The noun *prospect* also refers to presentation material in development projects. The verb *prospecting* refers exclusively to the search for metals, hydrocarbons and minerals, and the making of prospects for future mining and oil activity. A prospect opens knowledge gaps about its social, economic and environmental impacts. Thus, the perforated landscape exists at the discursive level as well as at the physical. Similar to how hegemonic mappings in certain circumstances are addressed by counter mapping, I suggest that disputed prospects may be addressed through what I term *counter prospecting*. I introduce counter prospecting as an experimental and interpretative praxis-based method that operates on two intersecting planes: It resists dominant and already given prospects, while on a plane of anticipation it reaches beyond these in a pro-spective exchange towards possible alternate futures.

Framings

> This is an image of my daughter. She is listening to the groundwater through the mining prospecting company's drill holes in our forest, says the Sámi nature photographer Tor Lundberg Tuorda. The young girl sits on the ground framed by a pair of survey stakes. Her right ear hovers just above two short iron tubes sticking up from the turf. Her concerned face made me acutely aware of threats to the groundwater.[4]

During the first decade of the millennium, a new generation of mining codes and strategies inspired by Canadian resource policy were in the making. As claimed by Alain Deneault and William Sacher,

> the new mining codes that Canada exported throughout the world – partly through the Canadian influence on agencies such as the IMF and the World Bank – all share one troubling characteristic: . . . public authorities are set up to 'solve' the thorny issue of the ancestral presence of indigenous peoples on these lands by subordinating it to the interests of Western corporations.[5]

In the Swedish mineral strategy of 2013, Sweden planned for a mineral boom, and 'the Government's opinion [was] that the needs of the mining industry must be satisfied quickly in order to make full use of the opportunities provided by the current boom in the industry'.[6] Echoing this notion, the Norwegian strategy for the mineral industry sector focussed on state funded mapping and 'an information strategy on mineral deposits in Norway directed towards Norwegian and foreign exploration and mining companies'.[7] The acquisition of mineral prospecting licenses in Norway increased from 1,112 square kilometres in 2010 to 18,663 in 2011, and the resulting prospecting license grid defines a possible new landscape of mineral exploitation in the remaining ranges of intact outfield landscapes.[8]

Today, prospectors extract geological core samples and leave behind holes in the ground. These are holes through which imaginaries pour and questions ripple. In northern Sweden, the media asks

Figure 9.1 The mineral prospectors have left the forest perforated, and a young girl listens to the ground water rippling below

Photo: Tor Lundberg Tuorda.

Figure 9.2 By the end of April, the reindeer graze on the ridges along the migration route to their calving grounds in the coastal mountains

Photo: the author.

questions about heavy metal leakage from drill holes, as the ground water is pouring out of tubes in low lying areas.[9] Still, the visible tubes, stakes, sticks and scars in the landscape yield unspoken territorial contexts: claims to the future landscape, expectations of prosperity, possible hazards, echoes of violence and expulsion of lifestyles.

Any planned usage change of outfields in Finnmark County should be assessed according to the Sámi Parliament's directive for outfield assessment, which '[aims] to safeguard the material basis for, and ensure further development of the Sámi culture'.[10] In the North-Sámi language, the landscape term *meahcci* conceptualizes these landscapes as providing for, and being central to, human life – landscapes '*where the nature resources are found*'.[11] There are many traditional harvesting activities of natural resources that together form the material base for Sámi culture and language. According to Associate Professor in Sámi literature science Harald Gaski, 'Sami descriptions of landscape can function as maps, in which are incorporated topography, geography and information as to which routes are best to take'.[12] This chapter focusses on the landscapes of reindeer pastoralism and the ways they are affected by mineral prospecting.

In *The Two Landscapes of Northern Norway*, philosopher Jacob Meløe describes the coastal fishermen's and the reindeer herders' landscapes as exemplary to understand how practices, landscape and language are interconnected. Based on Wittgenstein's language theory, he studied the praxis related to and the spatial implication of two words: the coastal fishermen's concept of a *natural harbour*, and the reindeer herders' concept of a *jassa*. The professional terminology in Sámi reindeer pastoralism is in Sámi language, and the philosopher relies on the herders to explain for him, not only that a jassa is a patch of snow with certain qualities that are important to the wellbeing of the reindeer herd during the summer, but also how the nomadic pastoral system works.

To understand a word, Meløe argues, the object that it refers to must be 'securely placed within its proper network of internal relations',[13] and 'within a given realm of human activities, or within a given practice, there is a network of implications between activities and activities, between activities and artifacts, between artifacts and their natural surroundings, and between artifacts and artifacts'.[14] The networks of implications constitute worlds and landscapes that yield concepts in the language: 'It is only the activity of sailing a boat too large for its crew to draw it ashore that will yield the concept of a landscape formation providing an adequate harbour'.[15] Reflecting elsewhere on how to understand how small children learn their mother tongue, Meløe argues that 'if [you] don't, already, have a rich understanding of the world . . . you are not able to extend the room where the words fit in place'.[16] Within the networks of landscape practices such as reindeer herding, a great diversity of approaches to knowledge can be found, even though it might appear as homogeneous from the outside. When discussing power-knowledges within reindeer herding research and management, social scientist and reindeer pastoralist Mikkel Nils Sara argues that 'the authority of science, representations in the form of diagrams, and precise quantitative analyses have a greater impact on the decision-making processes of mainstream society', while 'traditional Sámi reindeer herder knowledge, on the other hand, exists within the *siida* as its professional forum, where the practice-related approach to the problem at hand lays the foundation for what is viewed as relevant contributions for the situation'.[17]

Science-related forms of representation include maps, which, James Corner argues, have an agency to influence future landscapes:

> As expertly produced, measured representations, . . . maps are conventionally taken to be stable, accurate, indisputable mirrors of reality, providing the logical basis for future decision making as well as the means for later projecting a designed plan back onto the ground.[18]

In line with this critical perspective, geographer and critical cartographer Jeremy Crampton articulated a dialogue between cartography and critical human geography in order to 'give voice to cartography's nascent attempts to theorize representation and power relations'.[19] He perpetuated J.B.

Harvey's reading of cartographic history as an 'epistemic break' between a model of 'cartography as a neutral communication system, and one in which it is seen in a field of power relations; between maps as presentations of information, and exploratory mapping environments in which knowledge is constructed'.[20]

Counter moves

Doing architecture is a notoriously transdisciplinary project, both horizontally, between different disciplines, and vertically, between practical, administrative and scholarly approaches to development and design problems. Responding to the governmental preparations for a new Norwegian mineral strategy in 2012, the landscape architecture master studio *Fields of Exploration Limits of Exploitation* at the Oslo School of Architecture and Design sought to understand and to find a new vocabulary for what was going on.[21] We claimed that 'the perforated landscape impacts nature, places, and our lives'.[22] The studio discovered how a new version of the territory was being mapped, described and conceived in the light of global mining. Studio findings were included in the TV documentary *Gollegiisá – Skattkammeret* (The Treasure Shrine) by the Sámi film director Roger Manndal.[23] Thus, our studio mappings were charted in the discourse of mineral and landscape values in the Meahcci, the Sámi outfields.

As mentioned previously, Corner proposes to employ the power of maps as a liberating enterprise and draws inspiration from artistic appropriation of cartography: 'Unlike the scientific objectivism that guides most modern cartographers, artists have been more conscious of the essentially fictional status of maps and the power they possess for construing and constructing worlds'.[24] One example could be the Sámi artist Hans Ragnar Mathisen, who started a life-long project of making hand-drawn maps with Sámi place names in 1974, showing the presence of Sámi cultural landscapes.[25] Another example is the way in which the Tromsø-based artists Tanya Busse and Emilija Skarnulyte appropriate methods from mineral prospecting in the *Hollow Earth* project. Through aesthetic scrutiny of mining, they aim 'to understand the scale of which arctic geography is being transformed at the hands of resource extraction, metals and minerals'.[26]

The term *counter mapping* was coined by Nancy Peluso in 1995 and is associated with post-colonial and indigenous mapping practices. Counter mapping contests hegemonic map information and emphasizes traditional land rights that are subjugated or rendered invisible in state maps. Landscape architect Bieke Cattoor and geographer Chris Perkins have built on Corner's ideas and claim that 'architects employ cartographic knowledge differently than cartographers and provide *another counter mapping*'.[27] They suggest that by subverting cartographic norms, architectural atlases might 'escape from standard and accepted orthodoxies' by reimagining figure/ground relationships and employing hybrid forms of cartographic visualization, such as montages, juxtapositions and collages. What I call *counter prospecting* shares characteristics with counter mapping. I suggest, however, that *when* architectural appropriations of cartography give voice to subverted voices and contribute to counter mapping, it might enable counter prospecting because of the prospective intentions and capacities in architectural working methods.

Mixing methods

Jacob Meløe suggests that the meaning of an object can be investigated by 'situat[ing] yourself within the practice that this object belongs to, and then investigat[ing] the object and its contribution to that practice'.[28] I have located my study in qualitative inquiry and chart performative, ethnographic methods, alongside tools of landscape representation from the architectural disciplines. Fieldwork is an important aspect of my study; as ethnographer Anna Tsing writes, '*immersion* works because we are forced to enter other ways of life – that is to become social – before we have any idea what we are learning'.[29]

I suggested previously that in order to chart different forms of knowledge, one should place oneself in the voids of the perforated landscape and observe what circulates around them. Doing so involves a complex set of immersions with many arenas in which to learn to become social. It includes active participation and participant observation, interviews and conversations in addition to analyses of Government documents, media coverage and maps, as well as making my own sketches,

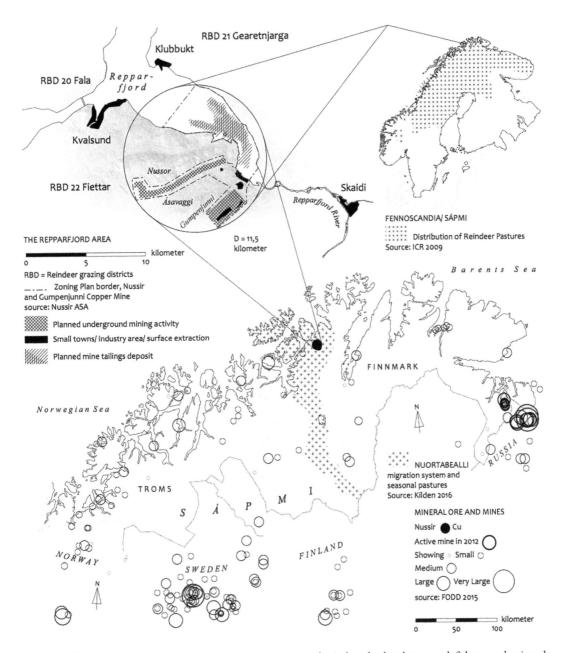

Figure 9.3 The Nussir copper mine prospect in the context of reindeer husbandry, coastal fishery and mineral prospecting

Map: the author.[31]

photos and maps. In these activities, I have pursued a multi-sited inquiry by focusing on the landscape from different perspectives, including those of members of coastal and pastoral communities, representatives from the mining industry, environmentalists and politicians, as well as researchers from other disciplines.

John Law insists that research is performative 'because realities are being made alongside representations of realities'.[30] In research involving mapping and design, this is indeed the case, as cartographies in turn construct ideas about the future landscape. Such ideas I have discussed thoroughly with my informants, whom I prefer to call conversation partners. I worked with various forms of representations: maps, collages, texts and diagrams, edited in *draft outfield atlases*. These I brought back to the field to discuss them critically with my conversation partners, whose landscapes I study, thus adding another layer of reflection to the research, which forms a knowledge base for *counter prospecting*. Counter prospecting implies a recursive practice, involving critically reworking my own tentative cartographies back into the empirical material that I analyse.

A window of opportunity

In 2005, a private Norwegian mineral prospecting company was established to exploit a copper ore at the base of the Nussor Mountain in Kvalsund Municipality. The company was seen favourably as a 'first mover' in realizing the objectives in the Norwegian Mineral Strategy, and the CEO of the Nussir Company, Øystein Rushfeldt, carefully sought opportunities to conduct dialogues with interested parties. Among geologists this area is known as the Repparfjord Tectonic Window. The gently curved bedrock is exposed, and rock layers from deep geologic time are accessible. Local occurrences of copper have been known here for a century. In the 1970s, the Mining Company Folldal Verk mined Gumpenjunni (the Wolf Ridge) and discharged mine tailings into the fjord Repparfjorden. After seven years, copper prices fell, and the mine was abandoned without any environmental remediation. Vast stretches of the fjord bottom were smothered. On the municipal land use map, the Gumpenjunni area remained coded in purple – an industrial area awaiting further exploitation.[32] In the same area, the interactive map from the Reindeer Husbandry Administration shows migration corridors and seasonal pastures. It clearly shows overlapping patterns of crucial functions in the pastoral system of the reindeer herding districts, 22 Fiettar and 20 Fálá.[33] In 2010, Kvalsund Municipality Council adopted Nussir ASA's scoping plan, which promised 150 jobs (including ripple effects). Thereafter followed years where Nussir ASA secured milestones in the planning process and further explorations of the copper ore. During this period, the reindeer walked every fall to the winter pastures in inner Finnmark, and returned every spring to the summer lands. In the pastoral community, children grew up and learned to participate and behave confidently in the *gárdi* every spring and fall.

Nussir ASA's zoning plan – accompanied by environmental impact assessments – was subject to a hearing in 2011. In March 2012, the final plan, which included references and comments to eighty-five consultation statements and two objections – from the Reindeer Husbandry Board and the Sámi Parliament – were presented to the Municipal Council. A third objection – from the Directorate for Fishery, Region North – was excluded by the municipal administration because it came in too late. On May 8, 2012, the Council approved the zoning plan and disregarded the objections. The County Governor of Finnmark then initiated an unsuccessful mediation between the Kvalsund Municipality and the Sámi Parliament, and the case was sent on to the Government for a final decision. In a letter of March 23, 2014, the Ministry of Local Government and Modernization emphasized the importance of exploiting mineral resources and the positive local economic ripple effects. The estimated annual revenue from the mine was at the time NOK 600–700 million, and the ministry approved the plan on the premise that the affected Reindeer Herding Districts and Nussir ASA came to an agreement on mitigating measures.[34] However, the planning process continued without such an agreement.

Bivdi, the Sea Sámi Coastal Fishermen's Association, the Fishermen's Association, the Institute of Marine Research and environmental NGOs all strongly discouraged sea disposal of tailings, a protest that gained global attention.[35] The prospect of a dead fjord has made media headlines all through the process, and the Young Friends of the Earth, Norway (Natur og Ungdom) prepared for civil disobedience if the plans were to go ahead. On December 8, 2016, the Norwegian Environment Agency (Miljødirektoratet) granted Nussir a permit, according to the Pollution Act, to operate two underground mines, annually producing fifty thousand tonnes of copper concentrate. From these mines, the company was allowed to annually discharge two million tonnes of mine tailings in the Repparfjord Fjord and establish a temporarily deposit on land of four hundred thousand tonnes of waste rock.[36] Friends of the Earth, Norway (Naturvernforbundet) and the Union of Norwegian Coastal Fishermen's Association (Kystfiskarlaget) claimed that the Norwegian praxis of depositing mine tailings in the fjords violated the European Union Water Framework Directive and have sent a joint complaint, together with ten other organizations, to the European Free Trade Association (EFTA) Surveillance Authority.[37]

Knowledge gaps

The impact assessment reviewed 'estimated impacts based on available information'.[38] Researchers from the Institute of Marine Research (IMR), as well as representatives from the seafood sector, claimed that knowledge about the impacts from the planned sea deposit were lacking on a range of issues. The applied methodology for valuating nature in the environmental impact assessment was taken from the Norwegian Public Roads Administration, and IMR questioned whether this methodology is suitable for assessing marine environments. Their consultation statement concludes:

> It is in many ways a very thorough assessment of the environmental impact of this applicant's prospect, and the impact assessment clearly shows that a fjord deposit will represent serious pollution of the fjord. It is, however worrying that the impact assessment writes down the value of marine life and marine resources and thus determines that the severe pollution have little consequence.[39]

At the time when the impact assessment was done, it was according to the Norwegian plan and building act sufficient to rely on *existing knowledge*. This has since been changed by the introduction of a second sentence in paragraph seven in the Directive for Impact Assessments, which was adopted in 2014: 'If such information about important matters is not available, new information shall be obtained as necessary'.[40]

For further contingency, the Norwegian Research Council initiated projects to explore the potential for mine tailings disposal in fjord systems. A state-of-the-art review made in one of these projects concludes that 'in most cases, submarine tailings disposal (STD) and sea mine tailings placement (DSTP) activities are taking place before sound scientific baseline information is available', and recommends 'the use of the precautionary approach when knowledge is too scarce to assess impacts'.[41]

Before commissioning the research on sea disposals of mining tailings mentioned previously, the Government insisted that mine tailings disposal in fjord systems – called *dumping* by the adversaries – was a sound environmental praxis. In the press release from the Ministry of Trade, Industry and Fishery about their recommendation of the Nussir mine to the Ministry of Local Government and Modernisation, the Minister of Fisheries, Elisabeth Aspaker, stated: 'Currently we have sufficient knowledge to conclude that sea disposal in Repparfjorden will not have unacceptable adverse impact on the seafood industry'.[42] Thus, it may be fair to claim that the Nussir prospect was pushed forward through a system perforated with knowledge gaps.

Supporting opportunities for the mineral industry, the Government encouraged a balanced co-existence between reindeer herding and mining.[43] In 2010, Nussir ASA came to an agreement with

Figure 9.4 A view from the coastal fishing boat in Repparfjorden towards the planned mine tailings deposit site
Photo: the author.

the Sámi Parliament on procedures for negotiations during the company's application process. 'If we succeed here, the whole world will follow', proclaimed the local paper, *Altaposten*, and cited Rush-feldt: 'As soon as we, here in Norway, have reached an agreement with the Sámi Parliament and the reindeer industry, others will follow'.[44] This attempt to get access to the reindeer pasture areas was not, however, agreed upon by the right holders – the affected reindeer owners and herders. In January 2016, the Sami Parliament Council announced that the agreement was no longer valid, but by then Nussir ASA had already received the discharge permit.

The Environmental Agency admitted in the discharge permit that the mine will negatively impact the livelihood of local fishermen and reindeer pastoralists, but stated that it was 'likely that the initiative [would] provide significant revenue for society'.[45] The Sámi Parliament questioned the assumption of societal benefits, and in 2016 they commissioned Vista Analyse to do the *first* social economic analysis of the Nussir case. This review found that neither documents from the environmental impact assessment nor complementary studies used in the decision-making process justified the strong belief in societal benefits held by the planning authorities. The report concluded that 'the uncertainty in the magnitude of both the benefits and the costs dictate that the precautionary principle should be emphasised in the management of the area'.[46]

The United Nations Convention on Civil and Political Rights, Article 27, states that 'minorities shall not be denied the right, in community with the other members of their group, to enjoy their own culture, to profess and practise their own religion, or to use their own language'.[47] In an attempt to justify the discharge permit relating to these rights, 'the Environmental Agency cannot see that the proposed initiative will affect religion or language particularly negatively'.[48] This sentence indicates a peculiar blindness to the close relations between culture and ways of life, landscape and language that I discuss in this chapter.

Voids

Proposing to change a landscape from being pastoral to becoming industrial is in itself an act of change, as the affected reindeer herders have to adjust their work days in order to attend to growing volumes of paperwork and hearings to ensure their interests. The then leader of the Sámi reindeer-herding district 22 Fiettar, Mikkel Nils Sara, explained it this way to the author Svein Lund: 'Nobody understands that already the plans and the threats they pose is an encroachment on reindeer pastoralism. It takes all the energy we should have used to develop our own practice'.[49]

The case shows that planning authorities have had a tendency to downplay pastureland encroachments in official documents. A former planner from Kvalsund Municipality recalls, 'it made no sense to me that the reindeer herders refuted the plan. We even gave them more grazing areas! Just look at the regulation map!'[50] He then explained that within the new borders of the zoning plan they had coloured the whole area – save of the processing plant and harbour area – in a green code, as LFNR – Agriculture, Outdoor life, Nature, and Reindeer Husbandry-areas. The zoning plan covers 37.6 km^2 on land and in the fjord, and the defined industrial area occupies only 0.4 km^2.[51] While this has given Nussir ASA a rhetoric point about placing a small footprint, such green painting of plans carries little credibility among the herders. Fiettar Reindeer Herding District commented in the consultation statement to the hearing in 2011 that for the LFNR coding to be a reality, the regulations of use of the area should at least be consistent with the LFNR functions.[52] The geographer Kathrine Ivsett Johnsen, who has analysed the political ecology of the decision-making process regarding Nussir, writes that, 'while the pastoralists claimed that the mining company and the municipality never understood their concerns, the decision-makers claimed that the pastoralists were not willing to discuss solutions for coexistence'.[53]

The pastoralists have, over years, adapted to a bit-by-bit encroachment of grazing land and now a number of infrastructural projects, including a 420-kW power line, are planned simultaneously. Mining will further 'reduce valuable land, threaten the balance between *spatial entities* [that the reindeer use under shifting weather and insect conditions], and create uncertainty about the future natural resource base that the traditional nomadic siidas and settled communities builds upon'.[54] The reindeer are constantly on the move in smaller and larger circles when they are in a seasonal pastureland; they eat while they walk. Since the reindeer themselves 'design' the pastures by choosing or avoiding areas, they are the ones setting the terms for co-existence. Males and older, experienced females can be tolerant towards human activity and prioritize access to food, while most female reindeer *avoid* disturbances, especially during the calving.[55] Thus, disturbances carve *a void* in the grazing movement of the reproductive herd.

Perforation

Nussir ASA made progress in the application process, and the Reindeer-herding Districts were preparing to fight for their rights in both national and international courts. Meanwhile, the prospecting of the copper ore continued through seasonal drilling campaigns every winter.

In October 2014 I observe and take notes at the research workshop Fate and Impact of Mine Tailings on Marine Arctic Ecosystems (MIKOS) at the Fram Centre in Tromsø. Everybody's attention is focused on the projector screen with a digital model of the Nussir ore body looking like a golden-metallic sheet of confectionary paper. 'This site has been drilled since 1985 – close to 200 drill holes at different depths,' says Øystein Rushfeldt. By now we know that the deposit is – he turns and points at the model – that 'piece of paper' going down into the mountainside. Thin lines representing the drill holes pierce through the lean form. He rotates it to the delight of the workshop attendants, and continues: 'Since The Geologic Survey of Norway is attending today, our great thanks to them and the state sponsored programme Minerals in North Norway.[56] They have had a lot of activity in this area. It is obviously a great help for our understanding of the region'.[57]

'The site' Rushfeldt referred to had been an 'untouched' mountain valley that is a calving land for reindeer. Nussir ASA's prospecting activity was a driver in the discourse as the media frequently reported their findings. In October 2014, High North News claimed that 'Nussir had become even bigger'.[58] Days earlier, the state broadcaster's regional service NRK Sápmi presented two critical feature articles about the terrain damages and disturbance in the valley called Ásavaggi. 'The tubes represent a hazard for the reindeer; they will not see them when they come running', Per Jonny Skum – reindeer owner, herder and board member of Fiettar – told the freelance journalist Bente Bjercke. 'Two iron tubes points up from the soil here too, [he] throws a small stone into the hole to check how deep it is'.[59]

In the pastures

Vaggi is a North-Sámi word that describes a broad *mountain valley without trees*. Ásavaggi rests between Nussor's steep east side and the stony and ragged massif of Gumpenjunni, the Wolf Ridge. The topography is particularly challenging in the western part of the Fiettar district, but Ásavaggi secures access to the different landscape entities. Reindeer walk all the time and circulate between different landscapes and snowscapes depending on insect and weather conditions. The economist Erik Reinert has compared the blocking of a migration passage to the removal of a stair in a house and the ensuing loss of access to entire floors.[60] Mikkel Nils Sara pedagogically explained to me the importance of timing with the early greening when the herd arrive at the spring–summer grazing lands: 'When the reindeer comes from the mountain pastures they must be slowly accustomed to the rich nutritious grasses, and it is detrimental to their wellbeing if they arrive too late'.[61]

The valley is a calving ground and preferred nursery for many of the female reindeer, who choose to spend their newborns' vulnerable first weeks there. The herders appreciate the reindeers' knowledge of the places they return to and recognize as home-places. 'A change of calving ground represents considerable disruption for both reindeer and herders, and is contested especially by fully grown females'.[62] In the autumn, the reindeer herders of Fiettar west depend on the reindeers' voluntary movement through Ásavaggi and around Gumpenjunni, so that they can congregate in Áisaroaivi by the southern district border. The herds from the neighbouring district, Fálá, then use the valley as rutting area and migration route southwards.

When Nussir introduced drilling activity in Ásavaggi, the reindeer owners adapted to the disturbance by moving their calf-marking fence from Ásavaggi to the Kvalsund Valley.[63] A smaller valley crosses the Nussor Mountain and connects the two places, but now even this passage, Solovaggi, is threatened by a hydropower project. When Rushfeldt in 2009 announced plans to mine the Nussir ore, the drilling activity had already caused controversies between the prospectors and the pastoralists. In 2007, the Reindeer Police – a special branch of the national police – prosecuted Nussir for environmental crimes, including illegal motorized traffic and damage to vegetation, and in 2008, Fiettar had to get a court rule to stop drilling activities that disturbed the calving.[64]

The first time I visited Fiettar, I came to the calf-marking fence in Kvalsunddalen. The temporary architecture of handheld ribbons and translucent fences followed the contours in the landscape and reminded me of Christo's Running Fence. One of the siida members explained to me that understanding and working *with* the reindeer's horizontal movement through the landscape is key to knowing where to place the temporary fences. The logic of prospecting, by contrast, is vertical.

In September 2015, I interviewed Øystein Rushfeldt at different locations of Nussir ASA, including the mountain valley they were prospecting. I sat in the backseat of the company's ATV four-wheeler. He had brought orange road marker sticks as temporary survey stakes to mark the GPS points for the next drilling campaign. We met the drill crew from Arctic Drilling, commissioned by Nussir ASA, on the way up. Ásavaggi lay before us like a green carpet after our climb up a high moraine on a trail that had grown deep from frequent use.

> Standing on the moraine, we hear a low frequency roar as if a storm is under way. I see two ATV bikes with skilled drivers in the distance. 'Could it be the guys from Artic Drilling?' Rushfelt wonders, 'they have nothing to do over there, no, it has to be someone from the reindeer-herding district, checking out what we are up to'.

Figure 9.5 Early in July 2013. Walking with the ribbon to make the reindeer move from the temporary corral to the gárdi where the calves are marked

Photo: the author.

Figure 9.6 The Mining Director marks the locations of the planned drill holes with vertical road sticks
Photo: the author.

He referred to the valley as the *prospecting site* and I referred to it as *the calving ground*.

> 'The Ása valley isn't such an important reindeer pasture as the researchers claim in the Environmental Impact Assessment', he says. 'We have been prospecting this valley for six years, and I have only seen a handful of reindeer. Take today: we haven't been here since January, and still there's no reindeer here'. In the upper part of the valley we stop by some muddy patches made by a drill rig. I notice a reindeer hoofprint in the mud. 'Look!' He sees it and answer, 'yes, one has been here. Here, at this very place they were supposed to be avoiding because of the machinery. What a shame!' I notice the humour in his voice, but just have to point out that 'there isn't any machinery here now'. I notice droppings almost every third meter, and comment that it looks like a lot of reindeer have roamed the valley. 'It's nice that they thrive here', he says, 'that it is possible for us to co-exist as we use to say'. I can't resist contesting: 'You just told me that you haven't been here since January'. 'Yes, this is the first trip. On the other hand, this drilling activity during some years will be just a moment in history. Now, could you do me a favour and contribute to the mining business by holding that red stick over there?' He laughs, 'I won't tell anybody!'[65]

That season, Arctic Drilling extracted ten thousand metres of core samples through thirty new holes.[66] In April 2016, when the reindeer returned to the calving grounds, the prospectors had, with a dispensation from Kvalsund Municipality, prolonged the operation. The herders detected a diesel spill and debris such as ropes, pipes and tarpaulin littered around: 'This is not good when it is a calving area'.[67]

At kitchen tables

While the outdoor activities provide rich experience and knowledge exchange that attach to strong memories, it is inside, at a table, that the conversations about the past events and discussions, including maps and images, take place. In autumn 2015, I brought a draft of my *Outfield Atlas* to the summer settlement in Áisaroaivi. It was a compilation of photos, maps, texts and diagrams I had made to learn by hand what I had been learning in the field. I felt more confident in the interview situation with something to show, but at the same time exposed to the pastoralists' evaluation of my work in progress. One of the women asked: 'What concepts about reindeer herding do you want people to get through your maps?' Since I was drawing cartographic representations while at the same time trying to learn about reindeer herding, I hadn't yet thought of the atlas as anything public, but her question was an important reminder of how cartographies define realities.

Everybody was waiting for the reindeer to arrive and prepared for intensive days. Fence work is an important arena for knowledge exchange between the herders and the other reindeer owners, their families and children. Occasionally, visitors from schools and kindergartens, journalists and researchers come to the *gárdi*. I also got to see other maps. One afternoon I visited the reindeer owner who had invited me to the fence work the previous year. A biologist stayed at her family's summer home. They worked with a joint research project about mitigation of loss to predators.

> There is coffee, milk, and chocolate on the kitchen table. The family's two Lapponian Herder dogs chill at the floor and the kids run to and from between children's TV and the chocolate dish. Their parents and the biologist evaluate plot patterns from the GPS bracelets that selected reindeer have carried during the summer. One hundred individual digital tracks provide indication about how the herd moves. The family father shows me eight tracks at the time on his screen. 'Look at this female', he says, pointing at a single track on the screen, 'she ran off directly to Ásavaggi to give birth, after our last stop at the spring-migration!' The machine eye had provided undisputable proof.

'I want our reindeer to be able to follow their own will', he says at another occasion. 'If the herds are forced to adapt too radically to human activity more of them might roam infields and gardens'.[68] The whereabouts of reindeer is a recurring theme both in the media and around kitchen tables. His brother said that he found it remarkable that people complain about reindeer when they are close to the settlements, but seem unable to observe them in the pasturelands.

> 'The reindeer observe the surroundings all the time and don't need to move much to be out of sight', he explained, 'but when they actually are there in plain sight, people still don't see them'. He scanned the landscape through the kitchen window; the herd was extraordinarily late this fall. It was a particularly good year for mushrooms, and the reindeer feasted on this treat instead of moving. Now, he said, they were arriving. I looked and looked and it amused him that I couldn't see them. At last I saw two moving white spots, and then, with this hint of the scale, I saw the whole lot of them coated in the various colours of the surroundings.

The two herders that the CEO of Nussir ASA and I had seen in Ásavaggi told me that they had been heading back to the road after checking up on the herd. 'You would have seen a group of close

Figure 9.7 The iron tubes are hard to see in bad weather. They represent a hazard for both reindeer and herders
Photo: the author.

to 200 reindeer if you had started out earlier and arrived there before us', said one of them. 'They moved away, you know, when we came'.[69] I recall a conversation with the prospector while we sat down by a pair of drill holes in the mountain valley.

> 'It is smart to keep the openings of the tubes covered in case someone in the future needs to access them', says Øystein Rushfeldt. He then removes the cover and throws in a small stone. It makes a long-drawn, tinkling sound as it falls and slides along the interior walls of the drill hole. At an unknown depth it breaks the groundwater surface with an echoing splash.
> 'Insane!' I whisper.
> I pick up another stone – a shard of a core sample – and hold it over the opening. When I drop it another howling sound swirls downward. When the shard is swallowed by the groundwater, the echo of the shattering water mirror seems to come back up through the hole until it slowly fades. 'A very interesting sound', the mining Director says, 'it sounds like science fiction'.[70]

Towards counter prospecting

Competing prospects in the *meahcci* – the outfields – imply a competition between the futures of different realms of landscape activities. In Ásavaggi, the continuum of reindeer herding in Fiettar and Fala competes with Nussir ASA's activity of prospecting for a future mine in the coastal mountain.

The prospects of the planned mine tailings waste disposal compete with the prospects of coastal fishery in the Repparfjord fjord.

The mining industry seeks advances and expansion of the extractive enterprises. An expanding mining industry has impacts that reach beyond the scope of any social and environmental impact assessment. The impacts continue to grow, both in advance of mining, during operation and after the eventual closure of mines as residual pollution continues to interact with the environment on land and in water. The Perforated Landscape is thus an extended landscape, a landscape where future excavations, artificial mountains and mudflats become ever expanding and interacting surfaces of unknown territories. The twofold objective of mineral prospecting – the search for minerals and the development of mining prospects – perforates the landscape both physically and discursively. In this chapter I have developed analytically the notion of 'the perforated landscape'.[71]

Through fieldwork experience informed theoretically by Meløe's analysis of the relations between landscape, praxis and language, critical moves within cartography and contemporary research on Sámi reindeer husbandry, I have pursued insight into how power structures in the discourse of disputed prospects both evolve from and impact the physical terrain. I have showed how Nussir ASA's prospecting activity, in addition to physically perforating the terrain, proposes interventions with unknown impacts, and thereby produces a possible future landscape perforated with knowledge gaps. Researchers, environmental NGOs, artists, journalists, indigenous rights advocates, stakeholders and interested parties in turn detect and scrutinize these knowledge gaps.

When *the perforated landscape* includes both discursive and physical dimensions, the term can be engaged analytically to explore the complexity and the dynamics at work in planned and prospected landscapes. As counter mapping previously has shown, resisting hegemonic mapping practices involves the making of counter maps to reveal what is invisible in the debate and enable an analysis of how power structures in the discourse of contested landscapes partly manifest through the use of maps. It is not sufficient, however, to analyse. When questioning hegemonic prospects, forward-directed concepts are needed. It must be possible to imagine other futures. We have seen in the case of Nussir that the copper mine prospect, through supportive policy and media attention, has been privileged as the main prospect of Kvalsund. The planning authorities have rendered the pastoral and coastal fishery community's plans for the future less important. I propose counter prospecting as a method to endow the voids in the perforated landscape with renewed interpretations of landscape values, to put forward alternative prospects and highlight subverted prospects of future landscapes. Counter prospecting implies exploring how knowledge gaps in the perforated landscape – as attention gravitates towards them – demand engagement with multiple landscape perspectives in the negotiation of the future north.

The planning authorities admitted that they approved an encroachment in the reindeer herding districts, but they did not acknowledge what the loss of the pastoral landscape implies: Neither the loss of the reindeers' intimate knowledge of their summer ranges, nor the pastoral relation to this particular place. The decision does not reflect the fact that the encroachments challenge the continuum of knowledge production within reindeer pastoralism and in turn affect Sámi language and culture. As Meløe instructs us, 'our everyday language . . . is in every bit of its existence tied to our everyday world, with its diverse activities, artefacts, weather, terrain, etc.'.[72] The experience-based knowledge about how reindeer use the place-specific landscape capacities under changing weather and insect conditions exist in the Sámi language only. 'In a world where there is no reindeer herding, there is nothing to yield the concept of a JASSA'.[73] It is clear that Ásavaggi, and the landscapes the valley connects, will be denuded of its capacity to renew the Sámi pastoral terminology if reindeer herding ceases to develop within it. The mode of knowledge building in the pastoral community is both traditional and forward-looking as it seeks both continuity and development. Further, as Sara reminds us in *Land Usage and Siida Autonomy*, there are power relations to be considered between different ways of knowing also within reindeer pastoralism.[74] As a pro-spective exchange towards alternate prospects, counter prospecting must be done in close interaction with both those who do the actual work of translating cultural concepts from the inside, and the performers of traditional landscape practices.

In this chapter, I have zoomed in on the scale of reindeer hoof prints and the diameter of core sample drill holes. Both are smaller than the palm of a hand, but in their aggregation, they define large territories. A young girl in the forest listened to the sound of groundwater through a cylinder of recently cut rock. In the calving land, a reindeer herder measured, by listening to the sound of a falling stone, how deep the hole is that frames the discourse of his herd's future. The director of the prospecting company listened to an alien sound from below that appeared to him as science fiction. The *perforated landscape* draws attention to its own future. The subterranean echoes of anticipated blasts remain in the soundscape of Ásavaggi. The cylindrical boreholes are like wormholes for the echoes to take on time travels. Down through the strata of geological time, and up to the discourse on the future landscape. The voids in prospected sites bear witness of both extracted matter and extracted agency as the negotiation of the future landscape is externalized from the landscape itself. The exploiters hold the core samples and claim to own the core questions. Counter prospecting returns to the landscape and its people the core questions about the direction of future development and the methods by which to answer them.

Notes

1 Field notes from Áisaroaive, September 2015.
2 James Corner, "The Agency of Mapping: Speculation, Critique and Invention," in *Mappings*, ed. Denis Cosgrove (London: Reaction Books, 1999), 213.
3 Sápmi is the North-Sámi name of northern Fennoscandia.
4 Notes from guest lecture by Tor Lundberg Tuorda at Tromsø Academy for Landscape and Territorial studies, November 2013.
5 Alain Deneault and William Sacher, *Imperial Canada inc. Legal Heaven of Choice for the World Mining Industries* (Vancouver, British Colombia: Talonbooks 2012), 40.
6 Government offices of Sweden, *Sweden's Mineral Strategy* (Ministry of Enterprise, Energy and Communications, 2013), 36.
7 Government of Norway, *Strategy for the Mineral Industry* (Ministry of Trade and Industry, 2013), 40.
8 The Geological Survey of Norway: *Mineralressurser i Norge 2011. Mineralstatistikk og bergindustriberetning/ Mineral Resources in Norway, Statistics* (NGU 2012), 28.
9 Urban Viklund, "Kvarglömda Järnrör vid Gamla Borrhål Läcker/ Left-behind Iron Tubes by Old Drillholes are Leaking," *Lokaltidningen Vilhelmina*, April 13, 2016.
10 My translation from 'Sametingets retningslinjer for utmarksvurdering (. . .) sikre naturgrunnlaget for og sikre videre utvikling av samisk kultur'. "Sametingets retningslinjer for utmarksvurdering § 1," Lovdata, accessed January 20, 2016, https://lovdata.no/dokument/SF/forskrift/2007-06-11-738#KAPITTEL_1.
11 Audhild Schanche, "Meahcci, den Samiske utmarka/ Meahcci, The Sámi Outfields," in *Samiske Landskap og Agenda 21. Kultur, næring, miljøvern og demokrati/ Sámi Landscapes and Agenda 21, Industry, Environmental Conservation, and Democracy, Diedut 1/2002*, ed. Svanhild Andersen (Guovdageaidnu: Sámi Instituhtta, 2002), 163. See also Stine Rybråten, "This Is Not Wilderness. This Is Where We Live" (PhD diss. University of Oslo, 2013), 81, about the English word *Outfield*, the Norwegian *Utmark* and the Sámi *Meahcci*.
12 Harald Gaski, "Indigenous Interdisciplinary Internationalism: The Modern Sami Experience, with Emphasis on Literature," in *Circumpolar Ethnicity and Identity*, ed. Takashi Irimoto and Takako Yamada (SENRI Ethnological Studies, 2004), 372.
13 Jakob Meløe, "Words and Objects," in *Working Papers No. 5*, edited by Henry Paul and Utaker Arild (Bergen UiB, 1992), 140.
14 Ibid., 136.

15 Jacob Meløe, "The Two Landscapes of Northern Norway," *Inquiry: An Interdisciplinary Journal of Philosophy* 31, no. 3 (1988): 392.

16 My translation from 'hvis man ikke allerede får en ganske rik forståelse for verden, . . . så får du ikke bygget ut det rommet som ord skal falle på plass innenfor'. Jacob Meløe, *About Practical Knowledge*, Lecture, September 27, Bodø 2006. Published online May 15, 2015. Accessed May 10, 2016. www.youtube.com/watch?v=suCPQ9f9SW0.

17 Mikkel Nils Sara, "Siida and Traditional Sámi Reindeer Herding Knowledge," *The Northern Review* 30 (2009): 166. Within the context of reindeer pastoralism, a siida is a group of reindeer herding families that work together to manage their herds.

18 Corner, "The Agency of Mapping," 215.

19 Jeremy Crampton, "Maps as Social Constructions: Power, Communication and Visualization," *Progress in Human Geography* 25, no. 2 (2001): 691.

20 Ibid., 691.

21 Knut Eirik Dahl, Espen Røyseland, Øystein Rø, and Kjerstin Uhre, "Fields of Exploration, Limits of Exploitation," (Landscape architecture studio, Oslo School of Architecture and Design, 2012), http://fieldsofexploration. blogspot.no/.

22 Thomas Vermes, "Stort arkitektprosjekt avdekker gruveindustriens nye makt/A Big Architecture Project Reveals the New Power of the Mining Industry," *ABC Nyheter*, April 12, 2012. Accessed May 15, 2016. www.abcnyheter. no/nyheter/2012/04/12/149381/stort-arkitektprosjekt-avdekker-gruveindustriens-nye-makt.

23 Roger Manndal, "Gollegiisá – Skattkammeret/The Treasure Shrine," TV Documentary on NRK Sápmi, 2013. www.nrk.no/video/PS★99127. Accessed January 13, 2014.

24 Corner, "The Agency of Mapping," 215.

25 See Kjerstin Uhre, "Sápmi and the Fennoscandian Shield, On and Off the Map," *Moving Worlds* 15, no. 2 (2015): 85.

26 Tanya Busse and Emilija Skarnulyte, *Hollow Earth* (Tromsø: Tromsø Academy of Contemporary Art, 2013), 3.

27 Bieke Cattoor and Chris Perkins, "Re-Cartographies of Landscape: New Narratives in Architectural Atlases." *The Cartographic Journal* 51, no. 2 (2014): 168.

28 Meløe, "The Two Landscapes of Northern Norway," 394.

29 Anna Tsing, "More-Than-Human-Sociality: A Call for Critical Description," in *Anthropology and Nature*, edited by Kirsten Hastrup (London: Routledge, 2013), 31.

30 John Law, *Making a Mess with Method* (Lancaster: The Centre for Science Studies, 2003), 7.

31 Sources to the map retrieved at Nussir.no, 2010, Kilden.no, and the Fennoscandian Ore Deposit Database.

32 Kvalsund Municipality Council's Area Plan 2004–2016, map dated April 27, 2004.

33 Kilden.nibio.no

34 Ministry of Local Government and Modernisation, "Kvalsund kommune – innsigelse til reguleringsplan for Nussir og Ulveryggen. Brev av 26.03.2014 fra Kommunal og Moderniseringsdepartementet til Fylkesmannen i Finnmark/ Kvalsund Municipality – objection to zoning plan for Nussir and the Wolf Ridge". Letter of March 26, 2014 from Ministry of Local Government and Regional Development to the County Governor of Finnmark. 2014.

35 Terry Odendahl, Roy Young, and Gary Wockner, "Why Is Mine Waste Being Dumped Directly into the Ocean?" *Eco Watch*, March 4, 2016. Accessed May 15, 2016. https://ramumine.wordpress.com/2016/03/07/why-is-mine-waste-being-dumped-directly-into-the-ocean/.

36 Miljødirektoratet. Oversendelse av tillatelse til virksomhet etter forurensningsloven – Nussir ASA. Brev av 8.12.2015 fra Miljødirektoratet til Nussir ASA/ MD 2015. Submission of permission for activities under the Pollution Control Act – Nussir ASA. Letter of 12.8.2015 from the Environment Directorate to Nussir ASA, page 3 and 9.

37 Naturvernforbundet, "New complaint and addition to the complaint: Norway has failed to comply with the Water Framework Directive. Letter of December 22, 2015 from the Friends of the Earth Norway to the EFTA Surveillance Authority," Accessed February 10, 2016. http://naturvernforbundet.no/finnmark/gruvedrift/klager-pa-dumping-i-repparfjorden-article34555-2023.html.

38 The Ministry of Local Government and Modernisation, "Regulations on Environmental Impact Assessment for plans pursuant to the Norwegian Planning and Building Act pursuant to the Act of June 27, 2008 No. 71." Accessed February 10, 2016. www.regjeringen.no/contentassets/f25837cb4dd045738e091f093ab06ccc/regulations_environmental_impact_assessment_for_plans.pdf, section 7.

39 My translation from 'Det er på mange måter utført en meget grundig vurdering av miljøkonsekvensene av dette omsøkte tiltaket, og konsekvensutredningen viser tydelig at et fjorddeponi vil representere en alvorlig forurensing av fjorden. Men det er foruroligende at konsekvensutredningen nedskriver verdien av marin natur og marine ressurser og på den måten kommer frem til at alvorlig forurensing får liten konsekvens'. Jan Helge Fosså et al.,

Institute of Marine Research, "Høring – reguleringsplan med konsekvensutredninger for planlagt gruvedrift i Nussir og Ulveryggen i Kvalsund Kommune/ Consultation statement – zoning plan impact assessment for planned mining operation at Nussir and Wolf Ridge in Kvalsund Municipality," September 15, 2011.

40 Ibid., section 7.

41 Eva Ramirez-Lodra et al., "Submarine and Deep-Sea Mine Tailing Placements: A Review of Current Practices, Environmental Issues, Natural Analogs and Knowledge Gaps in Norway and Internationally," *Marine Pollution Bulletin* 97, nos. 1–2 (2015): 18.

42 My translation from 'Vi har i dag god nok kunnskap til å kunne konkludere med at sjødeponi i Repparfjorden ikke vil ha uakseptable negative konsekvenser for sjømatnæringen'. "NFD anbefaler gruvedrift i Repparfjorden Pressemelding no17/ NDF recommend mining in Repparfjorden, press release number 17, February 13, 2014. Accessed April 5, 2014, https://www.regjeringen.no/no/aktuelt/anbefaler-gruvedrift-i-repparfjorden/id751034/

43 See Ivar Bjørklund, "Fra formynder til forhandler: Om inngrep, konsekvensanalyser og 'balansert sameksistens/ From Guardian to Negotiator: About Encroachments, Impact Assessments, and 'Balanced Co-Existence'," in *Samisk reindrift, norske myter/ Sámi Pastoralism, Norwegian Myths*, ed. Tor A. Benjaminsen et al. (Bergen: Fagbokforlaget 2016), 185.

44 Kita Eilertsen, "Lykkes vi kommer hele verden etter/ If We Succeed, The World Will Follow." (Altaposten, November 4 2010). Accessed October 10, 2014. www.altaposten.no/lokalt/nyheter/article403895.ece. My translation from 'Når vi i Norge har fått til en avtale med Sametinget og reindrifta, vil andre komme etter, sier Rushfeldt'.

45 Miljødirektoratet, 2016, 46; my translation from 'lagt særlig vekt på at det er næringspolitiske føringer for å utvinne mineraler i Norge og at det er sannsynlig at tiltaket vil gi betydelige inntekter for samfunnet'.

46 My translation from 'Usikkerheten i størrelsen på både nytte og kostnader tilsier at føre-var prinsippet bør vektlegges i forvaltningen av området.' Karin Ibenholt et al., "Gruvedrift ved Repparfjorden gjennomgang av utredninger om samfunnsmessige konsekvenser/ Mining by Repparfjorden Fjord review of assessments of societal impacts," *Vista Analyse, Report* 2016/26, 5. Accessed February 10, 2016. www.vista-analyse.no/no/publikasjoner/gruvedrift-ved-repparfjorden-gjennomgang-av-utredninger-om-samfunnsmessige-konsekvenser/

47 United Nations 1976, *Convention on Civil and Political Rights*, § 27

48 My translation from 'Miljødirektoratet kan ikke se at det omsøkte tiltaket vil påvirke religion eller språk negativt i særlig grad'. Miljødirektoratet, 2016.

49 My translation from 'Ingen ser at bare planane og trugsmåla allereie er eit inngrep i drifta vår. Det tar alle kreftene vi skulle brukt til å utvikle drifta'. Svein Lund, *Gull, gråstein og grums/ Gold, Stone and Mud* (Kautokeino, Davvi Girji 2015), 93.

50 Interview with former planner at Kvalsund Municipality, Tromsø, December 2015.

51 See T. A. Didriksen, *Nussir ASA. Reguleringsplan med konsekvensutredning for planlagt gruvedrift i Nussir og Ulveryggen i Kvalsund kommune/ Zoning Plan with Impact Assessments of planned mining in Nussir and the Wolf Ridge in Kvalsund Municipality*. Sweco-rapport nr. 02, 03.06.2011.

52 Mikkel Nils Sara, *Høringsuttalelse angående reguleringsplan med konsekvensutredninger for planlagt gruvedrift i Nussir og Ulveryggen i Kvalsund kommune/* Consultation statement regarding a zoning plan impact assessment for planned mining operation at Nussir and Wolf Ridge in Kvalsund Municipality, September 7, 2011.

53 Kathrine Ivsett Johnsen, "Land-Use Conflicts Between Reindeer Husbandry and Mineral Extraction in Finnmark, Norway: Contested Rationalities and the Politics of Belonging," *Polar Geography* 39, no. 1 (2016): 72.

54 My translation [and comment] from 'Fordi det reduserer verdifulle arealer, truer balansen mellom arealenhetene og skaper usikkerhet om det fremtidige grunnlaget som tradisjonelle nomadiske siidaer og bofaste lokalsamfunn er bygd på'. Sara, *Høringsuttalelse*.

55 Ingunn Ims Vistnes and Christian Nelleman, "Foreslått utbygging av Nussir gruver i reinbeitedistrikt 22 Fiettar – Konsekvenser for reindriften i 22 Fiettar og 20 Fálá/ Proposed Development of Nussir mines in reindeer grazing district 22 Fiettar – impact on reindeer pastoralism in 22 Fiettar and 20 Fálá," *Norut Report 2011: 2* (2011): 9.

56 The Norwegian Government allocated funds of NOK 100 million for the period 2011–2014 to the Geological Survey of Norway (NGU) to improve coverage of basic geological information relevant to assessment of the mineral potential in the three northernmost counties. See Government, *Strategy for the Mineral Industry*, 40.

57 Field notes, Tromsø, October 14, 2014.

58 Linda Storholm, "Nussir har blitt enda større/ Nussir has Become Even Bigger," *Highnorthnews*, September 30, 2014. Accessed October 15, 2014.

59 My translation from 'Rørene kan være farlige for reinen, den vil ikke se dem når den kommer løpende. (. . .) To jernrør stikker opp fra jorda her, også. Per Johnny Skum kaster en liten stein ned i hullet for å sjekke hvor dypt det er'. Bente Bjercke, "Kilometervis med kjørespor etter Nussir i Kvalsund/ Kilometres of

tracks after Nussir in Kvalsund," *NRK Sápmi*, October 3. Accessed October 10, 2014. www.nrk.no/sapmi/nussir-lager-dype-kjorespor-1.11964659
60 Erik Reinert, "Mineraler og rein/Minerals and Reindeer," *Klassekampen*, October 8, 2012.
61 Interview at the Sámi University, Diehtosiida in Kautokeino June 2013.
62 Mikkel Nils Sara, "Siida and Traditional Sámi Reindeer Herding Knowledge," *The Northern Review* 30 (2009): 151.
63 Magne Kveseth, "Frykter gruve-konsekvensene/ Fearing Impacts from Mining," *Altaposten*, November 14, 2009.
64 Vistnes, and Nelleman, Impact assessment, 10.
65 Ásavaggi September 7, 2015.
66 Halftan Carstens: "Han fant Malmen/ He found the Ore." *GEO365.no*, May 2016. Accessed 24 March, 2017, www.geo365.no/bergindustri/han-fant-malmen/
67 Erik Hind Sveen, "Reindriftsutøver oppdaget Nussirs oljesøl/ Reindeer herder detects Nussir's Oil Spill," *NRK Finnmark*, May 6, 2016. Accessed May 6, 2016, www.nrk.no/finnmark/reindriftsutover-oppdaget-nussirs-oljesol-1.12934193.
68 Áisaroaivi, September 2015.
69 Áisaroaivi, September 2015.
70 Ásavaggi, September 7, 2015.
71 Knut Eirik Dahl et al., *Fields of Exploration Limits of Exploitation*, 2012.
72 Meløe, "Words and Objects," 132.
73 Meløe, "The Two Landscapes," 388 (original use of capital letters).
74 Sara, "Siida and Traditional Sámi Reindeer Herding Knowledge," 165.

Bibliography

Bjercke, Bente. "Kilometervis med kjørespor etter Nussir i Kvalsund/ Kilometres of tracks after Nussir in Kvalsund." *NRK Sápmi*, October 3, 2014. Accessed October 10, 2014. www.nrk.no/sapmi/nussir-lager-dype-kjorespor-1.11964659.

Bjørklund, Ivar. "Fra formynder til forhandler: Om inngrep, konsekvensanalyser og 'balansert sameksistens/ From Guardian to Negotiator: About Encroachments, Impact Assessments, and 'Balanced Co-Existence'." In *Samisk reindrift, norske myter/ Sámi Pastoralism, Norwegian Myths*, edited by Tor A. Benjaminsen, Inger Marie Gaup Eira, and Mikkel Nils Sara, 177–93. Bergen: Fagbokforlaget, 2016.

Carstens, Halftan. "Han fant Malmen/ He found the Ore." *GEO365.no*, May 2016. Accessed March 24, 2017. www.geo365.no/bergindustri/han-fant-malmen/.

Cattoor, Bieke, and Chris Perkins. "Re-Cartographies of Landscape: New Narratives in Architectural Atlases." *The Cartographic Journal* 51, no. 2 (2014): 166–78.

Corner, James. "The Agency of Mapping: Speculation, Critique and Invention." In *Mappings*, edited by Denis Cosgrove, 213–52. London: Reaction Books, 1999.

Crampton, Jeremy W. "Maps as Social Constructions: Power, Communication and Visualization." *Progress in Human Geography* 25 (2001): 235–52. Accessed July 8, 2015. http://phg.sagepub.com.

Dahl, Knut Eirik, Kjerstin Uhre, Espen Røyseland, and Øystein Rø. "Fields of Exploration, Limits of Exploitation, 2012. (Landscape architecture studio, Oslo School of Architecture and Design)." http://fieldsofexploration.blogspot.no/

Deneault, Alain, and William Sacher. *Imperial Canada inc. Legal Heaven of Choice for the World Mining Industries*. Vancouver: Talonbooks, 2012.

Didriksen, T. A. *Nussir ASA. Reguleringsplan med konsekvensutredning for planlagt gruvedrift i Nussir og Ulveryggen i Kvalsund kommune/ Zoning Plan with Impact Assessments of planned mining in Nussir and the Wolf Ridge in Kvalsund Municipality*. Sweco-rapport nr. 02, March 06, 2011.

Eilertsen, Kita. "Lykkes vi kommer hele verden etter/ If We Succeed, the World Will Follow." *Altaposten*, November 4 2010. Accessed October 10, 2014. www.altaposten.no/lokalt/nyheter/article403895.ece.

Fosså, Jan Helge, Lars Asplin, Jan Aure, Sonnich Meier, and Terje van der Mehren. "Høring – reguleringsplan med konsekvensutredninger for planlagt gruvedrift i Nussir og Ulveryggen i Kvalsund Kommune/ Consultation statement – zoning plan impact assessment for planned mining operation at Nussir and Wolf Ridge in Kvalsund Municipality." Institute of Marine Research. September 15, 2011.

Gaski, Harald. "Indigenous Interdisciplinary Internationalism: The Modern Sami Experience, with Emphasis on Literature." In *Circumpolar Ethnicity and Identity*, edited by Takashi Irimoto and Takako Yamada, 371–87. SENRI Ethnological Studies, 2004.

Government offices of Sweden. *Sweden's Mineral Strategy.* Swedish Ministry of Enterprise, Energy and Communications, 2013. Accessed October 4, 2016. www.government.se/contentassets/78bb6c6324bf43158d7c153e bf2a4611/swedens-minerals-strategy.-for-sustainable-use-of-swedens-mineral-resources-that-creates-growth-throughout-the-country-complete-version.

Holm, Arne O. "Nærmere Kobberdrift." *High North News*, 14 February 2014. Accessed 5 April 2014. www.high northnews.com/naermere-kobberdrift/.

Ibenholt, Karin, Ingeborg Rasmussen, and John Magne Skjelvik. "Gruvedrift ved Repparfjorden gjennomgang av utredninger om samfunnsmessige konsekvenser/ Mining by Repparfjorden Fjord review of assessments of societal impacts." *Vista Analyse, Report,* 2016/26. www.vista-analyse.no/no/publikasjoner/gruvedrift-ved-repparfjorden-gjennomgang-av-utredninger-om-samfunnsmessige-konsekvenser/.

Johnsen, Kathrine Ivsett. "Land-Use Conflicts Between Reindeer Husbandry and Mineral Extraction in Finnmark, Norway: Contested Rationalities and the Politics of Belonging." *Polar Geography* 39, no. 1 (2016): 58–79.

Kvalsund Municipality Council's area plan 2004–2016, map dated April 27, 2004.

Kveseth, Magne. "Frykter gruve-konsekvensene/Fearing impacts from mining." *Altaposten*, November 14, 2009.

Law, John. *Making a Mess with Method.* Lancaster: The Lancaster University Centre for Science Studies, 2003.

Lund, Svein. *Gull, gråstein og grums/ Gold, Stone and Mud.* Kautokeino: Davvi Girji, 2015.

Lovdata. "Sametingets retningslinjer for utmarksvurdering/the Sami Parliaments Guidelines for Outfield Assessment," 2007. Accessed January 20, 2016. https://lovdata.no/dokument/SF/forskrift/2007-06-11-738#KAPITTEL_1.

Manndal, Roger. "Gollegiisá – Skattkammeret/the Treasure Shrine, TV Documentary." *NRK Sápmi*, 2013. Accessed January 13, 2014. www.nrk.no/video/PS★99127.

Miljødirektoratet, Oversendelse av tillatelse til virksomhet etter forurensningsloven – Nussir ASA. Brev av 8.12.2015 fra Miljødirektoratet til Nussir ASA/ MD 2015. Submission of permission for activities under the Pollution Control Act – Nussir ASA. Letter of December 8, 2015 from the Environment Directorate to Nussir ASA.

Meløe, Jacob. "The Two Landscapes of Northern Norway." *Inquiry: An Interdisciplinary Journal of Philosophy* 31, no. 3 (1988): 387–401.

Meløe, Jacob. "Words and Objects." In *Working Papers No. 5*, edited by Henry Paul and Utaker Arild, 109–41. Bergen: University of Bergen Wittgenstein Archives, 1992. http://wittgensteinrepository.org/agora-wab/issue/view/191.

Meløe, Jacob. "Om Praktisk Kunnskap. Del 2/2/ About Practical Knowledge. Part 2/2." Lecture, September 27, Bodø, 2006. Published May 15, 2015. Accessed May 10, 2016. www.youtube.com/watch?v=suCPQ9f9SW0.

Ministry of Local Government and Modernisation. "Kvalsund kommune – innsigelse til reguleringsplan for Nussir og Ulveryggen. Brev av 26.03.2014 fra Kommunal og Moderniseringsdepartementet til Fylkesmannen i Finnmark/ Kvalsund Municipality – objection to zoning plan for Nussir and the Wolf Ridge." *Letter of March 26, 2014 from the Ministry of Local Government and Regional Development to the County Governor of Finnmark.* www.regjeringen. no/no/dokumenter/Kvalsund-kommune – innsigelse-til-reguleringsplan-for-Nussir-og-Ulveryggen/id753930/

Ministry of Local Government and Modernisation. "Regulations on Environmental Impact Assessment for plans pursuant to the Norwegian Planning and Building Act pursuant to the Act of 27 June 2008 No. 71, section 7." Accessed February 10, 2016. www.regjeringen.no/contentassets/f25837cb4dd045738e091f093ab06ccc/regula tions_environmental_impact_assessment_for_plans.pdf.

Ministry of Trade and Industry, NHD. *Strategy for the Mineral Industry.* Oslo: Norwegian Ministry of Trade and Industry, 2013. Accessed May 1, 2013. www.regjeringen.no/contentassets/3fe548d142cd496ebb7230a54e71ae1a/strategyforthemineralindustry_2013.pdf

Naturvernforbundet. "New Complaint and Addition to the Complaint: Norway Has Failed to Comply with the Water Framework Directive." Letter of 22.12.2015 from the Friends of the Earth Norway to the EFTA Surveillance Authority." Accessed February 10, 2016. http://naturvernforbundet.no/finnmark/gruvedrift/klager-pa-dumping-i-repparfjorden-article34555-2023.html.

Novikova, Ksenia. "Gruveselskap har ingen avtale med reindriften/ the Mining Company Has No Deal with the Reindeer Pastoralists." *NRK Finnmark*, December 10, 2015. Accessed December 10, 2015. www.nrk.no/finnmark/nussir-har-ingen-avtale-med-reindriften-1.12696696

Oldendal, Terry, Roy Young, and Gary Wockner. "Why Is Mine Waste Being Dumped Directly into the Ocean?" *Eco watch*, 4 March 2016. Accessed May 15, 2016. https://ramumine.wordpress.com/2016/03/07/why-is-mine-waste-being-dumped-directly-into-the-ocean/.

Ramirez-Llodraa, Eva, et al. "Submarine and Deep-Sea Mine Tailing Placements: A Review of Current Practices, Environmental Issues, Natural Analogs and Knowledge Gaps in Norway and Internationally." *Marine Pollution Bulletin* 97, nos. 1–2 (15 August 2015): 13–35.

Reinert, Erik. "Mineraler og rein/ Minerals and Reindeer" *Klassekampen*, October 8. 2012.

Rybråten, Stine. "This Is Not Wilderness. This Is Where We Live." PhD diss. University of Oslo, 2013.

Sara, Mikkel Nils. "Siida and Traditional Sámi Reindeer Herding Knowledge." *The Northern Review* 30 (2009): 153–78.

Sara, Mikkel Nils. "Land Usage and Siida Autonomy." *Arctic Review on Law and Politics* 2, no. 2 (2011): 138–58. Accessed February 10, 2016. https://arcticreview.no/index.php/arctic/article/view/25.

Sara, Mikkel Nils. "Høringsuttalelse angående reguleringsplan med konsekvensutredninger for planlagt gruvedrift i Nussir og Ulveryggen i Kvalsund Kommune/ Consultation statement regarding a zoning plan impact assessment for planned mining operation at Nussir and Wolf Ridge in Kvalsund Municipality," September 7, 2011.

Schanche, Audhild. "Meahcci, den Samiske utmarka/ Meahcci, The Sámi Outfields." In *Samiske Landskap og Agenda 21. Kultur, næring, miljøvern og demokrati/ Sámi Landscapes and Agenda 21, Industry, environmental conservation, and Democracy*, edited by Svanhild Andersen, 156–70. Guovdageaidnu: Sámi Instituhtta, 2002.

Skarnulyte, Emilija, and Tanya Busse. *Hollow Earth*. Tromsø: Tromsø Academy of Contemporary Art, 2013.

Storholm, Linda. "Nussir har blitt enda større/ Nussir has Become Even Bigger." *Highnorthnews*, September 30, 2014. Accessed October 15, 2014. www.highnorthnews.com/nussir-har-blitt-enda-storre/

Sveen, Erik Hind. "Reindriftsutøver oppdaget Nussirs oljesøl/ Reindeer herder detects Nussir's Oil Spill." *NRK Finnmark*. 5 June 2016. Accessed June 5, 2016. www.nrk.no/finnmark/reindriftsutover-oppdaget-nussirs-oljesol-1.12934193

Tsing, Anna. "More-Than-Human-Sociality, A Call for Critical Description." In *Anthropology and Nature*, edited by Kirsten Hastrup, 27–42. London: Routledge, 2013.

Uhre, Kjerstin. "Sápmi and the Fennoscandian Shield, On and Off the Map." In *The Postcolonial Arctic*, edited by Graham Huggan and Roger Norum, 81–92. Leeds: Moving Worlds, 2015.

United Nations. *Convention on Civil and Political Rights*, 1976, §27. www.ohchr.org/en/professionalinterest/pages/ccpr.aspx.

Vermes, Thomas. "Stort arkitektprosjekt avdekker gruveindustriens nye makt/A Big Architecture Project Reveals the New Power of the Mining Industry." *ABC Nyheter*. April 12, 2012. Accessed May 15, 2016. www.abcnyheter.no/nyheter/2012/04/12/149381/stort-arkitektprosjekt-avdekker-gruveindustriens-nye-makt.

Viklund, Urban. "Kvarglömda järnrör vid gamla borrhål läcker/ Left-behind Iron Tubes by Old Drillholes Are Leaking." *Lokaltidningen Vilhelmina*, April 13, 2016.

Vistnes, Ingunn Ims, and Christian Nelleman. "Foreslått utbygging av Nussir gruver i reinbeitedistrikt 22 Fiettar – Konsekvenser for reindriften i 22 Fiettar og 20 Fálá. Norut Rapport 2011:2/ Proposed Development of Nussir mines in reindeer grazing district 22 Fiettar – impact on reindeer pastoralism in 22 Fiettar and 20 Fálá." *Norut Report 2011: 2*.

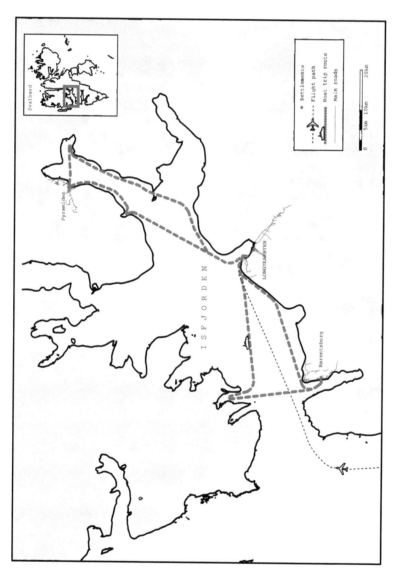

Map 4 A contested space of policy and industry. Svalbard has been exploited industrially for centuries, but is now transitioning to a post-industrial existence. This challenges the territorial status of this archipelago which is open to industry from everyone and currently has both Russian and Norwegian settlements. The fragile ecologies of the archipelago are threatened by chemical pollution, invasive species, oceanic plastic, tourism. Here, scientists monitor the effects of climate change on the natural environment, while its effects are experienced directly as increased landscape hazards in the settlements (map: Eimear Tynan).

10 Branding ice

Contemporary public art in the Arctic

William L. Fox

The Global Seed Vault

Ask people in the United States where to find Svalbard on the globe, and mostly you receive a blank stare in return, or perhaps a tentative "somewhere in the Arctic?" But if you mention that the Global Seed Vault (GSV) is in the Svalbard archipelago, they smile and nod and say "Oh, sure . . ." and then add "Have you been there?" It is telling that a conservation facility is now as or more prominent in our national mental map of Svalbard than the polar bears, glaciers, and Northern Lights (Aurora Borealis) that have served as its traditional talismans for the last 150 or so years. It's not that those earlier tropes have disappeared, but that another layer has been added. This evolution in Arctic metaphors, in particular as seen in comparison with public art throughout Norway, is what has brought me to Svalbard as I attempt to understand rapidly changing "brandscapes" in the Arctic.

I have been traveling intermittently for three years as a visiting researcher with the Future North team to towns such as Tromsø and Vardø, as well as touring northern Norway, to witness the intentional and unintentional deployment of art in this process. The diverse interests of the group serve well the goal of investigating the Circumpolar North as a regional laboratory for "studying the future landscapes of production, infrastructure, excavation, and environmental change."[1]

And that is why I find myself standing in front of the GSV on an afternoon in late May keeping out an eye for any patch of white in the snowy landscape that suddenly decides to stand up and move about, usually the first thing you notice when there's a polar bear nearby. Longyearbyen residents number about 2,000 people at any one time, while more than 3,000 polar bears roam the archipelago. You are supposed to be either armed or accompanied by an armed guide when outside the city limits, which we are – but none of us has a firearm, and several of us are scattered around the facility.

The Global Seed Vault is a more modest architectural proposition than you might suppose, given its title. Opened in February 2008 after almost two years of construction, the GSV was tunneled into the scree and moraine of the Platåfjell (Plateau Mountain) just south of the Longyearbyen airport on Spitsbergen, the largest of the nine islands in the Svalbard archipelago. Longyearbyen is the largest community in Svalbard, the islands being located between the 74th and 81st North parallels and about 600 nautical miles from the North Pole. The GSV then is located approximately halfway between the north coast of the Norwegian mainland and the North Pole, and is about 1,300 kilometers above the Arctic Circle. On this day in late May, it is cool enough standing outside it to wear a parka, hat, and gloves.

The Vault consists of a 100-meter tunnel at the end of which is a freezer vault divided into three rooms with enough space to hold 4.5 billion seed samples from around the world, or about twice the number of estimated genetically differentiated plant species. Constructed 130 meters above sea level, it is meant to be a backup for the world's existing seed banks, and keep safe the world's seed stock from rising sea levels produced by global warming as well as other disasters. It is a small irony that a few hundred meters to the east is the head frame of the played-out Trønder coal mine, one of the minor but myriad sources of fossil fuel that has accelerated global warming, a change prompting the creation of the Vault.

Svalbard is a place where land ends and the Arctic Ocean stretches uninterrupted to the North Pole, but it is also literally at the front line of climate change, the Arctic is warming twice as fast as other regions.[2] For this, and a complex set of economic and strategic reasons, Svalbard has become a geopolitical location of intense interest to scientists, as well as to the resource companies and, strategically, the militaries of multiple countries. It doesn't hurt Norway's strategic interests to publicly host the GSV on Spitsbergen, given that Russia also claims a territorial foothold in the archipelago.

Recognition of the Global Seed Vault by my countrymen is linked to the intense worldwide media attention given global change, a news story large enough to brand a science facility. The facade of the installation, designed as a prow to shed snow drifts and which the Norwegian government's property manager stated "will become a Svalbard landmark,"[3] is divided into three vertical segments that sit within a concrete frame: a stainless steel doorway at the bottom, a large ventilation grill above it, and a decorative panel at the top. The securely locked door sounds solid when I thump it, and the grill, which fronts the ventilation and compressor room, emits periodic mechanical roars as the

Figure 10.1 The Svalbard Global Seed Vault. Artwork: "Perpetual Repercussion" by Dyveke Sanne, 2008
Photo: Mari Tefre/Svalbard Globale frøhvelv.

cooling units inside kick on and off. The thick safety glass panel that covers multiple shards of steel in the aesthetic component at the top reflects light back in a constantly shifting pattern.

This last segment is a public art project integrated into the conservation facility that was commissioned by Public Art Norway and designed by Norwegian artist Dyveke Sanne. During the long winter night, the steel triangles, which actually extend back and over the top of the entire cantilevered entrance and include mirrors and prisms, are lit by fiber-optic cables. The point, according to Statsbygg, the public entity property manager, is that the art would make the facility visible year-round and from a distance, and express the "eternal values" of the conservation project. That may be, but from the nighttime photographs I have seen, the embellishment also mimics the Northern Lights.

The prow of the Global Seed Vault has, indeed, become a Svalbard landmark as promised. People come from all over the world to stand here, outside the vault with no chance of being admitted, just to marvel in front of what our colleague Peter Hemmersam calls a "pre-post-apocalyptic" piece of architecture and art. It used to be that people came to Svalbard, in particular Spitsbergen and its largest town, in order to mine coal, hunt the local game, and to admire the glaciers, peaks, and fjords. Nowadays they come to study permafrost and climate change, photograph polar bears, and to predictably admire the glaciers, peaks, and fjords. But what used to be remote and everlasting is now readily accessible and endangered. Tourists once sailed to Svalbard to stand at the edge of the world; now they fly in to stand at a receding ice front. Both are often considered sublime experiences, beautiful and terrifying in equal measures, if for different reasons.

White bears, black coal

Part of the appeal of the Arctic is that danger will accompany your pilgrimage, as in "nothing risked, nothing gained." Although classified as a marine mammal because of the time they spend in, on, or near the ocean, polar bears are also named as the largest land-based carnivore on the planet, growing up to eight feet in length, weighing as much as 680 kg, and running almost thirty kilometers an hour. The apex ursine predators have attacked and killed humans foolish enough to wander unprotected through their territory here. *Ursus maritimus* has been deployed since the nineteenth century as one of the single most important marketing tools used to attract tourists to Svalbard, which sees upwards of 55,000 visitors per year. One of the first sights to greet tourists flying into Longyearbyen is a large taxidermized bear standing atop the luggage carousel. Isbjørn, or "ice bears," are also the emblematic, charismatic megafauna for climate change.

Longyearbyen sits on the wide Advent Fjord, which is typically unfrozen in recent years, but nonetheless has blue bergy bits floating in it this late May, brought in by currents that sweep them from around the eastern side of the island. Much further away, across the larger Isfjorden, precipitous white peaks are separated by broad glaciers flowing out to the water. The town is heated by large, insulated hot water pipes propped atop the permafrost with other infrastructure lines, such as electrical cables, and there is not a tree in sight. The charms of polar towns, which can resemble the back end of a ski resort crossed with a mining operation, are not for everyone, albeit they are communities of people proud to thrive in harsh places.

Svalbard and other reaches of the far North, as with the far South of the Antarctic, are inhabitable only by virtue of importing massive amounts of external resources. There is no archaeological evidence to support the possibility that early indigenous hunters reached, much less settled Svalbard, and although it has been proposed that Norse sailors may have discovered the archipelago as early as 1194, the first confirmed sighting was by the Dutch explorer Willem Barents in 1596. Whalers established camps on the islands early in the 1600s, but it wasn't even verified until 1615 that Spitsbergen was an island and not connected to Greenland, more than 800 km to the east. The first permanent settlement was attempted in 1872 based on phosphate mining, and although there were already small coal mines being worked late that century, it took the American John Longyear's creation of the Arctic Coal Company in 1906 for a town to take hold. It was sustainable only because it earned revenue from mining a resource needed in Europe, and it remains the northernmost settlement of more than 1,000 people anywhere in the world.[4]

But the coal seams that brought permanent settlement to Svalbard are playing out and worldwide the use of coal is declining in favor of cleaner fuels. Multiple remnants of coal mining infrastructure remain visible around the town, however, the most obvious of which are the hundreds of wooden supports for the tramways once used to transport coal from the mines to the port. The conveyances traversed every ridge in sight from the seven major mines developed during the twentieth century. The only active coal operation in Longyearbyen is now a single mine providing fuel for the local power plant, the stack of which towers on the edge of town as its tallest structure.

Like many scenic European locales at the beginning of the twenty-first century, including the Swiss Alps and the mountains and fjords of Norway, Svalbard is transitioning from a role within the industrial production chain to one based on scientific research and the experiential economy, which is to say tourism. It now serves as the entry point into this part of the Arctic – and to that empty interior. It sits on the edge of the alluring, and that is how it is being marketed within the brandscape of the Arctic, which is predicated on the experience of wilderness. This is complicated, as both the physical reality of wilderness is changing, as well as our perception of it. The globe warms and the glaciers retreat, which makes the interior more accessible, even as the polar bears may decline with the simultaneous loss of sea ice as a habitat. The attraction of Svalbard is both the wilderness and the desire to see it before all the ice is gone.

An excursion through the brandscape of Norway

"Brandscape" is a term coined by anthropologist John Sherry in 1986, when he was investigating Nike Town Chicago and realized that the retail space had become a legitimate place.[5] The term refers to a process that consumers undergo as they navigate the blizzard of commercial identities to which we're exposed every day, and to what happens when brands embedded in buildings actually shape the built environment. Sherry initially defined brandscape as "a material and symbolic environment that consumers build with marketplace products, images and messages, that they invest with local meaning, and whose totemic significance largely shapes the adaptation consumers make to the modern world."

This idea was more comprehensively developed in 2010 by architect and critic Anna Kingman who postulated that the built environment is increasingly an experience that we consume as a marketed product, and that architecture as an aesthetic discipline is used to manage those experiences.[6] Times Square in New York City, the Ginza in Tokyo, and the Las Vegas Strip are extreme cases. One of the dangers in these intense brandscapes is that they can become parodies of reality, filled with architectural and experiential simulacra, in Guy Debord's terms.[7]

Australian landscape architect Nicole Porter describes how brandscaping has been applied to park systems and even entire countries, noting that "all landscapes . . . are increasingly subject to a coherent and comprehensive system of identity construction," and how such branding is undertaken through everything from the design of signage and brochures, as well as landscapes themselves.[8] In the past, signs and promotional materials sought to provide practical information to the traveler; landscape branding uses those devices to monetize the identity of place. As Greg Richards noted in 2011, "Creative resources are now regularly employed to generate more distinctive identities, offering regions and cities a symbolic edge in an increasingly crowded marketplace."[9] The Swiss ethnographer Suzanne Chappaz-Wirthher recently proposed that the deployment of public art in the Alps was part of the marketing of Switzerland as a brandscape to tourists and residents alike, an effort to overlay the existing perception of a sublime retreat with sophisticated cultural amenities.[10]

Norway has a longstanding and extensive history of branding its landscapes as sites for experiential consumption, and part of the relationship of tourists to both Norway and the Arctic arises from the art histories of both places, in particular that of circumpolar exploration and landscape painting, which have long been intertwined. Nineteenth-century tourism in both Norway and then the Arctic was based on a search for remote, unique landscapes that harbored authentic, unspoiled cultures.[11] Artists from England and Europe visited and painted Norway and the Arctic, which helped to shape the desires and expectations of visitors during the eighteenth and nineteenth centuries.[12] Cold climates and uncrowded places were also perceived to be healthy tonics in response to the increasingly industrialized cities. The relationship continues today with tourists seeking to experience sublime nature and artists attracted by mountain and polar environments.

One of the most extensive contemporary examples of brandscaping is the Norwegian Tourist Route Project, which is placing architectural and artistic framing devices along a network of eighteen scenic routes. The scenic climaxes of Norway include vertiginous fjords and waterfalls, snowcapped peaks and glaciers, and fractal coastlines peppered with islands. The most outstanding examples are threaded together by the National Tourist Routes, a program designed by the Norwegian Public Roads Administration to provide safe and aesthetic amenities for experiencing the landscape, thus increasing tourism and local economic activity. Begun in 1994 and anticipated to be finished sometime around 2023, each of the routes within the more than two-thousand-kilometer network will include artworks as well as the architectural enhancements of rest stops, viewing platforms, restrooms, and interpretive facilities.[13] As Janike Kampevold Larsen points out, touring the Norwegian landscape is a tradition that dates back to the late 1770s, and the Tourist Routes are a recent example of how cultural interventions in nature have been used to define and commodify landscapes.[14]

Among the most northern of the routes and artworks in the program are those in Vardø, a historical fishing community on Norway's northeastern coast. The town sits on a butterfly-shaped mound of rock above the Arctic Circle at 70°N, and rises only sixty meters above the Barents Sea. Russia is clearly visible from the top of the island only fifty kilometers to the southeast.

Vardø is still a regionally important fishing town, but industrial-scale trawling, market factors, and aging harbor facilities have led to a depopulation of the town during the last thirty years. The Tourist Routes first art project was built here, the *Steilneset Memorial* designed by Swiss architect and Pritzker Architecture Prize winner Peter Zumthor, who worked with the late French-American artist Louise Bourgeois to include her last sculpture. The project memorializes the seventeenth-century burning at the stake of ninety-one men and women suspected of being witches in the state of Finnmark. The memorial was originally conceived to be part of the national Millennium celebration in 2000, then adopted by the Tourist Routes art program and finished in 2011.[15]

The Future North team visited the memorial on a blustery afternoon in January 2014, the twilight sky and sea both shades of gray, our feet breaking through a crust of snow into the powder underneath. I chose to start my experience by walking up the long ramp at the eastern end of the main structure, entering a treated canvas bladder suspended from an enlarged version of a traditional wooden fish drying rack. Inside, I walked on an oak plank floor along a dimly lit corridor, the entire 27-meter-long structure vibrating both to the wind gusts outside and my footsteps. Every few feet, there were an asymmetrically placed window, light bulb, and individual plaque commemorated one of the victims.

At the other end of the memorial I exited onto another ramp that led to the open entrance of a smoked-glass and metal frame cube whose tall sides stopped several inches short of reaching the ground, allowing the wind and snow and cold to penetrate. Inside was the installation by Bourgeois, *The Damned, The Possessed, The Beloved* − a steel chair standing inside a concrete cone, the seat of

Figure 10.2 Steilneset Memorial in Vardø; Architect: Peter Zumthor, 2011

Photo: Asbjørn Nilsen/Statens vegvesen.

which is penetrated by four large gas jet flames. Seven large circular and slightly converse mirrors reflected and distorted the scene and myself within it. If the effect of the long building was one of somber historical reflection on the seventeenth-century burning of people accused of witchery, the glass cube was furious, immediate, and haunted.

The *Steilneset Memorial* is both site- and place-specific.[16] It is a striking and isolated structure perched on the coast, where its triangular wooden framework alludes to the fishing industry, and is thus anchored firmly to local history and vernacular building styles, while at the same time bringing international architectural and art vocabularies into play. The memorial is a clear indication that the area is shifting a large part of its identity from that of a site of production to one of experiential consumption, in this case public art as a walk-in history lesson.

Jan Andresen, the Director of the Tourist Route Section of the Norwegian Public Routes Administration provides an elegant formulation for the difference between the amenities and the artworks: "While the functional architecture along the tourist routes will primarily enhance travelers' experience of the landscape, the art will generate reflections and tell other stories than those that are immediately visible."[17] While most of the Tourist Route amenities serve to act as frames or focal points within the landscape, directing the attention of travelers to natural features, the artworks address the history, hopes, and desires of people.

The Tourist Routes are not the only places one finds new public art in the Norwegian landscape. In 1988 Anne Katrine Dolmen and Aaslaug Vaa proposed the Artscape Nordland sculpture project,

Figure 10.3 "The Damned, The Possessed and The Beloved", installation by Louise Bourgeois, 2011
Photo: Jarle Wæhler/Statens vegvesen.

which would place works outside by internationally prominent artists in a region-wide collection. Curator Maaretta Jaukkuri reports that the unofficial motto for the program "has been Michael Heizer's statement from the early seventies . . . that a work of art is not put in a place, instead it is the place".[18] The program has so far commissioned thirty-six sculptures throughout Nordland county, the aesthetics ranging from a figurative stone sculpture by Antony Gormley standing in a harbor to an abstract monolith along the shoreline by Tony Cragg. Furthermore, this private program served as an inspiration for the public Tourist Route projects.

Norway has successfully utilized art and architecture in an open effort for centuries to stimulate touristic curiosity and, more recently, to help create experiences that are unique to the country. What is of interest to me now, in Svalbard, is whether or not the same arc of intentional development present on mainland Norway is true at the leading edge of tourism further north at a time when sovereignty is no longer upheld through industry.

Longyearbyen's shift from coal mining toward increased tourism and scientific research is an opportunity to shift the identity of the community, and would thus be the logical foundation for its branding efforts. Both tourism and research have historical roots available for inspiration. Scientists have been visiting Svalbard aboard efforts to reach the North Pole since the Russian Čičagov Expedition in the 1760s. Beginning in the nineteenth century, and increasingly ever since, expeditions to and permanent field stations on Svalbard from multiple countries have been assessing ocean currents, the Aurora Borealis, permafrost, and Arctic biology and botany.[19] Adventurous travelers began visiting the islands in the 1820s and actual tourism started in the 1890s with the establishment of cruises from Tromsø. Development of the local tourism trade can be traced through the history of postcards and their local postal cancellations, which demonstrate that people were eager to have authenticated their presence at whaling stations, early polar exploration sites, and science field camps, and to view polar bears and glaciers calving into fjords.[20]

The Arctic in general, but Svalbard in particular, offers an ironic field site for the study of this progression in the deployment of tropes used to form its brandscape. Svalbard in the past has never hosted permanent occupants, so its layers of visual culture are relatively thin. But the environmental conditions that prevented settlement also preserve gestures committed on the landscape. Evidence of human presence is on almost every shoreline, from old hunting shacks to piles of whale bones, and even a cairn in the shape of a swastika set out by Germans in 1939. The evidence runs counter to the popular tourism trope that you're skating along the unknown edge of the world.[21]

The newest tag line for Svalbard is "Next to the North Pole," which appears on postcards, tote bags, t-shirts – all the usual logo placements used to promote the branding of a landscape. The polar regions in general – but because of geographical proximity to Europe, more especially the Arctic – have, since at least the eighteenth century, been posited as a pure, white, primeval *terra nullius* uncorrupted by human presence. None of that is completely true: the Arctic has been populated for thousands of years and pollution is sometimes unseen, but ever present. But one of the most widely bought tourist guides for the region, *Lonely Planet Norway*, contains within its first two paragraphs about Svalbard a list of time-honored descriptors: "dreamed, wondrous, dramatic, vast, forbidding, elemental, endless, perpetual, deeper, wilderness, and epic." The list is capped with the declaration that to travel to Svalbard is to cross the "remote frontier of the mind."[22] If traveling to Svalbard is no longer the terrestrial equivalent to reaching outer space, at least it can be marketed as a conceptual outland.

A tour of the art

Peter Hemmersam always leads our group on transects through the places we research in the circumpolar North, and on investigative walks that follow straight lines drawn on a map of the sites that cut rigorously through physical neighborhoods and document evidence of change over time. His line through Longyearbyen, which started not far from the airport, was no exception, crossing both geological and historical strata. The art and architecture encountered along the way were revealing. What makes Peter's transects so interesting is that, while they cross roads, they don't follow them. They disrupt the patterns designed to move people efficiently, and the narratives created to present the history and character of a place to residents and tourists alike. What's more, in Longyearbyen there are basically only two major roads, one that follows the seashore and fronts the town, and another at right angles that carries the town inland and up the valley to its terminus. Peter's transect cuts across both roads, making evident how changes in the extreme environment of Svalbard, the unique geography and history of Longyearbyen, and the relatively abrupt end to local coal mining have opened up new narratives, some of which conflict and compete with one another.

Place identity is formed through the adoption of the natural and cultural features that are perceived to be unique: local fauna, flora, and geomorphological features, but also community founders and foundational industries. Vision is the dominant human sensory apparatus, so it is no surprise that marketing co-opts these elements into the placement of public visual art to brand place. The gravitational pull of the Arctic is anchored in the ancient idea that by going over the edge of the world you will learn something about yourself, human nature, the cosmos – whatever your particular quest – that is otherwise unavailable knowledge. But, as with any proper edge of the world, there must be dragons just over the horizon, or in the case of Svalbard, polar bears. And there are polar bears here on everything that is flat enough to hold an image. It is appropriate that Peter started our traverse of the town at the sign between town and the airport indicating that polar bears are present. I counted nine representations of them between where we started and ended our walk through town, including one very handsome life-sized stainless steel version by sculptor Julian Warren and his son Thomas that stands outside the office of polar filmmaker Jason Robert. The issue of polar bears is complicated by the fact that, while they are dangerous, they are also protected as an endangered species. The guides who accompany tourists outside the town limits are not just seeking to protect the humans, but also the bears, and only at last resort is one to be shot.

The Northern Lights remain as another ubiquitous marketing theme, symbolic of circumpolar environments from Fairbanks to Tromsø. Locals tell us that five years ago Tromsø was dead in January; now the town is booked for the month with tourists from around the world looking for the lights. The same phenomenon started in Longyearbyen three years ago, and if it's not a polar bear on a t-shirt or calendar or poster, it's a preternaturally lurid depiction of the aurora. The public artwork on the Global Seed Vault evokes both daylight glinting of ice and the celestial lights during the dark, a clever stratagem given the location.

Svalbard also offers an intentional public art iconography that arises from coal mining, which is an industry-based brand in contrast with polar bears and Northern Lights. Two sculptures of miners stand in the middle of town where the street has been closed to vehicles, creating a pedestrian commons surrounded by everything from shopping and dining to the hospital and movie theater. The first public paean to mining that we found was a wooden sculpture, *The Miner* by noted Norwegian sculpturer Kristian Kvakland. A tall standing frame holds the carved horizontal figure of a miner drilling a coal seam, wedged between claustrophobic layers of wood. Created in 1993, you have to crane your neck back to take in the figure above your head, a nice manipulation by the artist that creates a physical allusion within the viewer to the cramped working condition of the miners.

I was at first surprised to find a memorial made of wood, until I was reminded that in the Arctic wood is a favored building material precisely because it does not decay as it would in lower latitudes. About a block downhill from the wooden sculpture was a more traditional, heroically styled, and slightly larger-than-life bronze sculpture of the same title by Tore Bjørn Skjølsvik, which was cast in 1999. I can imagine that the miners weren't satisfied with the first representation of their labors, and desired a more swaggering presence in town. Nonetheless, the conventional statue adds a requisite gravitas to anchor coal in the history of Longyearbyen.

The most contemporary identity marker in town is a colorful, two-story tower made of plastic waste harvested from the shoreline, a project that has had extensive exposure online. It's not located where tourists are likely to run across it, but when we walked down to the shore to investigate, we found a door into the tiny first-floor; inside, a couple of college students were sharing a beer. The ceiling is clear and you can peer upwards to continue the visual inventory of plastic items. This found-object architecture, *Ocean Hope*, was created in 2014 by Solveig Egelund, one in her series of *Håbet* ("Hope") sculptures that she has assembled at other sites, such as Lista in southern Norway.[23] By turning junk into habitat, she seeks to foster hope in the face of environmental challenges. The placement of the work is a vivid illustration of the growing global issue of plastic pollution in the oceans, but its location away from the main road is also emblematic of how we tend to ignore the dark side of globalization and its encroachment on what we pretend are pristine, untouched wildernesses.

These dual identities of Longyearbyen, and by extension Svalbard, one based on preserving natural resources (polar bears and clear skies for viewing northern lights), the other based on resource extraction, can be reconciled as components in an economic progression: coal is no longer as lucrative as tourism. They can also be viewed as occupying different points along the evolution of environmental ethics: coal extraction and burning harms the environment, while passive viewing of natural resources preserves it.

Figure 10.4 "Ocean Hope," sculpture by Solveig Egeland, 2014
Photo: Solveig Egeland.

Ocean Hope belongs to the most prominent category of contemporary art representing Svalbard, artworks that mark the spread of the human footprint, and the often negative effects of industry and people upon the environment. Such work is also the predominant form of contemporary art for which Svalbard has become known. If the Arctic is being marketed as a last frontier to tourists, it is also a terrain where artists have increasingly sought to understand and comment upon global change. But most of these artworks don't actually reside in the archipelago.

Research and the art of disappearance

Climate change is a subject of widespread public discourse that arises from the collection and analysis of data collected by scientists, who are often field researchers measuring changes in oceanic and atmospheric temperature and chemistry, the growth and loss of sea and glacial ice, and the effects all of this has on biodiversity and changes in habitat. Because climate change is happening so quickly in the Arctic, the media has widely promulgated related photographic and video images, including James Balog's extensive time-lapse photographs from his Extreme Ice Survey that have been featured in a popular video, *Chasing Ice*, and the book *Ice: Portraits of Vanishing Glaciers*.[24] Tourists not only visit Svalbard to see wildlife and landscape, but to visit this frontier of climate science where environmental changes are affecting the animals and land. During interviews with Longyearbyen residents and business leaders, we are told that scientific research is also part of the community's identity that they wish to capitalize on.

Among the most well-known contemporary photographic images of Svalbard are the winter views of research installations in Spitsbergen taken by Norwegian artist Christian Houge. His series "Arctic Technology & the Global Seedvault [sic]," made from 2000–2014, is a striking catalogue of radio telescopes, antennae fields, and the Seed Vault, the installations glowing faintly in the frigid winter twilight to create a technological sublime: the facilities are severely handsome and slightly menacing constructions built and operating in extreme conditions. Not only are the photographs shown and collected by art museums worldwide, but the Future North team noted that, when we visited the atmospheric studies telemetry field photographed by Houge, the instrument field laid out in a minimalist geometrical formation could be mistaken for a land art project.

The publicly visible research identity of Svalbard is also promulgated through architecture, for example in the shape of the Svalbard Science Center, a university and research facility built in 2005. Peter Hemmersam notes how the building gently resonates with the multiplex shape of the Ropeway Hub built above the town in 1957, a strictly utilitarian piece of architecture that served as a terminal for the coal-bearing aerial trams traversing various mountainsides from mines to town. Both buildings are strong geometrical solids elevated on stilts that also echo the form of the surrounding mountains.[25]

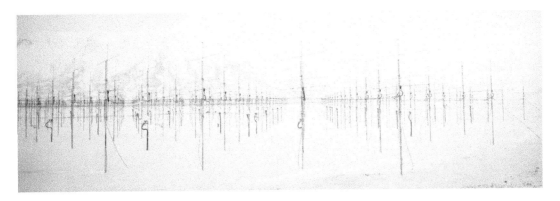

Figure 10.5 Christian Houge, *Antenna Forest* (2000), digital C-print, ed. 4/7, 100 x 300 cm

Source: Collection of the Nevada Museum of Art, The Altered Landscape, Carol Franc Buck Collection.

The research identity also appears in the interior design of the headquarters for the Svalbard Satellite Station atop Plateau Mountain, a facility that is the largest single collection point for downloading satellite data on the planet. It is one of only two stations on the planet that can tracks signals from all existing orbits of polar satellites. If you look at, listen to, or read a weather report anywhere on Earth, it is likely that the data came through the Longyearbyen facility. The Station is normally closed to the public, but we're lucky enough to be here on the first open house day in eight years. Hundreds of people show up, and it's clear that this would make a fine addition to the brandscape of Longyearbyen, given the strong sculptural presence that the radar domes create on the mountain plateau.

The private and governmental clients of "SvalSat" are received in the administrative building, which stands as an angular block amidst the thirty or so white radomes of the station. The main staircase of the building features suspended glass treads through which abstracted and quite handsome satellite views of the Earth are visible. As explained by station personnel, the artwork is used to sell visitors a sense of the global importance of the station: they are meant to feel that they are not only on the edge of the world, but also on the edge of space.

Scientific research also prominently underpins much of the art about Svalbard that's not visible permanently in Longyearbyen or in the archipelago. British photographer and sailor David Buckland, when confronted with oceanographic models predicting catastrophic changes in Arctic

Figure 10.6 The Great White Sale
Photo: David Buckland/Amy Balkin, 2008.

currents, wondered why scientists were having trouble convincing the public that climate change was a real and present danger, and caused at least in part by anthropic causes. In 2001 he founded Cape Farewell in the belief that a cultural response to climate change was necessary in order to transform scientific data into policy and action. Since then, he has led eight voyages to the archipelago with more than 350 artists, scientists, and writers.[26] Buckland's efforts have inspired numerous other "art & science" voyages to Svalbard, most notably the annual Arctic Circle expeditions and the 2012 Expedition Svalbard.[27]

Most of the art produced during these voyages follows the 1970s environmentalist "no-trace" camping practice that was later codified into the credo of "Leave No Trace".[28] The excerpts from Gretel Ehrlich's book *The Future of Ice* that David Buckland projected onto ice, then photographed, are a powerful example of an ephemeral practice, as well as a metaphorical branding of the Arctic. Ehrlich wrote her book after sailing on the 2003 Cape Farewell voyage, and Buckland projected words from it over the next two years.[29] The quotes – "The Great White Sale" and "Discounting the Future" among them – are condemnations of global consumption in the face of scientific evidence of climate change. The photograph of his projection of "Burning Ice" provided the title and cover image for the book about the project that was published in 2006, and has since come to emblematize both Cape Farewell and climate change in the Arctic.

An extensive project that exemplifies the art made in and around Svalbard, but not left in the archipelago, was initiated by British artist Alex Hartley, who sailed with Buckland in 2004. While circumnavigating the northern reaches of Spitsbergen, they located an island about the size of a football

Figure 10.7 Nowhere Island, installation by Alex Hartley, 2012
Courtesy of Victoria Miro Gallery.

field that had recently emerged as a glacier was retreating. Hartley was accorded the privilege of being the first human to set foot on the newly revealed land, which was eventually named Nyskjæret.[30]

Hartley returned to the island in 2011 and moved some of its glacial till onto a barge, a representative portion that he named *Nowhereisland*. Upon a moment's reflection the name quickly dissolves into various and ironic complementarities: nowhere island, now here is land, no where is land. Once this polyvalent piece of property was towed into international waters, Hartley declared it an independent nation. He then proceeded to have it tugged to Weymouth, England and then around the southeast coast to end in Bristol, a 2,000-mile-long journey that concluded when the island was dismantled and given away in pieces. Hartley's intervention was an intentional land art performance, and he cites, among other works, the floating island barge envisioned by Robert Smithson, Tania Kovats's *Meadow* barge from 2007, and the Francis Alÿs performance piece of 2002, *When Faith Moves Mountains*, as precursors. But it is also political theater that highlights global warming, international competition for natural resources, and the fickle nature of national boundaries. Before *Nowhereisland* was dispersed, it had attracted 23,003 people from 135 countries to sign up as citizens, who wrote the constitution for *Nowhereisland*, which in its first iteration consisted of 100 principles and conditions. The project remains online, a provocation about the nature of migration and global change.[31]

Into the Anthroposcenic

The sociologist John Urry defined the "tourist gaze" as a dangerous tendency for visitors to seek out predetermined sights deemed authentic, and to view them as elements disassociated from their local historical and physical contexts.[32] Visitors to the Global Seed Vault who view it solely as what Peter declared "pre-post-apocalyptic" architecture are turning the facility into one of those isolated elements in danger of becoming a cliché. I would submit that the artists whom Cape Farewell and other programs are bringing to Svalbard are, for the most part, creating a countervailing force against the assumptions of the tourist gaze. As Sibyl Ormin put it: "Artists often use reversal or exaggeration to create an intensified 'gaze' on touristic destinations or touristic images."[33] And, in fact, Janike Kampevold Larsen points out, a similarly disruptive dynamic is at work along the Tourist Routes, the artworks serving to deconstruct the framing devices of the architectural amenities, a point alluded to in conversation with her by artist/curator Svein Rønning.[34]

An example is the 2012 art project *Rock on Top of Another Rock* by Swiss artists Peter Fischli and David Weiss, a four-meter-tall cairn sited 1,389 meters above sea level on a mountain plateau accessible by car only during summer. At first glance, the stack of two gigantic boulders can appear completely natural, the size of the rocks seemingly beyond the ability of humans to manipulate. A twin work was installed in London's Kensington Gardens a year later, the two cairns linking the countries, a wry acknowledgment that English tourists have been finding their way to Norway for centuries. But, more importantly, the dual placement of the work also notes that nature and culture are not as clearly separated as tourists might assume.

Atmospheric chemist and Nobel laureate Paul Crutzen famously proposed in 2000 that we had left behind the Holocene, or Recent Era for the Anthropocene, the Human Era. He suggested that we take the 1790s as the starting point for this transition, when the consumption of fossil fuels increased enough to leave behind a global stratum of carbon. His notion of the Anthropocene has been further defined by Crutzen's colleagues to have three stages: a slow rise in atmospheric and oceanic chemical traces and temperatures from the 1790s until the 1950s; then a "Great Acceleration" that continues today as consumption and its effects undergo explosive growth; and, a third stage starting in the 1990s that marks public awareness of our role and agency in the changes.[35]

I have been proposing for the last decade that landscape art, as broadly defined, can be seen as evolving since the 1790s in response to these three stages, an "art of the Anthropocene". During the first stage, artists were part of the great scientific and cultural cataloging of the world, painting and then photographing land features, animals, even entire ecosystems. Famous examples are the South American paintings of Frederic Church, which are based closely upon Alexander von Humboldt's travels with the botanical illustrator Aimee Bonpland across the continent from 1799–1804. During the Second Stage, artists increasingly documented the human footprint across the landscape, most notably in the work of the New Topographic photographers and subsequent landscape artists since the 1960s, who sought to capture human alteration of the landscape.[36]

During this second stage, and starting roughly in the 1960s, a group of Land Artists, including Michael Heizer, Nancy Holt, and Patricia Johanson in the United States, and Richard Long and Agnes Denes and many others in Europe and South America, began to physically alter the landscape in order to create artworks. Their work proclaimed that art could be as powerful a presence in the landscape as dams or highway systems. I would argue that this is at least in part a response to the increasingly inescapable human footprint around the world. If we perceive there to be no longer almost any part of the planet that is free of visible human presence, even if it is simply a contrail 10,000 meters above us, then we counter that loss of "pure nature" with human interventions in nature. Art, too, could have a footprint.

The timing of earthworks, or Land Art projects, was poignant in that it coincided with the rise of the "Leave No Trace" landscape ethic adopted by the environmental community. The earthworks themselves, often made in relatively inaccessible and inhospitable locations where the artists could work cheaply – Michael Heizer's *Double Negative* (1969–1970) in Nevada and Robert Smithton's *Spiral Jetty* (1970) in Utah, for example – have since become tourist destinations, and helped kick off a worldwide movement that is so extensive that it has guidebooks to what has become known as "destination art".[37] The journeys to these often remote sites express the desires of the art tourist to experience both the sublime remoteness[38] and the artistic response, a complex series of aesthetic feedback loops of which both the architectural and art projects along the Tourist Routes take full advantage.

In the third stage of the Anthropocene, hundreds of artists worldwide have adopted the physical vocabulary and methods of the Land Artists to create sometimes long-term, sometimes ephemeral gestures in the landscape, works that address directly the issues and effects of global change. These projects, pioneered by the work of artists such as Joseph Beuys, Helen and Newton Harrison, and Agnes Denes, are not so much about the aesthetic dialogues with the histories of art and landscape architecture as they are an engagement with nature and the effects of the human footprint on it.[39] Eco-artworks, which often offer a critique of the petroleum-based economy of the Anthropocene, such as the tower of plastic by Solveig Egelund, can also be direct interventions seeking to mitigate the deleterious environmental effects of that economy.

The evolutionary path for the art of Svalbard, as well as its public art, follows the history of the town, an arc that is found throughout the Euro-American world. The first artists to visit remote locations – often traveling alongside explorers, scientists, and colonial military personnel – cataloged the natural world. What they pictured most often were emblematic landscapes and wildlife; their landscape paintings, and then photographs, were often made from vantage points that become scenic overlooks, and their images of megafauna such as polar bears persist today. Resource exploitation often followed exploration, and public art in particular was often created to memorialize the accomplishments of industry. In Longyearbyen, that would be the coal mining.

If the Arctic is one of the most rapidly changing parts of the globe in response to anthropic change, then it seems appropriate that the deployment of public art at high latitudes would have demonstrated a change from the "Leave No Trace" attitude of the early environmental movement to "Make a Mark." It is, for example, at times impossible to distinguish between the aligned stones made and photographed by artist Richard Long from the cairns and wind breaks and casual rock arrangements made by hikers in the backcountry or along the Norwegian tourist routes, for that matter. Cairns – stones which used to be stacked for wayfaring or to commemorate a significant site – are now such a prolific visual nuisance along trails that subsequent hikers will obliterate them, seeking to keep the scenery new for others, as well as preserving the habitat for insects and small reptiles and mammals that use local stones for cover. But who can blame the cairn makers when public entities such as national and local governments are funding the erection of works such as *Rock on Top of Another Rock* by Fischli and Weiss?

What we have called *nature* was once much more extensive than the human-altered landscape. Now the opposite is true. Humans have not only arguably been the most powerful geomorphological force on Earth for the last 1,000 years,[40] since 1950 we've literally shifted the Earth's axis and altered our planet's rotational periodicity through the storage of water in the Northern Hemisphere.[41] To co-opt the terms used by Neil Smith,[42] the "first nature" of the untouched world has been superseded by "second nature," in which the world has been converted into a landscape of commodification. Furthermore, that commodification has now shifted from an economics of production to one of consumption.

Historian William Cronon would argue that it is impossible to separate the two natures, and in the same year that Chao published his paper about dams altering the course of the planet, Cronon pointed out that humans have long mistaken the garden for being domestic and the wilderness as being wild. He reconciles both as parts of nature.[43] To complicate matters, other contemporary theorists would claim both are part of culture. The French anthropologist Phillipe Descola went so far in 2005 to argue that our very definitions of "wild" and "domestic" are incorrect: the distinction between the two arose only during the Neolithic. Given that the hunter-gatherers of the Paleolithic ranged widely to stalk and forage, the landscape was simultaneously a wilderness and a garden for the majority of our species' existence.[44]

The architectural framing devices of the Norwegian Tourist Routes don't insist that you are looking at wilderness, simply that you look at what is in front of you – all of which is now altered continuously by anthropic action. The artworks along the routes more often remind us to be critical about what we think we're observing; together, the art and architecture serve as an interface between the viewer and the viewed, even as they lessen the distance between what we may have once considered to be the domestic and the wild. We can trace the progress of this evolving modification of landscape as it tracks northward from the roadside amenities and art on the west coast of Norway to Vardø and onto Longyearbyen. What it demonstrates is that the art of the Anthropocene is morphing into the Anthroposcenic. It no longer makes sense to put nature and culture into a dialectic opposition with one another, but rather to accept that culture is an adaptive behavior arising from nature. And nowhere is that becoming more evident than in the evolving public art of the Arctic.

The public art and the brandscape of Svalbard are nowhere near mature enough to be as self-critical as the art and architecture projects along the Tourist Routes on the mainland of Norway. The history of Longyearbyen still has one foot in resource extraction, and the community is understandably more concerned with economic and political survival than on creating discourse about post-irony in the future landscapes of the Arctic. If the identity of the community appears to have progressed from coal mining to eco-tourism, it is striking to note how the deliberate integration of public art into the roof of the Global Seed Vault, a science and conservation facility, manages to become strategic branding in the geopolitical arena. If the Russians, for example, wish to use sites in Svalbard as bases from which to conduct oil drilling or to facilitate circumpolar shipping, the Norwegians can claim a higher moral ground based on sustainability and resilience in the face of climate change. During our time here I've not seen any art that couldn't be co-opted into that argument.

Inside/out

Standing before the entrance to the Global Seed Vault, having apparently escaped the attention of any polar bears, the slowly declining light of a late May afternoon prompts us to think about returning to town. It's important to remember that parsing brandscapes from wilderness areas is an imprecise exercise. While the allure of Svalbard has been sold to tourists since the nineteenth century, and businesspeople such as John Longyear sought to attract buyers for his coal, the global marketing of the archipelago only began when tourism began to be seen as the primary way of keeping Longyearbyen alive as a post-industrial economy, and as a strategic Norwegian outpost in the rapidly opening Arctic waters. A brandscape here is nascent, not fully formed, and certainly not yet buttressed up with a national art and architecture program, as with the Tourist Route program. But Svalbard at present is a perfect opportunity to watch the process of this acculturation.

As part of that, the Future North team is off to the newly established Kunsthall Svalbard, a branch of the Northern Norway Art Museum in Tromsø, which is showing the premier of *Glacier*, a video by American artist Joan Jonas. Then uphill at the Galleri Svalbard we will examine paintings and photographs by local artists ranging over several decades, as well as one of the most stunning collections of Arctic maps in existence, which is appropriate for a place that self-identifies as the edge of the world. If we've been looking at art outside, now we'll be looking at landscape inside. It's not without a smile that we acknowledge how our presence here, and what we write about Svalbard, is integral to how human cognition constantly pulls the world inside out of itself.

Notes

1 Janike Kampevold Larsen, Peter Hemmersam, and Andrew Morrison, "Future North" (Research Application, Oslo School of Architecture and Design, 2012): 1.

2 An overview of Arctic warming was accessed on August 15, 2015 at: http://earthobservatory.nasa.gov/IOTD/view.php?id=81214.

3 This quote and all specifications about the Global Seed Vault were taken from the webpage of Statsbygg, the public sector organization that "is the Norwegian government's key adviser in construction and property issues, a building commissioner, property manager and property developer … Statsbygg is intended to be an active instrument in implementing political objectives in relation to the environment, architecture, aesthetics, innovative user solutions and health and safety at the workplace, based on the ministries' priorities". Statsbygg, *Svalbard globale frøhvelv/Svalbard Global Seed Vault* (Oslo: Statsbygg, 2008), 3, www.statsbygg.no/files/publikasjoner/ferdigmeldinger/671_SvalbardFrohvelv.pdf.

4 Bjørn Fossli Johansen, ed. *Cruise Handbook for Svalbard* (Tromsø: Norwegian Polar Institute, 2011). This is a compact and thorough history and guidebook for the archipelago.

5 John F. Sherry, "The Soul of the Company Store: Nike Town Chicago and the Emplaced Brandscape," in *Servicescapes: The Concept of Place in Contemporary Markets*, ed. John F. Sherry, Jr. (Chicago: NTC Business Books, 1998): 109–50.

6 Anna Kingman, *Brandscapes: Architecture in the Experience Economy* (Cambridge: MIT Press, 2010).

7 Guy Debord, *La Société du spectacle* (Paris: Buchet-Chastel, 1967). See also Umberto Eco, *Travels in Hyperreality* (New York: Harcourt Brace & Company, 1983).

8 Nicole Porter, *Landscape and Branding: The Promotion and Production of Place* (London: Routledge, 2016): 32–33.

9 Greg Richards, "Creativity and Tourism," *Annals of Tourism Research* 38, no. 4 (2011): 1230.

10 Suzanne Chappaz-Wirthner, "Site-Specific Actions in Brandscape Valais," (presentation at the seminar "Ars Contemporaenus Alpinus: A critical approach to site-specific art in natural environments," Sierre, l'Ecole cantonale d'art du Valais, September 17, 2014). www.ecav.ch/fr/projets/sous-projet/suzanne-chappaz-wirthner-570.

11 John Urry, *The Tourist Gaze* (London: Sage, 2002), 20.

12 Knut Ljøgodt, "Wild Nature: Swiss and Norwegian Landscape Painting," and Magne Malmanger, "The Alps: An Iconographic Theme in Romantic Art," in *Den Ville Natur: Sveitsisk og Norsk Romantikk*, ed. Anne Aaserud and Knut Ljøgodt (Tromsø: Nordnorsk Kunstmuseum, 2007).

13 Ken Schluctmann, *Architektur und Landschaft in Norwegen* (Ostfildern: Hatje Cantz.Verlag, 2014).

14 Nina Frang Høyum and Janike Kampevold Larsen, *Views: Norway Seen from the Road, 1723–2020* (Oslo: Forlaget Press, 2012).

15 Reidun Laura Andreassen and Liv Helene Willumsen, eds. *Steilneset Memorial: Art Architecture History* (Stamsund: Orkana, 2014).

16 Peter Hemmersam and Janike Kampevold Larsen, "Landscapes on Hold: The Norwegian and Russian Barents Sea Coast in the New North," *Critical North: Space, Theory, Nature* (Fairbanks: University of Alaska Press, 2017).

17 Line Ulekleiv, ed. *Peter Fischl/David Weiss: Rock on Top of Another Rock* (Oslo: Forlaget Press, 2012), 7.

18 Maaretta Jaukkuri, ed. *Artscape Nordland* (Oslo: Forlaget Press, 2001), 19.

19 Thor B. Arlov, "The Discovery and Early Exploration of Svalbard. Some Historiographical Notes," *Acta Borealia: A Nordic Journal of Circumpolar Societies* 22 (2005): 3–19.

20 John T. Reilly, *Greetings from Spitsbergen: Tourists at the Eternal Ice 1827–1914* (Trondheim: Tapir Academic Press, 2009).

21 Urban Wråkberg, "Re-Photography in Northern Interdisciplinary Field Studies: Cultural Testing Grounds Around Ice Fjord, Svalbard," in *Expedition Svalbard: Lost Views on the Shorelines of Economy*, ed. Tyrone Martinsson, Gunilla Knape, and Hans Helberg (Göttingen: Steidl, 2015).

22 Anthony Hamm, Stuart Butler, and Miles Raddis, *Lonely Planet Norway* (Footscray: Lonely Planet Publications Ltd., 2011), 348.

23 Solveig Egeland, "Ocean Hope," www.oceanhope.no.

24 James Balog and Terry Tempest Williams, *Ice: Portraits of Vanishing Ice* (New York: Rizzoli, 2012).

25 Peter Hemmersam, "The Contorted Architecture of Geopolitics," *Future North*, June 09, 2015, www.oculs.no/projects/future-north/news/?post_id=4078.

26 David Buckland, Ali MacGilp, and Sion Parkinson, eds. *Burning Ice: Art & Climate Change* (London: Cape Farewell, 2006). See also: www.capefarewell.com.

27 "The Arctic Circle," Accessed September 4, 2015, www.thearcticcircle.org; Tyrone Martinsson et al., *Expedition Svalbard: Lost Views on the Shorelines of Economy* (New York: Steidl, 2015).

28 Jeffrey L. Marion and Scott E. Reid, "Development of the U.S. Leave No Trace Program: An Historical Perspective," published January 2001, accessed September 5, 2015, http://web.archive.org/web/20060718233839/www.lnt.org/about/history.html.

29 Barbara Matilsky, *Vanishing Ice: Alpine and Polar Landscapes in Art, 1775–2012* (Bellingham: Whatcom Museum, 2013).

30 "Nowhere Island," accessed September 19, 2015, http://nowhereisland.org.

31 William L. Fox, "The Art of Svalbard, May 23–June 1, 2015," *Center for Art + Environment Blog*, July 26, 2015. www.nevadaart.org/the-art-of-svalbard-may-23-june-1-2015/.

32 John Urry, *The Tourist Gaze, Leisure and Travel in Contemporary Societies* (London: Sage Publications, 1990, revised 2002).

33 Sibylle Omlin, "The Critical Turn in Tourism Studies: Working with Images Between Seduction, Deconstruction and Critique," in *Tourists Like Us: Critical Tourism and Contemporary Art*, ed. Federica Martini and Vytautus Michelkevicius (Sierra, Switzerland: Ecole Cantonal d'Art du Valias, 2013).

34 Janike Kampevold Larsen, "Global Tourism Practices as Living Heritage: Viewing the Norwegian Tourist Routes Project," *Future Anterior: Journal of Historic Preservation, History, Theory, and Criticism* 9, no. 1 (2012).

35 Will Steffen, Paul J. Crutzen, and John R. McNeill, "The Anthropocene: Are Humans Now Overwhelming the Great Forces of Nature?" *Ambio* 36, no. 8 (2007).

36 Peter Pool, ed., *The Altered Landscape* (Reno: The University of Nevada Press, 1999). See also Ann Wolfe, ed., *The Altered Landscape: Photographs of a Changing Environment* (New York: Skira Rizzoli, 2011).

37 Amy Dempsey, *Destination Art* (Berkeley: University of California Press, 2006).

38 Federica Martini, "Romantic Frequent Flyers," in *Tourists Like Us Critical Tourism and Contemporary Art*, ed. Federica Martini and Vytautas Mickelkevičius (Vilnius: University of Vilnius Press, 2013).

39 Heather Davids and Etienne Turpin, "Art & Death: Lives Between the Fifth Assessment & the Sixth Extinction," in *Art in the Anthropocene: Encounters Among Aesthetics, Politics, Environments and Epistemologies*, ed. Heather Davis and Etienne Turpin (London: Open Humanities Press, 2015).

40 Roger LeB. Hooke, "On the History of Humans as Geomorphic Agents," *Geology* 28, no. 9 (2000). See also Bruce H. Wilkinson, "Humans as Geologic Agents: A Deep-Time Perspective," *Geology* 33, no. 3 (2005).

41 Benjamin Fong Chao, "Anthropogenic Impact on Global Geodynamics Due to Reservoir Water Impoundment," *Geophysical Research Letters* 22, no. 24 (1995).

42 Neil Smith, *Uneven Development: Nature, Capital and the Production of Space* (Oxford: Blackwell, 1984).

43 William Cronon, "The Trouble with Wilderness: Or, Getting Back to the Wrong Nature," in *Uncommon Ground: Rethinking the Human Place in Nature*, ed. William Cronon (New York: W. W. Norton & Co., 1995).

44 Phillipe Descola, *Beyond Nature and Culture* (Chicago: University of Chicago Press, 2005).

Bibliography

Andreassen, Reidun Laura, and Liv Helene Willumsen, eds. *Steilneset Memorial: Art Architecture History*. Stamsund: Orkana, 2014.

Arctic Circle. Accessed September 4, 2015. www.thearcticcircle.org/#.

Arlov, Thor B. "The Discovery and Early Exploration of Svalbard. Some Historiographical Notes." *Acta Borealia: A Nordic Journal of Circumpolar Societies* 22 (2005): 3–19.

Balog, James and Terry Tempest Williams. *Ice: Portraits of Vanishing Ice*. New York: Rizzoli, 2012.

Buckland, David, Ali MacGilp, and Sion Parkinson, eds. *Burning Ice: Art & Climate Change*. London: Cape Farewell, 2006.

Cape Farewell. Accessed September 15, 2015. www.capefarewell.com.

Chao, Benjamin Fong. "Anthropogenic Impact on Global Geodynamics Due to Reservoir Water Impoundment." *Geophysical Research Letters* 22, no. 24 (1995): 3529–32.

Chappaz-Wirthher, Suzanne. "Site-Specific Actions in Brandscape Valais." Presentation at "*Ars Contemporaenus Alpinus*: A critical approach to site-specific art in natural environments" at l'Ecole cantonale d'art du Valais, Sierre, Switzerland, September 17, 2014.

Cronon, William. "The Trouble with Wilderness: Or, Getting Back to the Wrong Nature." In *Uncommon Ground: Rethinking the Human Place in Nature*, edited by William Cronon, 69–90. New York: W. W. Norton & Co., 1995.

Davids, Heather and Etienne Turpin, "Art & Death: Lives Between the Fifth Assessment & the Sixth Extinction." In *Art in the Anthropocene: Encounters Among Aesthetics, Politics, Environments and Epistemologies*, edited by Heather Davis and Etienne Turpin, 3–29. London: Open Humanities Press, 2015.

Debord, Guy. *La Société du spectacle*. Paris: Buchet-Chastel, 1967.

Dempsey, Amy. *Destination Art*. Berkeley: University of California Press, 2006.

Descola, Phillipe. *Beyond Nature and Culture*. Chicago: University of Chicago Press, 2005.

Eco, Umberto. *Travels in Hyperreality*. New York: Harcourt Brace & Company, 1983.

Fox, William L. "The Art of Svalbard, May 23-June 1, 2015." *Center for Art + Environment Blog*. Last modified July 26, 2015. www.nevadaart.org/the-art-of-svalbard-may-23-june-1-2015/.

Hamm, Anthony, Stuart Butler, and Miles Raddis. *Lonely Planet Norway*. Footscray: Lonely Planet Publications Ltd., 2011.

Hemmersam, Peter and Janike Kampevold Larsen. "The Norwegian and Russian Barents Sea Coast in the New North." In *Critical North: Space, Theory, Nature*, edited by Keven Maier and Sarah Jaquette Ray. Anchorage: University of Alaska Press, 2017.

Hemmersam, Peter. "The Contorted Architecture of Geopolitics." *Future North*. Last modified June 09, 2015. www.oculs.no/projects/future-north/news/?post_id=4078.

Hooke, Roger LeB. "On the History of Humans as Geomorphic Agents." *Geology* 28, no. 9 (2000): 843–46.

Høyum, Nina Frang and Janike Kampevold Larsen. *Views: Norway Seen from the Road, 1723–2020*. Oslo: Forlaget Press, 2012.

Jaukkuri, Maaretta, ed. *Artscape Nordland*. Oslo: Forlaget Press, 2001.

Johansen, Bjorn Fossili, ed. *Cruise Handbook for Svalbard*. Tromsø: Norwegian Polar Institute, 2011.

Kingman, Anna. *Brandscapes: Architecture in the Experience Economy*. Cambridge: MIT Press, 2010.

Larsen, Janike Kampevold. "Global Tourism Practices as Living Heritage: Viewing the Norwegian Tourist Routes Project." *Future Anterior: Journal of Historic Preservation, History, Theory, and Criticism* 9, no. 1 (2012): 67–87.

Larsen, Janike Kampevold, Peter Hemmersam, and Andrew Morrison. "Future North." Research Application, Oslo School of Architecture and Design, 2012.

Ljøgodt, Knut. "Wild Nature: Swiss and Norwegian Landscape Painting." In *Den Ville Natur: Sveitsisk og Norsk Romantikk*, edited by Anne Aaserud and Knut Ljøgodt, 149–54. Tromsø: Nordnorsk Kunstmuseum, 2007.

Malmanger, Magne. "The Alps: An Iconographic Theme in Romantic Art." In *Den Ville Natur: Sveitsisk og Norsk Romantikk*, edited by Anne Aaserud and Knut Ljøgodt, 155–61. Tromsø: Nordnorsk Kunstmuseum, 2007.

Marion, Jeffrey L., and Scott Reid. "Development of the United States Leave No Trace Programme: A Historical Perspective." *Enjoyment and Understanding of the Natural Heritage*, edited by M. B. Usher, 81–92. Edinburgh: Scottish Natural Heritage, 2001.

Martini, Federica. "Romantic Frequent Flyers." In *Tourists Like Us: Critical Tourism and Contemporary Art*, edited by Federica Martini and Vytautus Michelkevicius, 71–86. Sierr: Ecole Cantonal d'Art du Valias, 2013.

Martinsson, Tyrone, Gunilla Knape, and Hans Hedberg, eds. *Expedition Svalbard: Lost Views on the Shoreline of Economy*. New York: Steidl. 2015.

Matilsky, Barbara. *Vanishing Ice: Alpine and Polar Landscapes in Art, 1775–2012*. Bellingham: Whatcom Museum, 2013.

Nowhere Island. Accessed September 19, 2015. http://nowhereisland.org.

Ocean Hope. Accessed September 19, 2015. www.oceanhope.no.

Omlin, Sibylle. "The Critical Turn in Tourism Studies: Working with Images Between Seduction, Deconstruction and Critique." In *Tourists Like Us: Critical Tourism and Contemporary Art*, edited by Federica Martini and Vytautus Michelkevicius, 29–40. Sierra: Ecole Cantonal d'Art du Valias, 2013.

Pool, Peter, ed. *The Altered Landscape*. Reno: University of Nevada Press, 1999.

Porter, Nicole. *Landscape and Branding: The Promotion and Production of Place*. London: Routledge, 2016.

Reilly, John T. *Greetings from Spitsbergen: Tourists at the Eternal Ice 1827–1914*. Trondheim: Tapir Academic Press, 2009.

Richards, Greg. "Creativity and Tourism." *Annals of Tourism Research* 38, no. 4 (2011) 1225–53.

Schluctmann, Ken. *Architektur und Landschaft in Norwegen*. Ostfildern: Hatje Cantz. Verlag, 2014.

Sherry, John F. "The Soul of the Company Store: Nike Town Chicago and the Emplaced Brandscape." In *Servicescapes: The Concept of Place in Contemporary Markets*, edited by John F. Sherry, Jr., 109–50. Chicago: NTC Business Books, 1998.

Smith, Neil. *Uneven Development: Nature, Capital and the Production of Spac*. Oxford: Blackwell, 1984.

Statsbygg. *Svalbard globale frøhvelv / Svalbard Global Seed Vault*. Oslo: Statsbygg, 2008. Accessed April 29, 2016. www.statsbygg.no/files/publikasjoner/ferdigmeldinger/671_SvalbardFrohvelv.pdf.

Steffen, Will, Paul J. Crutzen, and John R. McNeill. "The Anthropocene: Are Humans Now Overwhelming the Great Forces of Nature?" *Ambio* 36, no. 8 (2007): 614–21.

Ulekleiv, Line ed. *Peter Fischl / David Weiss: Rock on Top of Another Rock*. Oslo: Forlaget Press, 2012.

Urry, John. *The Tourist Gaze, Leisure and Travel in Contemporary Societies*. London: Sage Publications, 1990, revised 2002.

Wilkinson, Bruce H. "Humans as Geologic Agents: A Deep-Time Perspective." *Geology* 33, no. 3 (2005), 161–64.

Wolfe, Ann, ed. *The Altered Landscape: Photographs of a Changing Environment*. New York: Skira Rizzoli, 2011.

Wråkberg, Urban. "Re-Photography in Northern Interdisciplinary Field Studies: Cultural Testing Grounds Around Ice Fjord, Svalbard." In *Expedition Svalbard: Lost Views on the Shorelines of Economy*, edited by Tyrone Martinsson, Gunilla Knape, and Hans Helberg, 157–69. Göttingen: Steidl, 2015.

11 Place as progressive optic

Reflecting on conceptualisations of place through a study of Greenlandic infrastructures

Susan Jayne Carruth

Infrastructure – the means of supporting and structuring the basic needs of society – lies at the core of Greenland's development, implicated in the fields of tourism, resource extraction, ecology, trade, health, education, and industry. Infrastructures deeply affect where and how people live in Greenland, enabling, disabling, and shaping how the land is experienced and inhabited. This is, of course, true of everywhere, but in Greenland the sparse, scattered population; harsh climate; vulnerability to climatic change; and fragile politico-economic conditions render infrastructure particularly pertinent both at the national level and at the local scale, transforming everyday lives, landscapes, and places. Furthermore, infrastructures in Greenland, which rely upon informal, microscale, ephemeral, networked operations just as much as official civil engineering, fundamentally challenge what we understand infrastructure to be. Studying infrastructure in Greenland, therefore, necessitates the study of its landscapes and places. This chapter suggests that place is a particularly important optic for the consideration of Greenlandic infrastructures, and furthermore enquires whether such a place-framed understanding of Greenlandic infrastructures might not also enrich contemporary conceptualisations of place.

Contemporary popular framings of Greenland, especially from the outside, tend to fall into two ostensibly antagonistic camps. On one side stand those that focus upon the economic opportunities conferred by transforming bio-geo-cryo-climatic conditions. On the other hand, there are those that are concerned with preserving the beauty of Greenland's sublime 'nature' and the integrity of its fragile ecological systems. While superficially polarized, these dominant perspectives have a conceptual commons: *landscape*.

In the former, the Greenlandic landscape is viewed as a mappable, consumable territory ripening for economic exploitation. Greenland is thus seen as an infrastructural opportunity for the production, distribution, and consumption of natural resources. In the latter, the Greenlandic landscape is conceived of as one of the few remaining 'wildernesses', and thereby as a site of international heritage deserving of protection. Acting as a kind of scenic litmus paper, the Greenlandic landscape from this perspective is suffering at the hands of anthropogenic climate change triggered by the exploitation of vast quantities of natural resources via global infrastructures. Landscape, and in particular landscape's interrelationship with infrastructure, is the protagonist in both these perspectives of Greenland, despite their opposing frameworks and ideologies.

Yet, within these discourses another concept is conspicuous through its absence – *place*. Place, hindered perhaps by atavistic connotations of sentimentalism, parochialism, and apparently negligible relevance in the face of the challenges of climatic change, tends to be overshadowed by landscape in Greenland. In the twentieth century, the notion of 'place' was the subject of foundational works by Michel De Certeau, Michel Foucault, and Yi-Fu Tuan, amongst others.[1] But with some notable exceptions, the concept of place within contemporary cultural geography and urbanism has not been subject to the same interdisciplinary critical and conceptual overhaul that landscape has benefitted from in recent decades.[2] Landscape, not place, has been the dominant framing of Greenland. This domination of place by landscape – both in urbanism/geography and specifically in the Arctic – is however challenged by the deeply entangled relationship between place and infrastructures in Greenland, and the manner in which such places in the Greenland (and other Arctic territories) are networked with larger infrastructures that are complicit in transforming climatic change and the international geopolitical turn to the 'North'.

Drawing on my recently completed PhD thesis, this chapter argues not only for the legitimacy of studying infrastructures in Greenland through the optic of place as well as landscape, but also, crucially, how the study of infrastructures in Greenland can unfold new perspectives on place with relevance outside of the Arctic. Its central points are, firstly, that a redefined concept of place, following the lead of Doreen Massey and Kenneth Olwig, lends an apt theoretical framing for understanding infrastructures in Greenland that pays needed attention to the everyday lives, spaces, and cultures of Greenlanders, while simultaneously connecting these with larger infrastructural systems.[3] Secondly, it suggests that a study of Greenland's highly visible, informal, networked, microscale infrastructures sheds new light on such conceptualisations of place, which are relevant beyond Greenland and the Arctic.

The chapter is split into three sections. The first section differentiates landscape from place through three concrete examples in Greenland – Greenpeace's relationship with Greenland, the history of Ivittuut/Grønnedal, and speculative architectural concepts for a new airport in Nuuk – that each illustrate the problematic neglect of place in Greenland's development discourses. After briefly sketching transforming conceptualisations of landscape over recent decades, the second section draws attention to the work of British social scientist and geographer Doreen Massey and American landscape geographer Kenneth Olwig as important exceptions to the often still largely conservative contemporary conceptualisations of place still found in urbanism and geography. Finally, the third section employs five examples of infrastructures in Greenland – over-ground water grids; transforming harbours; informal 'canine' infrastructures that support dog sleds; off-grid solar panels and rubbish dumps – which illustrate both the significance of infrastructures in the construction of place, and how such seemingly localised infrastructure-places are in fact condensed 'moments' within larger infrastructural networks. I conclude the chapter by arguing that place, just as much as landscape, is a productive and relevant optic for the understanding of Greenlandic infrastructures, particularly in the context of climatic change, as it effectively uncovers a density of associations and experiences contained and condensed within a particular set of relationships in space and time – and these are both 'local' effects and atmospheres and larger, 'global' systems and resonances. Finally, I suggest that this empirical study reciprocally enriches and progresses conceptualisations of place due to the highly visible nature of Greenland's informal, diffuse infrastructures and their intertwinement with the production of place at multiple scales.

Forgotten place

What distinguishes place from landscape? While it is certainly not my intention to emphatically separate or dichotomise place and landscape – they are undoubtedly fuzzily and messily overlapping – there are meaningful distinctions in connotation. There is a complex and extended history of investigation into defining and distinguishing place from landscape, and it is outside the scope of this chapter to comprehensively review such discussions.[4] One could suggest that place is distinguished by having 'meaning', named on a map and filled with social and cultural significance. But landscape is not a purely materialist framing, and conventional understandings of landscape as purely a large-scale concern, and place as specific bounded 'points' within such landscapes are no longer valid, as discussed in the next section of this chapter in more detail. In light of this, rather than a theoretical analysis, three Greenlandic examples – the long-span history of Greenpeace's campaigns in Greenland, the story of Ivittutt/Grønnedal's changing status within the context of a localised setting, and a recent speculative proposal for a new airport in Nuuk, the capital of Greenland, by BIG and Tegnestuen Nuuk – are described to nuance the distinction within the specific, concrete context of Greenland.

The tendency towards considering infrastructure through the optic of landscape at the expense of place is unfolded at a large scale, and over a long timeframe, through the often-fraught relations between Greenland and Greenpeace. Greenpeace has long-standing interests in the Arctic and has mounted campaigns against the hunting industry in the Arctic, particularly in relation to the international fur trade. When Greenpeace, however, famously protested against seal pup hunting in the 1970s, and lobbied, successfully, for its banning, it arguably did not distinguish sufficiently between the hunting methods used by the Inuit compared to commercial hunters from other regions, nor did it comprehend the importance of seal hunting as a livelihood and its cultural-economical role in Inuit communities. The ban led to the Greenlandic Inuit, as well as other Arctic communities, losing what was a significant and depended-upon income, and was a major contributor to a steep rise in widespread poverty, which led in turn to a substantial reliance upon government benefits and a dismantling of the identity, livelihoods, and cultural status of many Greenlandic men. The suicide rate for young men rose along with other social pathologies that continue to reverberate.[5] Furthermore, and ironically, Greenpeace would only condone 'true' Inuit hunting, i.e. hunting without guns, motorboats, and snowmobiles; the Danish colonialist restructuring of Greenland had obliged Greenlanders to move away from scattered settlements near hunting grounds to larger towns, thereby necessitating such new technologies.[6] Still today Greenpeace receives a rather cool welcome in Greenland and their recent campaigns and protests over hydrocarbon extraction in the Arctic have once again angered some Greenlanders who feel that there remains a lack of sensitivity to Greenlandic and Arctic human lives and livelihoods, and that such campaigns effectively question the sovereignty of the Greenlanders over their own land.[7] Greenpeace's 'Save the Arctic' campaign in Greenland illustrates the intersections and reverberations between 'traditional' colonialism, 'Eco-Colonialism', and the potential 'Corporate-Colonialism' of international gas and oil companies. Landscape, considered both as an ecological system and sublime beauty, has played the leading role in this intersectional colonial narrative. Concurrently, a lack of attention has been paid to places and the people, daily life, and culture that such places are produced by and with: the 'resolution' of such discourse is simply not high enough to capture these apparent 'details' of place.

If this example illustrates tensions between infrastructure, landscape, and place on the large scale, the story of Ivittuut/Grønnedal, in the south west of Greenland, demonstrates them within a localised setting. The now-abandoned twin settlements of Ivittuut/Grønnedal were once an important resource for the Allies in World War II. Ivittuut was one of the very few sites in the world with a significant cryolite deposit; a key mineral in the manufacture of aluminium at that time. The Ivittuut mine was opened in 1854, and as technology changed in aluminium production in the 1920s, the demand for cryolite boomed and a mining village grew around the mine brining wealth to the area. There came even greater demand for Ivittuut's cryolite during World War II to supply aeroplane manufacturing and other military equipment, and at its peak the mine was exporting up to 85,000 tonnes in 1942.[8] Once Denmark was occupied by Germany in April 1940 there was a fear of the German Nazis capturing this important resource and consequently the American military set up a base just a few kilometres from the mine – Grønnedal.[9] This military base prompted the first, and practically last, road to be constructed between two settlements in Greenland; a minor piece of infrastructure but a major innovation in this context. Post-war, the strategic importance of Ivittuut fell as alternative methods for smelting aluminium were invented, exacerbated by the depletion of Ivittuut's cryolite in the 1980s. Furthermore, in 1951 the USA handed Grønnedal base over to the

Danish military, which then occupied it until 2012.[10] These connected shifts in military and resource status have led to a dwindling and then 'closing' of both Ivittuut and Grønnedal. Today Ivittuut and Grønnedal stand almost empty and are crumbling as unheralded sites of global industrial and military heritage – unique places that each condense a moment of global history and connect Greenland to larger geopolitical narratives. This story exemplifies many of the hallmarks of the global tapping of Arctic resources – a quick move in, and a quick move out, with little responsibility for the longer and wider view. And, unlike in other regions, in the Arctic there is no immediate repopulation of such places, nor a critical mass of people even aware of such changes. The wealth of the area has disappeared and the structures and infrastructures lie stagnant; a relic of the dangers of short-term thinking about Arctic infrastructures. This heightened example of the colonialist military-industrial complex working in Greenland is a caricature of infrastructure-as-landscape-as-resource. There has been almost no consideration of Ivittuut/Grønnedal as a place – of the lived everyday experience of the people that resided there, the connection between these infrastructures and other spaces and places, or Greenlandic cultural and spatial heritage. Instead Greenland's physical landscape and the usefulness of its resources have been the only optic.

Finally, these tendencies towards the deprioritisation of place are also detectable within speculative proposals for the future of Greenland. In 2012 'Conditions' journal published a special edition on the theme of 'Possible Greenland', featuring a wide selection of essays, interviews, and proposals, many of which pertain to infrastructure.[11] The schemes in 'Possible Greenland' are largely thoughtful and ambitious, but many of them focus on international connections above all else. An example of this is BIG Architects and Tegnestuen Nuuk's proposal 'Air + Port' – a hybrid transport hub on an island close to Nuuk that would facilitate greater tourism and global trade through increased accessibility.[12] Addressing the expensive and complicated air and ocean transport infrastructures in Greenland, Air + Port is compelling, exciting, and both conceptually and graphically strong, replete with well-crafted diagrams, maps, and seductive images. The ambition to make Greenland more accessible is backed-up with infographics concerning opening-up shipping routes and activating global transport networks. There is, however, a high degree of distantiation. Catering for export, tourism, and industrial flows and processes is prioritised, and while this is a needed perspective, the attention to the everyday lives and places and cultures of Greenlanders is absent. The proposal engages with the specific physical landscape of the region, but there is a tendency to negate the sociocultural landscape and deprioritise the grounded, experiential dimensions of such an infrastructural proposal. This is evident in the choice of visuals, which concentrate upon 'aeroplane-eye' views and long shots across dramatic 'empty' landscapes. Clearly this project is a conceptual provocation, not a developed solution, but nowhere is Nuuk, the journey from the city to the airport or people, workers, and connections with local places evident in this conceptual diagram.[13]

Regardless of 'greater cause' or good intentions, and regardless of the scale of perception, it is vast, 'wild' landscapes and large territorial strategies that dominate in these varied narratives: 'Big Nature' pushes human settlements, and the everyday lives of people living there, out of the limelight. Place and landscape are not opposite terms, yet in the examples mentioned they are polarised as place is supressed by a highly cartographic, 'from-afar' interpretation of landscape.[14] It therefore seems imperative to assert that Greenland is as much about places as it is about landscapes: as much about the realm of everyday culture and local people, the 'close up' and the experiential, as it is about the larger, longer, grander landscape.

Transforming conceptualisations of landscape and place

Across multiple disciplines in the social sciences, natural sciences and the humanities, the concept of landscape has undergone critical re-evaluation in recent decades, particularly within cultural geography and urbanism. Since the middle of the twentieth century, the concept of landscape within cultural geography has progressed through a series of shifts, the 'tentpoles' of which are marked via seminal texts and theories. In a departure from the earlier thinking of Carl Sauer and Lewis, W.J.T Mitchell's dissection of 'Landscape and Power' argued for the consideration of landscape as an instrument of power and agent of force rather than a scenic, neutral backdrop.[15] Concurrently, the work of Denis Cosgrove, often in collaboration with Stephen Daniels, represents 'New Cultural Geography' which argues for comprehending representations of landscape as iconographic and ideological.[16] More recently this was challenged by poststructuralist 'non-representational' and then 'more-than-representational' theories, which have sought to reframe landscape in terms of embodied experience and in multiple dimensions.[17] These factions within cultural geography do, however, all share a rejection of the notion of landscape as scenery, and an acceptance that landscapes are human constructions and natural/cultural hybrids: these are points which are also reflected within landscape architecture and urbanism, particularly in the rise of Landscape Urbanism, which conceives of the city and nature as one hybrid landscape.[18] Together these theoretical and conceptual developments within the spatial disciplines of the humanities have progressed the conceptualisation of landscape from one of static, neutral backdrop to a transforming and transformative natural-cultural hybrid.[19]

Place, however, has not received the same progressive, far-reaching overhaul as landscape has over recent decades. Frequently (with exceptions as noted later) the concept of place is still framed as a static node, an enduring character, or the outcome of the process of landscape within landscape theory.[20] This mirrors the common usages of place; the prosaic notion of a bounded location. Christian Norberg-Schulz's fixed, singular 'genius loci' interpretation of place remains on many architectural and urbanism reading lists as the defining text on place, however his fixed notion of place has been questioned by some thinkers within the spatial disciplines.[21] A good example is architectural historian Mari Hvattum's argument that 'genius loci' is used to justify expressions of nationality/nationalism, and in response to this she instead proposes 'place as action', suggesting that seeing place as a space of action rather than static form might enable the 'local' to remain significant in a globalised era.[22] Hvattum's dynamic and fluid interpretation of place dovetails with the research of two scholars from within cultural geography that this section will focus on: Kenneth Olwig and Doreen Massey. Both Massey and Olwig – via their respective primary subjects of space and landscape – have written extensively on place, and read together, their work offers an alternative, progressive understanding of place, that convinces of the interrelationship between the large scale and the microscale, the international and the local. This shared attention to the significance of multiscalar thinking in the production of place provides a particularly appropriate and relevant framing for re-engaging with place and infrastructure in contemporary Greenland.

Kenneth Olwig has written persuasively on how landscapes are shaped by, and participate in the shaping of, politics, culture, and the connection between landscape and place. He notes that the rise of globalisation and transnational organisations has led to an increased interest in regions, that is, an understanding of place and 'placeness' that goes beyond nation-state borders.[23] Olwig discusses how landscapes on a regional scale are the result of sociocultural processes and that such regions are not a 'monolithic unity of environment and culture determined by nature, but rather *the place of a polity* constituted through human law and custom'.[24] This designation of region as place is further expanded in 'Choros, Chora and the Question of Landscape'. Drawing on the Greek roots of chorography, Olwig states that, conventionally, geography relates to the global scale, topography relates to a locale, and chorography sits in the middle as 'the study of a region as place'.[25] This identification of a region (rather than just a town or settlement) as a place is significant, and Olwig does this because he recognises that the idea of place is not merely reliant upon scale. Suggesting translating chora not directly as place but instead as 'place-region', Olwig argues that place is formed through a commonality of culture and politics, rather than a circumscribed settlement boundary.[26] Noting Jacques Derrida's writings, he argues that 'chora' referred to 'invested' and political places, not abstract space, and that Ancient Greek geographer Strabo saw chorography as concerned with the 'narration of conditions' over spatial geometry.[27] Therefore, a region can be a place, its edges defined by the reach of common culture and conditions. Olwig proceeds to discuss how one scale of place – the smaller topos – can influence, and build, a larger scale of place – a choros. Using the agora as an example, he describes how it is through the activities of the agora that a chora is formed, i.e. the agora – an ostensibly discrete and bounded place – influences and shapes the larger politicocultural place-region. Through his engagement with both the notion of connected scales of place and the role of customs and practices, Olwig presents an interpretation of place associated with dynamic relations rather than static characters or things. He blurs distinctions between place, landscape, and region, suggesting that it is not only landscape that acts upon place, but place that acts upon landscape.[28] This has particular bearing for Greenland and the Arctic as the Nordic notion of landscape specifically grows from cultural activities that regulate the land, rather than purely bio-geological processes.[29]

Doreen Massey's interrogation of the politics and practices of place and space provides a related interpretation of place allied with postmodern perspectives on (post)colonial theory, gender, race, and queer studies. In 1991 Massey published 'A Global Sense of Place' in 'Marxism Today', which was later republished in 'Space, Place and Gender'.[30] This short essay sets out Massey's case for an understanding of place that is 'extroverted', relational, and 'outwards looking' as a positive, progressive response to globalisation and the compression of time and space. While recognising that a sense of place and belonging are important to most of us, she unfolds the problematic reading of place as a closed, fixed, bounded entity in relation to nationalistic, romanticised, sanitised political perspectives. Massey challenges this dichotomous reading – where global/open is automatically progressive, and local/closed is regressive – by proposing that place is constructed as a constellation; an assemblage of relations that meet at a particular location. She suggests four criteria of this 'global sense of place'; places are not static but are always in progress; places are not bounded with insides and outsides, but always connected and linked to other places; places are constructed of multiplicities, that is, they are not formed by a single identity or heritage; and lastly, that each place's uniqueness comes from a specific mixture of larger relations. In 'For Space' Massey lengthily unfolds these concepts, eloquently suggesting that 'If space is rather a simultaneity of stories-so-far, then places are collections of those stories, articulations within the wider power-geometries of space' and 'Places (are) not as points or areas on maps, but (. . .) integrations of space and time'.[31] Her writing has heightened bearing for Greenland in its discussion of the intersectionality of place with colonialism and neoliberalism, and her understanding that even the most remote, geographically peripheral places are in conversation with other places and always in flux.

Massey and Olwig overlap convincingly on a revised, more progressive notion of place. Place, through their eyes, is no more static than landscape is: it, too, is about connections, dynamics, and the lives and practices of people at multiple scales. Place is not subservient to landscape but its equal, the two concepts working together to coproduce conditions in an iterative manner. And places are always networked with other places. I propose that such a reconceptualisation of place not only has great potential as an investigative framing within infrastructural discourses, and particularly so in Greenland, but that a study of infrastructures in Greenland can further progress conceptualisations of place, nuancing humanistic and social-sciences readings of what place means in a globalised context. The next and final section aims to demonstrate this argument through five existing Greenlandic infrastructures.

Place as a conceptual optic for Greenlandic infrastructures

Infrastructures, including energy, water, transport, and waste, are highly visible components in the urban realm of Greenland: due to permafrost and extremely hard rock substrates, they are most often above ground rather than buried, implemented without much concern for concealing their purpose, and in some cases located in prominent positions in a town or settlements rather than 'backstage'. This visual and spatial prominence renders these infrastructures part of daily life, criss-crossing with other civic programmes, spaces, and structures. Indeed, the fact that infrastructures in Greenland are part of the urban fabric in a particularly straightforward, direct manner only under-lines the role of infrastructures in the production of place. But however micro in scale or seemingly localised these infrastructures are, the challenges they address (or at times produce) connect Green-land to larger infrastructural discourses, bigger networks, and other places. Greenlandic infrastruc-tures are therefore, as Massey puts it, 'articulations', or 'collections of stories' of both localised and global narratives.

This section describes five examples of infrastructures in Greenland.[32] The infrastructures each pertain to differing technological systems and are commonly found within many areas of Greenland. Headlined with a photograph, the relationship between each infrastructure and place is described in two ways – firstly, how these infrastructures are related to practices, forms, atmospheres, and spaces within the town or settlement that they serve, and secondly how these infrastructural places relate to broader 'geometries' and global discourses on infrastructure. In this way the optic of place, inter-preted in a progressive manner, is used to unfold our understandings of infrastructure in Greenland, and vice versa, the empirical infrastructural studies sharpen our understandings of place.

Water – over-ground grids

Water grids and distribution mechanisms in Greenland are almost always run over-ground. Towns generally have developed water grids with running water indoors, all fed by surface laid pipes. A typical settlement in Greenland, however, does not have a formal water grid; instead a central water tank filled with treated water feeds a number of tap houses distributed strategically in the settlement. To collect water in settlements, residents either fill a container, or connect a hosepipe.[33] In most towns, metal-encased pipes zigzag over rocky terrain (Figure 11.1) so as to avoid drilling into and excavating hard rock. In settlements, rubber hosepipe often lies coiled outside of houses, brightly coloured, waiting to be connected to the bright blue tiny tap houses. Over-ground water grids in Greenland are therefore comprised of a mixture of fixed, insulated, heated pipes; rub-ber hosepipes; and human labour, and are highly visible and legible urban elements. Water infra-structures are visually and spatially present components of place and a constant reminder of the challenges Greenland faces to supply this essential infrastructure. And more than a spatiovisual distinction, it also invokes certain practices, cultural understandings, and a heightened awareness of the expense and energy involved in gaining clean drinking water. There is a certain irony here in a country whose landmass is approximately eighty per cent covered by ice-sheet, and inhabited purely on coastal fringes.

This irony extends further. While melting ice and the impact this has on global sea levels domi-nates headlines, the very same melting ice has multiple implications for Greenland, ranging from altering ecosystems which in turn impact socioeconomic conditions, to the appearance of entrepre-neurial proposals to package and sell Greenlandic glacier water. In 2015 a company in Greenland announced plans to export bottled water to China and Indonesia, harvested from icebergs in Disko Bay.[34] While ostensibly this could be an economic opportunity for Greenland, it elucidates the complex and often paradoxical nature of global 'water grids' – using hydrocarbon produced plastic bottles to export water released due to climatic change over huge distances, in order to serve over-seas luxury markets.[35] This stands in stark contrast to the difficulties and challenges Greenland faces in providing drinking water for its own citizens, made visible by the labour and energy intensive over-ground water grids.

Figure 11.1 An insulated, heated, metal-encased water pipe zigzagging across the facets of the rocky terrain in Nuuk
Photo: the author.

Marine transport – transforming harbours

Harbours are a very significant component of place in Greenland as well as a critical infrastructure. Due to terrain, geography, and climate there are no significant road-based transport infrastructures, and while air transport is common, it is marine transport that dominates. No matter how big or small, every settlement and town has a harbour, which acts as the main point of access and egress for food, fuel, consumer goods, and people. The harbours are also home to the major industry of Greenland – fishing – and the fishing company Royal Greenland has many factories throughout the country, from larger cities to isolated settlements. There are approximately 250 harbour facilities in Greenland, which range from large scale ports to small jetties and mooring points in settlements (Figure 11.2).[36]

While many generic harbour elements such as containers and cranes are present in Greenland, the harbours also have a particular atmosphere through a dense overlapping of programmes and forms. The Royal Greenland factories, always painted blue, act as a centre, billowing steam and activity. A vast number of small boats, owned and operated by local fishermen cluster around moorings or jetties, and the slaughter of fresh catch often takes place right beside the boats, bloodying the ice. Alongside fishing, several other activities are performed in harbours, from sailor accommodation, to grill bars, fuel stations, boat or canoe rentals, occasional tourist shops, and commuter boat services. The harbours are intensive places of action and the heart of a town or settlement.

Greenland's economy, and society, continues to be intrinsically dependent upon the fishing industry.[37] In settlements a large proportion of the community make a living through hunting and

Figure 11.2 Sisimiut's harbour in April. The Royal Greenland factory is visible on the far left of the picture. On the other side of the harbour is a mixture of grill bars, containers, fuelling points and outhouses.

Photo: the author.

fishing-related activities, supplemented by government subsidies. Income from fishing, however, is becoming less stable in part because of shifting fish stocks, and government subsidies cannot reach far enough. At the same time two other harbour programmes are growing, both provoked by climatic change. Tourism, including cruise-tourism, is increasing, in part motivated by 'last chance' and 'eco' tourists.[38] At the same time, climatic change is slowly opening up sea routes made accessible as summer sea ice melts in the Arctic. These materialising shipping routes hold great commercial interest, as they are keystones in the reduction of global transportation distances and the circumvention of less politically stable waters. In addition, in response to the centralisation policies of the twentieth-century and contemporary economic challenges, discussions continue regarding the relative social, cultural, and economic benefits and drawbacks of Greenland's dispersed settlement pattern, including the inter-settlement travel this entails. This complex combination of global and regional factors is resulting in transmuting harbours, which offer a snapshot of the multiscalar dynamics within a specific place.[39]

Land transport – canine infrastructure

Above the 'dogline', Greenlandic dogs are one of the most important forms of land transport infrastructure.[40] The dogs are not treated as pets but as working animals and they are an essential infrastructure as well as an important visual component and daily practice of place. One can smell and hear the dogs from the centre of a town or settlement, even though the dogs are usually located on the peripheries. This canine infrastructure is connected with multiple other infrastructures and systems – they feed on the scraps of the hunt, are housed in 'dog areas' complete with their own architectural and structural approach, and are also a tourist magnet. Every dog owner builds a dog shed or two for his or her animals; built from various offcuts and waste materials, reimagined and reappropriated (Figure 11.3), they have a particular informal aesthetic and atmosphere – no two are the same.

Due to climatic change, however, this land transport infrastructure is suffering. Warming has led to a shorter ice season, and consequently the hunting season has been reduced, limiting the livelihood of many.[41] And people are not the only victims: some hunters can no longer afford to keep their dogs fed all year around. The decline in the use of dogs is exacerbated due to the competition from snowmobiles, which can simply be parked during summer without requiring any upkeep and therefore offer a more flexible alternative to the care of dogs. Some hunters, in desperation, are therefore killing their dogs as they can no longer afford them.[42] These dog areas, and their self-built shelters, are consequently under threat due to climatic change. The fact that dog sheds are often constructed from used packaging material and other imported components is itself symbolic of this fragile and contested interdependence with the larger world despite the regional specificity of such canine infrastructures.

Figure 11.3 A dog shed in Sisimiut. The structure is made of scraps of waste materials lending an informal, 'scrapbook' atmosphere.

Photo: the author.

Energy – displacing hydrocarbons

Greenland's energy infrastructure is split in two: in the largest towns electricity is supplied via hydropower, equating to approximately seventy per cent of national electricity consumption. The remaining settlements and smaller towns remain, however, heavily dependent upon imported fossil fuels. This imported fuel is very expensive and so pilot schemes for alternatives such as wind, hydrogen, and solar energy have been carried out in recent years. In particular, there is a significant number of microscale solar energy schemes in various locations throughout Greenland – mostly private homes but also institutions.

These renewable alternatives to imported hydrocarbons produce particular spatiocultural conditions in the settlements and smaller towns. Solar panels – both photovoltaics and solar thermal collectors – are often casually retrofitted to existing structures. Due to the high latitude, the panels are generally affixed to vertical surfaces such as walls and balconies rather than roofs, so as to better capture the lower sun angle. Consequently, they appear almost like hung pictures: modular, and attached wherever there is blank wall (Figure 11.4). These forms of off-grid energy also give rise to particular ways of living, where energy usage is constantly monitored and limited and activities carefully choreographed. There are very few organised, utility scale arrays and so this ad hoc appearance communicates the self-sufficiency of Greenland's solar infrastructure.

Figure 11.4 A private house in Nuuk with a casually mounted solar panel on its façade. As vertical mounting is needed due to the low sun angle, the facades of homes and buildings are altered through these additions.

Photo: the author.

These micro-scale experiments and self-sufficiencies are in clear response to larger energy infrastructure discourses surrounding Greenland. Greenland is estimated to host considerable fossil fuel resources and many, although not all, Greenlanders are positive towards both hydrocarbon extraction and other mining activities as this could provide a means of socioeconomic development and, arguably, ultimately political independence from Denmark.[43] Oil exploration in Greenland dates back to the late 1970s, but five test drillings during the 1970s and 1990s were unsuccessful[44] and there has been slow progress since. In 2010 a Scottish company, Cairn Energy, successfully discovered hydrocarbons in Greenland and following this breakthrough the Greenlandic government awarded its first round of offshore exploration licenses in 2011. Following the initial positive results, however, more recent attempts have not been as successful with no commercially viable discoveries to date. The falling price of oil is now making oil extraction even less likely due to relative expense, and Norway's Statoil, France's GDF Suez, and Denmark's Dong Energy, returned their exploration licences last year.[45] This ultimately dampens Greenland's hopes for independence, which hoped to displace Danish subsidies through the revenue from successful oil and gas exploration. As such, these alternative infrastructures are becoming more important, particularly regards the survival of small settlements, and their visible proliferation is telling of wider energy landscapes.

Waste – dump/recycle

Greenland's waste infrastructure diverges into two approaches. On one hand, it suffers from an 'end-of-pipe' approach to its sewage. Grey and black water is commonly simply discharged untreated into the sea or onto unoccupied terrain, relying heavily upon the deep water and high tidal rate to transport it away from the coastline.[46] Similarly, incinerators burn what cannot be sent to landfill, and dumps are brimming with packaging materials: a symptom of the extremely high level of importation. There is, however, another side to Greenland's waste infrastructure – a pragmatic approach to reusing materials and components, often as found in the rubbish dumps. The dump is therefore also used as a resource (see Figure 11.5). Plywood, rebar, ICPs, pallets, and so on are all repurposed in the building of sheds and outbuildings, and to furnish repairs to buildings, decks, and vehicles. The dumps therefore are also a part of town, despite being on its periphery, and the clever reuse of materials and components contributes to a particular aesthetic atmosphere in the towns and settlements.

This divided waste infrastructure – the wasteful and the resourceful – relates to a number of larger infrastructural discourses. Greenland's dependence upon importation is instigated by geophysical limitations but is also a vestige of Greenland's colonial history. Combined, this necessitates a high level of consumption through transnational and domestic transportation, and is an amplification of a key symptom of globalisation seen across the world: as the gap between production and consumption grows, and flows of goods, information, and materials become increasingly entangled and opaque.[47] Conversely, the resourceful approach to waste through infrastructures that repurpose packaging and waste mirrors the rise of a global movement of 'hackathons' and upcycling, prompted by an awareness of resource scarcity and the social and environmental problems evidenced in disposable fashion, built-in obsolescence of consumer goods, and seemingly ever quickening cycles of consumption.[48] Although in Greenland this is of course not practiced as a part of an anti-consumption lifestyle, but rather a long-established means of survival in a difficult economic and social climate. Similar approaches can be witnessed in many global regions where financial scarcity, and distance from the 'core' (whether politically, geographically, or economically), promotes the diversion of conventional resources, from remote islands to marginalised subcultures within large cities.

Figure 11.5 Sisimiut's rubbish dump. The incinerator is to the right and, in front of it, different types of rubbish are categorised to enable easy sorting and, in many cases, repurposing.

Photo: the author.

Reflections

Contemplating the intersection between a Massey-ian/Olwig-ian approach to conceptualising place and the empirical case study of select infrastructures in Greenland, two significant points emerge: firstly, that Massey and Olwig's shared belief in an outwards-looking, progressive conceptualisation of place provides an apt, and much-needed, multiscalar lens for the study of Greenland's infrastructures. Secondly, that this progressive conceptualisation of place is not only supported by empirical study, but in fact is enriched by it: the multiscalar sociocultural, political, economic, and environmental associations of infrastructure, and its co-production of places locally and globally, are made explicit and magnified by Greenland's informal, networked, microscale, and, crucially, highly visible infrastructures.

To expand upon the first point, Kenneth Olwig and Doreen Massey's conceptualisations of place provide a more progressive perspective that frees place from parochialism, embeds it within larger discourses and networks, and links it with the critical re-evaluation of landscape. Using this understanding of place as an optic for studying infrastructures in Greenland uncovers a density of associations and experiences contained and condensed within a particular set of relationships in space and time – and these are both 'local' effects and atmospheres and larger, 'global' systems and resonances. As such the optic of place contributes to the erasure of persistent and unhelpful binaries within infrastructural discourses – between local/micro and global/macro – instead enabling these concepts to index one another by condensing vast, global, far-reaching narratives into visible, sensible, tangible places at the human scale. This has particular relevance for understanding, and addressing, the consequences and complexities of climatic change. Greenland faces significant challenges, overcoming decades as cryospheric, hydrological, and biological parameters shift. While these physical changes are of interest globally, the direct and indirect impacts they have on the social, cultural, and economic life of Greenland itself is also critical to understand and manage, for Greenlanders, the broader Arctic region and other peripheralised – whether geographically, socioeconomically, or both – regions globally. This is not straightforward, as for example the dual character of existing waste infrastructures demonstrates, whereby Greenland is at once the involuntary importer of global effects and outdated practices, while it simultaneously draws upon highly localized traditions and pragmatic spatial logics. It is this complexity that needs to be grasped and addressed, and place provides a meaningful framing for doing so. This is not to say that the concept of landscape is not useful for studying infrastructures in Greenland – it is – nor that the concept of place does not have the potential to be tipped back into an introverted, static, conservative perspective. But place, when approached in the way described in this chapter, is an equally valid lens for understanding that reveals infrastructure as a network that is both place and landscape, object and system, aesthetic and the ethic, technology, and culture. To study the places of infrastructures is to study microcosms of infrastructural complexity.

The empirical study of infrastructures in Greenland also renders aspects of this progressive conceptualisation of place particularly acute and clear: the operational logics of place – the role of infrastructures in the production of place – are highly visible in Greenland. Electricity conduits are visible urban elements rather than hidden subterranean magic. Water grids are visible as snaking lines over hard rock, relatively unobscured due to low-density development, and, in the case of settlements, water grids further demand the active participation of inhabitants. This leads to a highly visible, straight-forward picture of the ongoing, connected production of infrastructure and place, as well as creating a particular aesthetic and atmosphere where there is no clear separation between 'behind the scene' and 'on stage'. The contradictions of operational logics of place are also highly visible; whether through the split-personality approach to rubbish – a culture of reuse, yet a habit of untreated dumping of sewage – or the drive to capitalise upon the effects of climatic change on water and ice. The global-local dynamics of infrastructures and place, as expounded by Massey and Olwig, are magnified due to the very fact that Greenland is a 'remote' island in the North Atlantic, peripheralised from the 'core', yet intrinsically connected to larger global systems and structures, both mechanically through freighters and more subtly through resource chains and geopolitical geometries.

Appreciating the informal, networked, microscale nature of infrastructures in Greenland and their central role in sustaining place is particularly important in regards to the survival of small settlements and dispersed patterns of settlement that are seen in many vast and marginal territories in the Arctic and beyond. Such 'alternative' forms of infrastructure, which are so critically embedded in manifold dimensions of life and society in such regions, can serve to further progress our understanding of Olwig and Massey's reading of place, moving towards a conceptualisation that is at once local and global, multiple, and particular.

Notes

1 Michel Foucault and Jay Miskowiec, "Of Other Spaces," *Diacritics* 16, no. 1 (1986), doi:10.2307/464648; Yi-Fu Tuan, *Space and Place: The Perspective of Experience*, 5th or later Edition (Minneapolis: University of Minnesota Press, 2001); Michel de Certeau, *The Practice of Everyday Life*, trans. by Steven Rendall (Berkeley; London: University of California Press, 1984).

2 Doreen Massey, *For Space*, 1st edition (London; Thousand Oaks, CA: SAGE, 2005); Kenneth Olwig, "Choros, Chora and the Question of Landscape," in *Envisioning Landscapes, Making Worlds: Geography and the Humanities*, ed. Stephen Daniels et al. (London; New York: Routledge, 2011); Mari Hvattum, "Stedets Tyranni," *Arkitekten* 2

(2010). Both Massey and Olwig are further explored in this chapter. Landscape has been explored as a central framing by many in the spatial disciplines. Stephen Daniels et al., eds., *Envisioning Landscapes, Making Worlds: Geography and the Humanities* (London; New York: Routledge, 2011); Denis E. Cosgrove, *Social Formation and Symbolic Landscape* (Madison: University of Wisconsin Press, 1998); Denis Cosgrove and Stephen Daniels, eds., *The Iconography of Landscape: Essays on the Symbolic Representation, Design and Use of Past Environments*, New edition (Cambridge; New York: Cambridge University Press, 1989); Pierre Bélanger, "Landscape Infrastructure: Urbanism beyond Engineering" (PhD diss. Wageningen University, 2013); Rachel Z. DeLue and James Elkins, eds., *Landscape Theory* (London; New York: Routledge, 2008); Hayden Lorimer, "Cultural Geography: The Busyness of Being 'more-than-Representational',"* Progress in Human Geography* 29, no. 1 (2005); Nigel Thrift, *Non-Representational Theory: Space, Politics, Affect*, New edition (London; New York: Routledge, 2007); W. J. T. Mitchell, ed., *Landscape and Power*, 2nd edition (Chicago: University of Chicago Press, 2002).

3 Susan J. Carruth, "Infrastructural Urbanism That Learns from Place: Operationalising Meta Material Practices to Guide Renewable Energy Planning in Greenland" (PhD diss. Aarhus School of Architecture, 2015).

4 See notes 3 and 20 for a brief listing of important research in this field.

5 *Arctic Climate Impact Assessment* (Cambridge: Cambridge University Press, 2005), www.cambridge. org/us/academic/subjects/earth-and-environmental-science/climatology-and-climate-change/arctic-climate-impact-assessment-scientific-report.

6 Anthony Speca, "Greenshit Go Home! Greenpeace, Greenland and Green Colonialism in the Arctic," paper presented at University of Chichester, 27 March 2014. www.arcticpoliticaleconomy.com/wp-content/uploads/2014/09/20140327-Greenshit-go-home-Speca.pdf.

7 Ibid.

8 Charles Emmerson, *The Future History of the Arctic: How Climate, Resources and Geopolitics Are Reshaping the North and Why It Matters to the World* (London: Vintage, 2011).

9 What I refer to here as Grønnedal has in fact a more complicated nomenclature. The Americans referred to the settlement as 'Green Valley', however once the base changed hands from the USA to Denmark it became known as 'Grønnedal'. It also known by its Greenlandic name – 'Kangilinnguit'. This change in name reflects the shifting populations, politics, and interests of the settlement. What I refer to here as Ivittuut has also undergone a shift in name, as it was formerly known as Ivigtût.

10 In August 2011, the Danish Defence decided to shut the Grønnedal base and establish the Arctic Command in Nuuk in order to streamline its structure.

11 This edition also acted as the catalogue for the Danish Pavilion at the Venice Biennale

12 BIG and Tegnestuen Nuuk, "Air + Port," *Conditions* Possible Greenland, 11&12 (2012).

13 Carruth, *Infrastructural Urbanism That Learns from Place*.

14 Leena Cho and Matthew Jull reflect upon the varied interpretations of landscape in the Arctic through their categorisation of conceptualisations of the Arctic into the three strands of 'Climatic Apocalypse – Deterministic Landscape', 'Treasure Trove – Commodified Landscape', and 'Territorial Conquest – Vied Landscape'. This perceptively analysis reflects not only contemporary but historical global attitudes towards the Arctic. Leena Cho and Jull, Matthew, "Urbanized Arctic Landscapes: Critiques and Potentials from a Design Perspective," paper presented at Arctic Urban Sustainability Conference, Washington, DC, 2013, www.gwu.edu/~ieresgwu/programs/conference.cfm.

15 Mitchell, *Landscape and Power*.

16 Cosgrove and Daniels, *The Iconography of Landscape*; Daniels et al., *Envisioning Landscapes, Making Worlds*; Denis Cosgrove, ed., *Mappings* (London, England: Reaktion Books, 1999); Cosgrove, *Social Formation and Symbolic Landscape*.

17 Thrift, *Non-Representational Theory*; Lorimer, 'Cultural Geography'.

18 See Bélanger, "Landscape Infrastructure: Urbanism Beyond Engineering"; Elizabeth Mossop, "Landscapes of Infrastructure," in *The Landscape Urbanism Reader*, ed. Charles Waldheim (New York: Princeton Architectural Press, 2006).

19 Susan J. Carruth, "Developing Renewable Energy in Discontiguous Greenland: An Infrastructural Urbanism of 'material Practices'," *Journal of Landscape Architecture* 11, no. 1 (January 2, 2016) doi:10.1080/18626033.2016.1144686.

20 The artist Tacita Dean writes that 'place is to landscape as "identity" is to portraiture'. Tacita Dean and Jeremy Millar, *Place* (New York: Thames & Hudson, 2005), 12. suggesting that place is an emergent property, or perhaps objective, of landscape. Writing in W.T. Mitchell's seminal 'Landscape and Power', Robert Pogue Harrison states that 'place is defined by its boundaries, its intrinsic limits, its distinctly local "here" that remains fixed in space even as it perdures in time'. Robert Pogue Harrison, "Hic Jacet," in *Landscape and Power*, edited by W. J. T. Mitchell (Chicago: University of Chicago Press, 2002). Paul Groth and Todd Bressi suggest that 'landscape denotes the interaction of people and place' thereby implying that place is a material element somehow contained by

landscape; a part of the whole. Paul Groth and Todd Bressi in "Understanding Ordinary Landscapes," in *Understanding Ordinary Landscapes*, edited by Paul E. Groth and Todd W. Bressi (New Haven: Yale University Press, 1997), 1. And in 'Landscape Theory', Jessica Dubow explicitly asserts 'landscape as process rather than as place'. Jessica Dubow, "The Art Seminar: Conversation from June 17th 2006 at the Burren College of Art, Ballyvaughan, Ireland," in *Landscape Theory*, edited by Rachel Z. DeLue and James Elkins (London; New York: Routledge, 2008).

21 Christian Norberg-Schulz, *Genius Loci: Towards a Phenomenology of Architecture* (New York: Rizzoli International Publications, 1980).

22 Hvattum, "Stedets Tyranni", 24.

23 Michael Jones and Kenneth Olwig, eds., *Nordic Landscapes: Region and Belonging on the Northern Edge of Europe* (Chicago: University of Minnesota Press, 2008).

24 Ibid. xiii, my italics.

25 Olwig, "Choros, Chora and the Question of Landscape."

26 Ibid., 50.

27 Ibid., 49.

28 Carruth, "Infrastructural Urbanism That Learns from Place."

29 Maria Hellström Reimer, "Scapelands of the North – Roots, Rights, Routes." Paper presented at Latitud Norte: Ética y estética del habitar, Valencia, 2011.

30 Doreen Massey, *Space, Place and Gender* (Cambridge: Polity Press, 1994).

31 Massey, *For Space*, 130.

32 Drawing on textual analysis of government documents, journalistic and academic articles, as well as fieldwork carried out in Nuuk, Sisimiut, and Kapisillit in 2013 and 2014.

33 Carruth, "Developing Renewable Energy in Discontiguous Greenland."

34 Hector Martin, "Ice, Ice, Everywhere," *The Arctic Journal*, 24 January 2015, http://arcticjournal.com/business/1277/ice-ice-everywhere.

35 Mia Bennett, "Greenland: Barrels of Oil and Bottles of Water", *Cryopolitics*, January 29, 2015, https://cryopolitics.com/2015/01/29/greenland-barrels-of-oil-and-bottles-of-water/.

36 Statistics Greenland, *Greenland in Figures 2013* (Nuuk: Statistics Greenland, 2013), www.stat.gl/publ/kl/GF/2013/pdf/Greenland%20in%20Figures%202013.pdf.

37 Statistics Greenland, *Greenland in Figures 2015* (Nuuk: Statistics Greenland, 2015).

38 Sarah Woodhall, *Spring 2013 Visitor Survey Report: Greenland: A Full Report on International Tourists in Greenland* (Nuuk, Greenland: Visit Greenland, 2013), http://corporate.greenland.com/media/4653/visitor-survey-report_uk.pdf.

39 The pros and cons of continuing to support the decentralised structure versus returning to centralisation policies is complex and much debated. Kåre Hendriksen, "Grønlands Bygders Udviklingsdynamik: Udvikling Eller Afvikling? (2)," *Groenland (charlottenlund)*, 1 (2014).

40 It is estimated that there are more than 30,000 sledge dogs in Greenland, living in packs and transporting hunters into the open land. See David C. King, *Greenland (Cultures of the World)* (New York: Cavendish Square Publishing, 2009).

41 The shortening of the hunting season is also exacerbated by a number of other factors, including the dilution of the salt content of sea ice, which alters the melting point and behaviour of frozen landscapes.

42 Tim Folger, "How Melting Ice Changes One Country's Way of Life," *National Geographic*, October 2015, http://ngm.nationalgeographic.com/2015/11/climate-change/greenland-melting-away-text/.

43 Although just how feasible such a route to independence is contested. See Minik Rosing et al., *For the Benefit of Greenland: The Committee for Greenlandic Mineral Resources to the Benefit of Society* (Nuussuaq and Copenhagen: Ilisimatusarfik, University of Greenland and the University of Copenhagen, 2014).

44 Timo Koivurova and Kamrul Hossain, "Offshore Hydrocarbon: Current Policy Context in the Marine Arctic" (Arctic Transform and the Arctic Centre, 2008), http://arctic-transform.org/download/OffHydBP.pdf.

45 Stine Jacobsen, "Statoil Hands Back Three Greenland Exploration Licenses", *Thomson Reuters*, January 14, 2015, http://af.reuters.com/article/energyOilNews/idAFL6N0UT1Q620150114.

46 Ragnhildur Gunnarsdóttir, "Wastewater Treatment in Greenland" (PhD diss., Denmark's Technical University, 2012).

47 Carruth, "Developing Renewable Energy in Discontiguous Greenland."

48 Ibid.

Bibliography

Arctic Climate Impact Assessment. *Arctic Climate Impact Assessment*. Cambridge: Cambridge University Press, 2005. www.cambridge.org/us/academic/subjects/earth-and-environmental-science/climatology-and-climate-change/arctic-climate-impact-assessment-scientific-report.

Bélanger, Pierre. "Landscape Infrastructure: Urbanism Beyond Engineering." PhD diss., Wageningen University, 2013.

Bennett, Mia. "Greenland: Barrels of Oil and Bottles of Water." *Cryopolitics*, 29 January, 2015. https://cryopolitics.com/2015/01/29/greenland-barrels-of-oil-and-bottles-of-water/.

BIG, and Tegnestuen Nuuk. "Air + Port." *Conditions (Possible Greenland)*, 11 & 12 (2012).

Carruth, Susan J. "Developing Renewable Energy in Discontiguous Greenland: An Infrastructural Urbanism of 'material Practices'." *Journal of Landscape Architecture* 11, no. 1 (2016): 66–79. doi:10.1080/18626033.2016.1144686.

Carruth, Susan J. "Infrastructural Urbanism That Learns from Place: Operationalising Meta Material Practices to Guide Renewable Energy Planning in Greenland." PhD diss., Aarhus School of Architecture, 2015.

Cho, Leena, and Matthew Jull. "Urbanized Arctic Landscapes: Critiques and Potentials from a Design Perspective." Paper presented at Arctic Urban Sustainability Conference, Washington, DC, 2013. www.gwu.edu/~ieresgwu/programs/conference.cfm.

Cosgrove, Denis, ed. *Mappings*. London: Reaktion Books, 1999.

Cosgrove, Denis, and Stephen Daniels, eds. *The Iconography of Landscape: Essays on the Symbolic Representation, Design and Use of Past Environments*. Cambridge: Cambridge University Press, 1989.

Cosgrove, Denis E. *Social Formation and Symbolic Landscape*. Madison: University of Wisconsin Press, 1998.

Daniels, Stephen, Dydia DeLyser, J. Nicholas Entrikin, and Doug Richardson, eds. *Envisioning Landscapes, Making Worlds: Geography and the Humanities*. London: Routledge, 2011.

Dean, Tacita, and Jeremy Millar. *Place*. New York: Thames & Hudson, 2005.

de Certeau, Michel. *The Practice of Everyday Life*. Translated by Steven Rendall. Berkeley; London: University of California Press, 1984.

DeLue, Rachel Z. and James Elkins, eds. *Landscape Theory*. The Art Seminar 6. London: Routledge, 2008.

Dubow, Jessica. "The Art Seminar: Conversation from June 17th 2006 at the Burren College of Art, Ballyvaughan, Ireland." In *Landscape Theory*, edited by Rachel Z. DeLue and James Elkins, 87–156. London: Routledge, 2008.

Emmerson, Charles. *The Future History of the Arctic: How Climate, Resources and Geopolitics Are Reshaping the North and Why It Matters to the World*. London: Vintage, 2011.

Folger, Tim. "How Melting Ice Changes One Country's Way of Life." *National Geographic*, October (2015). http://ngm.nationalgeographic.com/2015/11/climate-change/greenland-melting-away-text/.

Foucault, Michel and Jay Miskowiec. "Of Other Spaces." *Diacritics* 16, no. 1 (1986): 22–27. doi:10.2307/464648.

Groth, Paul E. and Todd W. Bressi, eds. *Understanding Ordinary Landscapes*. New Haven: Yale University Press, 1997.

Gunnarsdóttir, Ragnhildur. "Wastewater Treatment in Greenland." PhD diss., Denmark's Technical University, 2012.

Harrison, Robert Pogue. "Hic Jacet." In *Landscape and Power*, edited by W. J. T. Mitchell, 349–64. Chicago: University of Chicago Press, 2002.

Hellström Reimer, Maria. "Scapelands of the North – Roots, Rights, Routes." Paper presented at Latitud Norte: Ética y estética del habitar, Valencia, 17–18, 2011.

Hendriksen, Kåre. "Grønlands Bygders Udviklingsdynamik: Udvikling Eller Afvikling?" *Groenland (charlottenlund)*, 1 (2014): 66–83.

Hvattum, Mari. "Stedets Tyranni." *Arkitekten*, 2 (2010): 33–44.

Jacobsen, Stine. "Statoil Hands Back Three Greenland Exploration Licenses." *Thomson Reuters*. 14 January 2015. http://af.reuters.com/article/energyOilNews/idAFL6N0UT1Q620150114.

Jones, Michael, and Kenneth Olwig, eds. *Nordic Landscapes: Region and Belonging on the Northern Edge of Europe*. Minneapolis: University of Minnesota Press, 2008.

King, David C. *Greenland (Cultures of the World)*. New York: Cavendish Square Publishing, 2009.

Koivurova, Timo, and Kamrul Hossain. "Offshore Hydrocarbon: Current Policy Context in the Marine Arctic." Arctic Transform and the Arctic Centre, 2008. http://arctic-transform.org/download/OffHydBP.pdf.

Lorimer, Hayden. "Cultural Geography: The Busyness of Being 'more-than-Representational'." *Progress in Human Geography* 29, no. 1 (2005): 83–94.

Martin, Hector. "Ice, Ice, Everywhere." *The Arctic Journal*, 24 January 2015. http://arcticjournal.com/business/1277/ice-ice-everywhere.

Massey, Doreen. *For Space*. London; Thousand Oaks: SAGE Publications Ltd, 2005.

Massey, Doreen. *Space, Place and Gender*. Cambridge: Polity Press, 1994.

Mitchell, W. J. T., ed. *Landscape and Power*. Chicago: University of Chicago Press, 2002.

Mossop, Elizabeth. "Landscapes of Infrastructure." In *The Landscape Urbanism Reader*, edited by Charles Waldheim, 163–78. New York: Princeton Architectural Press, 2006.

Norberg-Schulz, Christian. *Genius Loci: Towards a Phenomenology of Architecture*. New York: Rizzoli International Publications, 1980.

Olwig, Kenneth. "Choros, Chora and the Question of Landscape." In *Envisioning Landscapes, Making Worlds: Geography and the Humanities*, edited by Stephen Daniels, Dydia DeLyser, J. Nicholas Entrikin, and Doug Richardson, 44–54. London: Routledge, 2011.

Rosing, Minik, Rebekka Knudsen, Jens Heinrich, and Lars Rasmussen. "For the Benefit of Greenland: The Committee for Greenlandic Mineral Resources to the Benefit of Society." Nuussuaq and Copenhagen: Ilisimatusarfik, University of Greenland and the University of Copenhagen, 2014.

Speca, Anthony. "Greenshit Go Home! Greenpeace, Greenland and Green Colonialism in the Arctic." Paper presented at University of Chichester, 27 March 2014. www.arcticpoliticaleconomy.com/wp-content/uploads/2014/09/20140327-Greenshit-go-home-Speca.pdf.

Statistics Greenland. *Greenland in Figures 2013*. Nuuk: Statistics Greenland, 2013. www.stat.gl/publ/en/GF/2013/pdf/Greenland in Figures 2013.pdf.

Statistics Greenland. *Greenland in Figures 2015*. Nuuk: Statistics Greenland, March 2015. www.stat.gl/publ/en/GF/2015/pdf/Greenland in Figures 2015.pdf.

Thrift, Nigel. *Non-Representational Theory: Space, Politics, Affect*. London: Routledge, 2007.

Tuan, Yi-Fu. *Space and Place: The Perspective of Experience*. Minneapolis: University of Minnesota Press, 2001.

Waldheim, Charles. ed. *The Landscape Urbanism Reader*. New York: Princeton Architectural Press, 2006.

Woodhall, Sarah. "Spring 2013 Visitor Survey Report: Greenland: A Full Report on International Tourists in Greenland." Nuuk: Visit Greenland, 2013. http://corporate.greenland.com/media/4653/visitor-survey-report_uk.pdf.

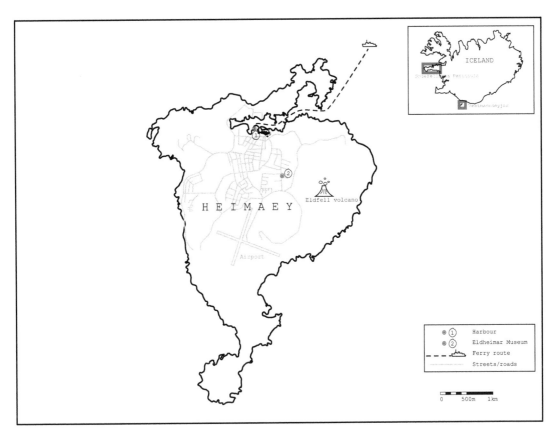

Map 5 The geologically active landscapes of Iceland feature untraditional landscape typologies displaying vast energy infrastructure, flexible infrastructure as well as the geologic landscape in the making. While a contemporary tourism industry is based on landscape appreciation. Circling the island and visiting the smaller Heimaey, the geologic forces, still so vividly at work on Iceland, presents itself to us both as an energy landscape and as a live pre-Anthropocene landscape generation yet untouched by human agency (map: Eimear Tynan).

12 Inhabiting change

Elizabeth Ellsworth and Jamie Kruse

Between 2012 and 2014, while we were researchers for the Future North project we arrived at a new way to formulate thoughts and actions that arose for us. To engage with ideas at the heart of Future North, we carried out a project titled Inhabiting Change at several of the research sites.[1] As we composed a series of linked field dispatches in performative response to what we experienced, the phrase "turning at the limits" grabbed our imaginations and we started to use it to capture some of the embodied, affective charge that the work was taking on for us.

In retrospect, we realize that the text-based artwork that emerged from our lived experience of the field research, entitled Turning at the Limits of the World, constitutes our primary research results. It can be read as "notes to self" about what it takes to turn into the Anthropocene – that is, to alter individual and shared senses of self, world, others, time, place, geologic materiality in ways that acknowledge and act in recognition of Anthropocene realities. Simultaneously, "turning at the limits" can be read as a list of "instructions" for others who might want to adapt and perform the turn it describes – namely, the refusal to continue on in habitual ways while disregarding or resisting material limits of life itself, and turning, instead, toward enactments of the difference that the limits propose. As an artwork, Turning at the Limits of the World also functions as something of a manifesto for living the Anthropocene.

We offer this particular research result, along with the related methodology and dispatches from Inhabiting Change, as more than representations, reports, or interpretations. We offer them as practices for performing embodied acceptance of limits. Ultimately, as a project of performative research, Inhabiting Change constitutes a gesture in response to the fact that the current material conditions of the Anthropocene are the outcomes of the breaching of ecological limits that constitute the human habitat on Planet Earth.

The following five dispatches are the results of two journeys to Iceland and one in the Norwegian North. They consist of images and accounts of how unprecedented intensities, scales, and speeds of contemporary change inflected our embodied, daily lives, imaginations, and interpretive acts as we traveled through and lived with various landscapes and social contexts in the far North. For the performative inhabitations, we sought out sites and times where forces of change are unfolding with particular intensity, palpability and exquisiteness. We activated built structures, landscapes, and ephemeral events as "field stations" or "apertures" for observing, sensing, documenting, and creatively responding to volatile forces of change that we found to be in play on a daily basis around us. We made observations about, and creative responses to, global pressures of change as they reshaped daily live experiences and meanings – our own, and the scholars, community members, students, and artists with whom we spoke.

We sent our field reports from Snaefellsnes Peninsula, Iceland; Heimaey, Iceland; Tromsø's Small Projects, a collaborative exhibition and artists' residency; Vardø's Steilneset Memorial; and Kirkenes, Norway. We offer them as dynamic tracings of the arrival of new futures of the North into widespread public cognizance. They are experiments in composing a hybrid voice of artist-researcher, one that we further hybridized by "collaborating" with nonhuman forces and materials. We did this by attempting to write across multiple, moving points of view as we worked and lived in the midst of the big, fast changes that currently are composing emerging futures north. We also attempted to invent this collaborative, hybrid voice by following Jane Bennett's challenge to "devise new procedures, techniques, regimes of perception that enable us to consult the nonhuman more closely, to respond and listen more carefully to their outbreaks, objections, tests, and propositions."[2] Our field practices of "inhabiting change" and then performatively responding to the resulting experiences are attempts to devise such new procedures and regimes of perception. We wanted to pick up and disclose "outbreaks," "objections," and "propositions" of nonhuman things that are now setting up futures north.

The images and dispatches put the diverse sites of our field research for Future North into relation to one another. Our methods of inhabiting change in collaboration with nonhuman things allowed us to move-with some of the local flows of material change reconfiguring life, landscape, and the Earth itself, and make something of the generative potentials they offer to designers, artists, and citizens.

We used the immediacy of new digital media to slow down, pay deep attention, move-with, and make-from-within edges and forces of change itself. We wagered that the lively alter-world – that is, that the strange and unfamiliar habitats both of and not of our own making which are now emerging in the Far North – would catch us and gesture back. Much of what we needed for this project was learned and invented along the way.

The performative approach that we took to our field research created an aesthetic perspective onto themes and concerns of the overall Future North project, specifically: change itself, futures that are in the midst of arriving, and how to frame, for broader audiences, field research data on changing landscapes.

The lived realities of our inhabitations in the field are evidence that the present moment in the history and future of northern landscapes is composed of the ramifying accumulation of moments, spaces, and times when humans acted in ways that ignored the limits of our habitat, and in ways that were out of tune our species' material, technological, cultural, and psychological finitudes.

Dispatches from inhabiting change

Eventscapes: Iceland

05.17.2012, 09:49

> "We know when we build a bridge it will not last."
> Pétur Matthíasson, Public Relations Manager, Icelandic Road Administration[3]

We just returned from Iceland, where we took part in the Landscape Journeys project sponsored by the Oslo School of Architecture and Design's Institute of Urbanism and Landscape and the Norwegian Research Council.[4] The project's mandate is to take up travel as methodological form and acknowledge the importance of developing research projects in response to landscapes – while actually moving through them. Fellow journeyers included Janike Kampevold Larsen, Mason White, Luis Callejas, Peter Hemmersam, Alessandra Ponte, and Giambattista Zaccariotto.[5] Our international contingent had an array of interests including glaciers, geothermal energy, landscapes, and the literary imagination, and "extreme" landscapes. smudge studio's particular focus was what we have come to call "streaming" landscapes of Iceland, and how we might address the Icelandic landscape as a concatenation of events occurring along various speeds, intensities, and temporal trajectories.[6]

Before leaving Reykjavík to conduct research on the road, our group met with Pétur Matthíasson, a representative from the Icelandic Road Administration.[7] He shared information about effects of glacial floods on the highway system in southern Iceland, with a focus on Iceland's Highway 1, also known as the Ring Road.[8] This 1,339-kilometer stretch of two lane roads and one-lane bridges opened in 1974. It is the lifeline for inhabitants outside of the capital. During our interview with Mr. Matthíasson following his presentation, he shared a map with us that illustrated particular stretches of roadway in Iceland at risk of being washed away when volcanic eruptions occur and set the jökulhlaups, or glacial outburst floods, into motion. At one point in our conversation he stated that "there's nothing you can do" in the face of some eruptions, and that "we wouldn't design [the bridges] without considering all that can happen in nature." For us, his words were a powerful demonstration of design limits that are accepted routinely and worked with creatively in Iceland.

We learned that glacial outburst floods can arrive within hours, days, or even months after an eruption. After a series of strong earthquakes in 2000,[9] bridge designs were upgraded to meet earthquakes force of 8.0 on the Richter Scale. This included deliberately designing "weak points" into Icelandic roads that allowed them to be washed away in small sections. Such intentional breakage relieves the pressure on the remaining roadway and bridges, often saving the bridges from being

Figure 12.1 Icelandic outwash
Photo: smudge studio 2012.

torn completely away by the massive floods. Our guest speaker told us that the Road Administration keeps a cache of 100–300 meters of "stock" bridge material on hand at all times, just in case.

After a day and a half in Reykjavík, we began our journey

In many ways, it's still the Pleistocene (1.8 million to 11,700 thousand years ago). And in many other ways, it's still the Cambrian (541 to 485 million years ago). At 66° north, Iceland is on a northern latitude where one would expect its landmass to be covered with ice. But due in part to the Gulf Stream that flows from the south and warms its environment, this island is in a state of constant thaw.[10] One-third of Iceland's 104,600 square kilometers is volcanically active. Massive glaciers, including the largest ice mass in Europe, loom in the mountains, and daily life along the southern edge of the country is shaped directly by glacial materials washing, flooding, eroding, and falling from the sides of volcanoes.

We had been prepared by our guest speaker to encounter Icelandic roadways that had been designed to be "responsive" to their surrounding forces. Nevertheless, the turbulence of the landscapes we encountered surprised us. We passed infrastructural and geological detritus from a jumble of previous glacial/volcanic events that were playing out still. At farm after farm, we saw workers driving large front-loaders, dredging outwash that sometimes stretched to the horizon. Volcanoes and glaciers sharply inflect life here on a continuous basis, even when no active large-scale "event" is apparently underway. From the viewpoint of road infrastructure, the last landscape event is still happening and the next is always approaching (on average, an eruption occurs in Iceland every five years).

Only an hour outside of Reykjavík we stopped at Markarfljot. In July 2010, a massive flood of glacier water arrived here, melted by the eruption of the Eyjafjallajökull volcano. As thousands of Europeans found themselves immobilized when the eruption caused air traffic cancellations, some people in Iceland were dealing with the extraordinary outpouring of thick ash and water mobilized by the volcano. According to Matthíasson, as the Eyjafjallajökull eruption began, the Road Administration estimated it would take 90 minutes for the flood to reach the ring road in the Markarfljot area. In response, a brave, local roads worker activated his machinery and broke a hole through the highway. This channeled water toward the sea and away from cultivated fields and took pressure off of the bridge. The bridge was spared, and despite the raging waters, only a small section of roadway had to be replaced.

During the last two days that we were in Iceland, we were based out of Vík í Mýrdal. Vík is the southernmost village in Iceland and boasts beautiful black basalt beaches. Vik is also notable for its proximity to Katla.[11] Katla is the volcano that scientists have warned could unleash devastating consequences world-wide.[12] In Vík and the farm communities nearby, residents practice periodic evacuation drills in preparation for local consequences.

While we were in Vík, no eruptions appeared imminent, but it was hard to deny our visceral senses of instability that marked our stay there. Katla's material presence ensured that we didn't easily forget its proximity – nor the reality that it is capable of massively rearranging Vík's landscape at a scale we can barely imagine.

In 2008, a group of researchers spent time in Vík and in the nearby farming community of Álftaver.[13] They interviewed citizens of these two communities to determine their perception of risk and their preparedness for future eruptions While residents' responses and senses of safety varied greatly, several farmers commented that there is more at risk today than during the last major eruption in 1918 because of their current dependence upon critical modern infrastructures such as electricity, utilities, and transportation: "Our life is based on our land and I sometimes wonder what I will do if an eruption takes place and everything is taken away from me! Will it all be over?"[14] Another farmer said, "what if the roads are blocked, no electricity, no phone connections, petrol and so on."[15]

The senses of precariousness that we experienced in Vík echoed those we felt previously in Tokyo, where, according to some scientists, there is a 70 percent risk of a major earthquake before 2016.[16] Both locations, and many others around the world, are in the midst of realizing the contemporary consequences of the ancient fact that when streaming geologic material realities collide with infrastructures, sometimes the human and the built will be massively out-scaled. A passage from Yasunari Kawabata's 1954 novel, *Sound of the Mountain*, suggests that the medium for such realizations can be the streaming landscape itself:

> It was a windless night. The moon was near full . . . Shingo wondered if he might have heard the sound of the sea. But no – it was the mountain. It was like the wind, far away, but with a depth like a rumbling of the earth. . . . The sound stopped, and he was suddenly afraid. A chill passed over him, as if he had been notified that death was approaching. He wanted to question himself, calmly and deliberately, to ask whether it had been the sound of the wind, the sound of the sea, or a sound in his ears. But he had heard no such sound, he was sure. He had heard the mountain.[17]

We sampled and creatively documented several material traces of streaming geology events that are still evolving. And we conducted an "imaginative investigation" into probable next events. We used the results of this process as inspirational data for the field note that we titled Moving Parameters. The starting point for Moving Parameters is a familiar practice that speaks directly to designers, architects, and regional planners, namely, the definition of design specifications.

Through the language of design specifications, we offer a set of experimental concepts and dynamic images that we believe address designers and architects from within the midst of streaming landscape events. We locate our field notations along infrastructures designed to meet the jökulhlaup's glacial outburst flood. There, we stage an imaginary encounter between the determination of design specifications and the extraordinary speeds and scales of geologic change in southern Iceland. We consider the question: What new ways of designing and planning in relation to landscape events might such an encounter make possible and thinkable?

In Moving Parameters, we arrive at a proposal of four experimental design specifications. We offer them in an effort to activate particular potentials within the imaginations of publics: the potentials to evolve new capacities to "see the movement" of Iceland's landscapes, and to see that movement as something that is present, at varying speeds, in all landscapes – including Norway's. Not only is movement present and ongoing in the landscape, it also presses upon the works and approaches of farmers and citizens, designers, and architects.

As an event, as a vibrant thing that modifies other actors through a series of actions, and as a condition of contemporary daily life, the geologic presents publics, engineers, architects, and planners with an ever-unfolding situation of changing parameters. Inspired by encounters with streaming landscapes in southern Iceland in May 2012, we offer four experimental design specifications to those engaged in imagining, building, and living within landscape streams.

Design specifications for streaming landscapes | landscape events

1   Design for "no zero." There is no zero, no moment of closing the cycle of material transformation or of returning land, materials, geomorphology, or ecosystem to what was "before." There is only an interminable downstream. Design for that.

2   Design for humans as infrastructure and infrastructure as human. In southern Iceland, we encountered landscape events that provoked us to see infrastructure as composed of both human and nonhuman actants "all the way down." Such a situation requires design specifications that

are able to flex for inevitable but unpredictable moments when landscape events require humans to physically assemble-with built infrastructures and "shepherd" or redirect and redesign nonhuman infrastructural actants at crucial moments – often on the go.

3   Design for and with the limits of design. Some landscape events out-scale human design capacities altogether. Rather than trying to overcome the limits of design through ever-more resilient or flexible materials or ever-more technically complex (and therefore more vulnerable) assemblages, what might happen when we take the limits of design to be generative? That is, when designers respond to the limits of the act of design itself by generating as-yet-unimagined public policies, social connections, narrative accounts of "limits," or aesthetic appeals? What new social or governmental processes, and what new transportation or infrastructural objects become thinkable and possible?

4   Design for time's materiality. Streaming landscapes are spatial-temporal events. In landscapes events, both time and space stream in nonlinear fashions. Design for the material realities and consequences of nonlinear time's agency in landscape events.

In the Far North, projects of infrastructure and energy extraction/production are underway that will alter ecologies and geologies for thousands of years to come – and in ways unknowable from here. What approaches to living with and within landscapes of energy become possible when publics, designers, engineers, and architects work from a palpable, even aesthetic, sense of the speeds, scales, intensities, and dynamics of northern regions' geologic streamings?

Turning at the limits of the world

06.01.2014, 13:30

When forces of fast and big change pass through us, when they alter our trajectories, moods, imaginations, and experiences of self, they reverberate sensorially in our bodies and minds. This happened to us at Öndverðarnes, the westernmost point of the Snæfellsnes Peninsula, Iceland, on May 30, 2014.

Our project in Inhabiting Change was to make aesthetic equivalents, rather than to make documentation, of some of the sensations of the arrival of change, bare change, and to trace some of the ways that inhabiting change altered our Western cultural awareness and our individual psychologies while doing fieldwork for Future North. In response to our experience of inhabiting the westernmost point of Iceland and addressing that place as an aperture onto change itself, we decided to report our "findings" as a text-based artwork. In retrospect, we realize that, in fact, we unintentionally and inadvertently performed the practice that we describe/instruct in this artwork when we were at Öndverðarnes.

This practice is something like taking a long exposure in photography. It allows minute impressions to accumulate, and it is capable of coaxing into awareness events of planetary change that are occurring beneath the thresholds of ordinary stimuli or cultural habits. Without necessarily intending it to be so, this practice has some of the qualities of ritual. It moves us out of the "ordinary" and attunes us to nonhuman dynamics that co-exist and co-act with our (human) present. It is not about rational communication, teaching, or prescribed change. Its enactment calls for a performance that is highly reflective, psychological, and individual. And at the same time, it requires us to project our imaginations and conceptualizations in ways that include, communicate, and connect with people other than just ourselves, who are often quite different from ourselves and often located far from our performances.

For us, proof of this practice is that it sets up an occasion in which we sense for ourselves some of the ways that we are highly entangled with dynamic and changing earth forces at cultural and

Figure 12.2 Öndverðarnes, Iceland
Photo: smudge studio, 2014.

material edges and limits. It sets up occasions for us to invite, respect, and turn towards the difference that such sensation presents to our daily habits and assumptions.

It is particularly important to us at this point that the practices we invent turn us toward our own entanglements within and exposures to big and fast change with something other than fear. Instead of trying to control, damper, ignore, or erase our compounding feelings of radical inter-involvement and raw exposure, this practice, which we are calling Turning at the Limits of the World, addresses such feelings as interminable, and as material traces of being alive on this planet. Many of us who are contemporary Western-encultured humans have difficulty recognizing, much less attuning to, or turning at our limits as individuals or even as a species. We made our best attempt to do all three, and arrived at the following practice-as-artwork:

> *Turning at the Limits of the World*
>
>> *Make a deliberate intention to meet forces that compose a limit of the world*
>> [Take this to be an obligation.]
>> *Before setting out, invent a practice for turning at the limits of the world*
>> [Once you arrive, you will not be able to think abstractly or act directly.]
>> *Set out for a geomorphological edge*
>> [Do research. Find a geomorphological edge where human and nonhuman forces converge and delimit one another. Define the route, timing, mode, support, and affordances of travel. Each of these will prove to be highly consequential.]

Arrive

[You will arrive for the first time only once. Feel the forces of this place for your own body/brain/mind – and not as represented in guidebooks, research, others' photos, habitual assumptions.]

Re-tune the media. Make them able to signal what the forces of edge-ness disclose here

[Choose or reinvent media on the spot – something capable of attuning to forces at play at this particular moment and edge. Use them to signal the limits that are being delivered at this place by geomorphological materials and human and nonhuman events right now. Use the media to signal what is being disclosed here.]

Locate the site-moment of going no further

[Collaborate with forces that are in the midst of making/remaking the limits of the world here. Sense how these forces indicate when and how they deliver limits here. Use the media and your body/brain/mind to sense when and why you will declare: just enough and turn.]

Pause there

[Co-exist with the zero and the infinity that is your declared turning point, your human+nonhuman limit. Experience a long exposure. Like a photograph, let impressions accumulate via any means you choose.]

Perform the turn

[This is not a defeat. Declaring "just enough" and turning at the limit is not (ever) "turning back." Rather, it is to inflect your own movement in response to addressing and being addressed by limits of this world. It is a heroic act to encounter and take in the limit of the zero – its "full stop." Your act of turning can be a ritual; it is most definitely a gesture of address to the strange stranger of the limit that is arriving here. Your turning is a highly consequential act. Not to turn would have significant consequences. To turn has significant consequences. The only thing that you cannot do here is to live in this zero/infinity. The turning point is not livable. It is an un-livable state because it is always and only The Transition itself. The turning point is the trans-siting, the transit, the trans-formation. It is Change Itself. You cannot inhabit or know the turning point. You can only pass through it. This passage is the work. It is the practice.]

Return a difference

[It is a heroic act to encounter and take in this fact of the physical universe: to turn is to generate and live a difference. To turn is to acknowledge a limit and live by that limit – but with a generative difference. To turn at the limits is to generate potential and open the future. To turn is to perform the winding up and the letting go into difference and surprise. What is returned by this practice is a difference. A difference is the gift that your turning offers back to the world. It is a bow to the fact that the world is a continuously unfolding configuration and reconfiguration. A bow is a turn, it is a waveform. With this practice, you bow to the difference that the limits of the world make in yourself and in the world. The gesture of turning-bowing returns you to the interlocking material reality that is you+the limits of the world.]

Pompeii of the North: Heimaey

06.05.2014, 10:51

In area, the home island is so small that it approximates Manhattan south of the Empire State building. The volume of material that came pouring out on Heimaey in 1973 would be enough to envelop New York's entire financial district, with only the tops of the World Trade Center sticking out like ski huts. The image is not as outlandish as it seems. A few miles west of Manhattan, the high ground of Montclair – of Glen Ridge, Great Notch, and Mountainside – is the product of a similar fissure eruption.[18]

On the southern edge of Iceland, in January of 1973, a 2.4-kilometer-long fissure opened and began an outpouring of lava that lasted until July. In addition to birthing an entirely new 225-meter-high volcano, the event buried over 300 homes and left millions of tons of tephra in its wake.

Our journey to Heimaey began with a short 30-minute ferry ride from Iceland's main island. Once off the boat, we hiked through a "house graveyard" where homes are still buried beneath 15-meter-deep tongues of solidified lava. From there, we walked on to a new museum, Eldheimar, built in remembrance of the events of 1973. It has been built around what is now its centerpiece: an excavation of a buried home.

On Heimaey, we performed a conceptual turn at the limits of the world. This is a place where people have adapted to massive, fast, unexpected change. In a matter of hours people had to abandon their homes. In many cases, they returned to homes buried in lava or ash. The lava flow was relatively slow, so as physical and material limits of the town were exceeded, local citizens were able to save their harbor and the community did not suffer a major loss of life. Because of this, residents had the time and the psychological capacity to inhabit the change that continued to take place across several months. This afforded active adaptation and creative response (such as spraying sea water on the lava to cool it in a successful effort to protect the harbor from destruction). Inhabitants could experience "catastrophic change" and turn towards it in part because a fortunate configuration of events confined experiences of loss to the loss of material things and not human lives. After 40 years, inhabitants narrate the life of their island in terms of "before" and "after" the eruption. People responded to the changes that engulfed Heimaey by recreating their town with new physical structures, infrastructures, and meanings.

What gave us pause on Heimaey was our deepened sense that human psychological limits greatly influence how we are able to meet change. If we have the opportunity to meet change from within our individual physical, psychological, and cognitive limits, there is a much better chance that we will

Figure 12.3 Outside the Eldheimar Museum, Heimaey Island, Iceland

Photo: smudge studio, 2014.

be capable of creatively living within and moving in accord with changing material realities. However, if many lives had been lost on Heimay, it's very likely this island would be a place of mourning and the new museum a memorial.

Yet, we all now live in a contemporary context that is in the process of exceeding limits that have composed many received material realities – psychological, social, environmental, cultural, economic, geological. Geophysical dynamics such as climate, seismic pressures, coastal erosion are churning out change without concern for humans or our built environments.

There's much that humans cannot control when it comes to the geologic (earthquakes, tsunamis, volcanic eruptions). We design ourselves into even more risk when we stake our own species' viability upon so many affordances poised beyond the limits of the world. Human activities set algorithms of change into motion long ago. Many of the systems now beginning to "tip" can't be stopped.

Unlike those who lived through the catastrophic changes that took place in Heimaey in 1973, those of us now awakening to the catastrophic changes arriving at our own doorsteps in the form of the local effects of global warming have had the luxury of being duly warned. Unfortunately, unlike the catastrophic forces that played out during the Heimaey eruption, the compounding forces of global change that are now co-mingling with one another and pressing upon the entire human species far exceed the physical, psychological, political, economic, and technological limits of individuals and cultures. The narratives of the islanders' experiences, efforts, and ultimate renewal told through the Eldheimar Museum's compelling exhibits are remarkable and inspiring. But the forces of change now pressing upon our species' habitat will require humans to invent radically different narratives than these. Contemporary Iceland and all of the Far North are now being turned at and by the limits of the human habitat on this planet.

Practice as passage | Steilneset memorial, Vardø

06.13.2014, 14:58

> He [a Sami from Vahranger by the name of Anders Poulsen] also said that when he lifts the rune drum high into the air, or when his son Christopher lifts the stone high into the air, they will get an answer, just as two persons do when they speak to each other . . . when asked, [he] replied that when he learnt the rune drum craft from his mother, it happened because he wanted to know how people were faring far away, whether they were enjoying good fortune, and he wanted to know whether travellers will be in luck, and he wanted to help people in distress, and with his art he wanted to do good, and his mother said that she would teach him such an art. He himself had not asked to learn. He was questioned further at length, and he abided by his previous confession and did not change it in any way, nor would he confess more about his activities than that this was an art of playing the drum with which he had done no harm. Thus, on the basis of what has been confessed, the following was decided. After diligent examination and due consideration of the nature of this case and of Anders Poulsen's lengthy confession, we have learnt how exhorted creatures, represented by figures on his rune drum, induce him to believe, at the Devil's whim, the acts and signs he asked about and looked for, acts and signs which according to him are indeed confirmed by events, and he states that he has learnt this craft from his mother and another Sami in his youth.
>
> (Finnmark Witchcraft Trial Court Proceedings)[19]

A variety of planetary equilibria made it possible for the human species to evolve. Human activities on a global scale are now pushing those equilibria to their tipping points. Our species continues to act in ways that challenge the limits of its own habitat on this planet. At the Steilneset Memorial, we brought our own psychological states and material embodiments to a site in which we met up with

Figure 12.4 Steilneset Memorial
Photo: smudge studio, 2014.

and addressed earth forces. Our wager was that our encounter with earth forces at the site would activate recognitions of our personal, cultural, and perhaps even species limits.

Outside the Steilneset Memorial in Vardø, Norway, we paused for the morning on a vibrant edge of a stunning geological formation.[20] The vertical tilt of black rock that torqued its way up and out of its previously horizontal stratum and twisted what was below into what is above, signaled dynamic forces that animate the earth's continuous folding, unfolding, and refolding. Citizens and administrator tried and burned Sami people and supposed witches in the 1600s near the monument and these rocks. According to the Guidebook distributed at the Steilneset Monument and produced by the Varanger Museum, one of the accused, Anders Pouelsen, was 100 years old. His case was deferred, but he was murdered with an axe while in legal custody. Perhaps the executioners feared not only the power of the Sami people and their media capable of connecting humans across time and space. A number of names given by local people to places and natural features in the area make references to hell or to the ends of the earth, testifying to the likelihood that they also feared the physical sensation that these rocks induce: the embodied sensation that the earth is an animate force.

The present situation of our species is that we live on a planet whose bio- and geo-systems are now moving farther from equilibrium. The process-methodology that we have been practicing in the past few weeks (an aesthetic/psychological/physical turning at the limits of the world) distances us, however gently, from twenty-first-century affordances. It positions our bodies where forces that

Figure 12.5 Rock outcropping adjacent to Steilneset Memorial
Photo: smudge studio, 2014.

configure and reconfigure the planet's material realities can reach us. When we step into the practice, we try to sense forces pressing upon us (geographic, geologic, cultural, historical, psychological, etc.) more immediately and barely than they do in everyday life. The practice gives us time and place to remember, acknowledge and respect the reality of the fact that humans cannot reign in or control the planetary changes that are now underway. The practice is ritual in the sense that it is an end unto itself. It is not about communication, teaching, or creating change other than within ourselves, while we are in the process. It is highly internal, psychological, and individual – while at the same time it is highly connective across human and nonhuman distance and difference. It is a humbling reminder and a provocation to know and be with human-animal sensations of exposure to our individual and species' limits – instead of attempting to tame, control, or ignore them.

What can be learned when we invite, respect, and turn towards the difference that these sensations present, and do so with something other than fear?

As contemporary Western-encultured humans, our imaginations and capacities as designers and citizens are dulled to embodied sense of limits. Turning at the Limits of the World is our attempt to turn our own practice towards a necessary difference.

We performed just such a turn on Friday, June 13, 2014, via passage through Peter Zumthor's Memory Hall; Louise Bourgeois' The Damned, The Possessed, The Beloved; as well as the Earth's adjacent shoreline in Vardø, Norway.

70°N | *There is no darkness here*

06.21.2014, 15:09

During our month of conducting research in Iceland, Vardø, Tromsø, Kirkenes, we've been subject to a wide range of changing circumstances and planetary forces. But one force has been constant – the force of light. Before visiting the Far North, we had naively thought that the "midnight sun" in the Norwegian cities would be bright and clear. We thought that our days here would be bathed in light similar to the light we know in New York City – only for longer periods of time. We had imagined that we would adapt to that difference easily.

We now understand that in the land of the midnight sun, things are a bit less familiar than we had imagined. Here, people talk about the extremes of light as if they were a weather system. The early summer days of our trip have been filled with rain and low hanging, overcast skies. Instead of brilliant clear light, for long strings of days we've experienced an undifferentiated continuum of grey skies. The light of 14:00 has been the same as that of 07:00, and that of 16:00 as well. And this day's light is the same as yesterday's and tomorrow's. It became incredibly challenging (without watches) to determine if time was passing at all, or even if we were tired. Decisions to sleep have come less out of exhaustion and more out of randomly realizing "what time it is" and then deciding to close the

Figure 12.6 Midnight sun in Tromsø

Photo: smudge studio, 2014.

curtains because of what the mechanical clock says, rather than what the body says. The effects have accumulated. They've resulted in a low-grade psychological and physical drain. We've shared many conversations about how to manage these effects, and no one seems to have come up with a way to work around what can't be stopped: no matter the curtains, blinds, or T-shirts over the eyes, the light leaks through, psychologically if not physically. Closed eyelids continue to detect that our evolved and geographically encultured limits regarding light and sleep are being exceeded.

As days ebbed by, we started to wonder about the toll this was taking on our brains/bodies. We were moving more slowly, or perhaps the world was? How might we alleviate this intensifying sensation of disorientation? It felt as if somehow a higher atmospheric pressure was impinging on our bodies in Tromsø and Vardø. We couldn't seem to pull out of our cognitive fog. Could there be such a condition as "latitude sickness?" If so, we had it, and it wasn't the first time.[21]

As the disorientation intensified, it deepened our realization that we were, indeed, in The North. Its vastness and mythic intensities were sinking into our bodies. Our bodies knew that we were far from the landscapes, ecologies, and daily weather that were so familiar as "home." Psychological vertigo set in. Our conceptions of time and our bodies' experiences of it, long and short scale, felt as though they were stretching and collapsing.

In Kirkenes, from within the resulting mental haze, we looked out across the Barents Sea and behind us across rock landscapes that were still rebounding from their dramatic crush by Pleistocene Ice. We sent our imaginations into the not-so-distance future, when these vast spaces around us will be punctuated by new systems of transiting the thawing Northeast Passage and new affordances (new railroad lines and harbors in the far north) designed to exploit the changes brought on by the warming climate.

For the life span of our species thus far, "the North" has demarcated the edge of the human world for a number of cultures. What lay beyond acted as a physical limit to human habitation. The force of light in the far north left us feeling as though we had come upon limits of our own capacities to inhabit this place. Sensing this relatively benign "local" limit helped us to sense that we also are connected to and embedded within much larger, even planetary limits.

In the far north, the national border between Norway and Russia is enforced by lookout towers and checkpoints even as they are lived more fluidly by local inhabitants. Temperatures are warming. The population and urban infrastructures are increasing. Harbors are being prepared now for the opening of the Northeast Passage and resulting new economic and trade "opportunities." Fishing industries that collapsed decades ago are starting up again as fish such as cod migrate to cooler northern waters. But as warming continues, the cod will likely keep heading even further north, leaving another round of economic bust in their wake.

It seems to us that as resources along these northern coasts and offshore are exploited, "the North" will become less of one sort of edge – known to some cultures as the Ultima Thule, the end of the world – as it becomes more of another sort of edge: the advancing edge of big, fast-moving regional changes (opening of new trade routes, migrations of people and animals northward as a result of climate change pressures, extraction of increasingly depleted natural resources, effects of the slowing of the Gulf Stream) that will ramify southwards and around the globe.

For our research for Future North, we attempted to bring an aesthetic framework to the task of addressing challenging and complex contexts of change in the North. As researchers who are passing through towns and landscapes that are home to local people, we realize now that for the past month, we have been living one, "local" Far North version of the massive change that is arriving with unequal speeds, forces, and consequences planet-wide, but from which no location or peoples are excluded.

In many ways, we are all strangers now, even in our own local landscapes. And our home landscapes are becoming strangers to us. Wild deviations from the familiar norms of our home places are becoming more frequent and straying farther and farther from known models for prediction. Compounding strangeness is not specific to the North, but it can be particularly fast and intense here.

As natives of the fortieth parallel, our contribution to the Future North project, perhaps, is to bring with us a perspectival distance and difference that comes from being from elsewhere – the elsewhere that is quickly becoming everywhere.

The local relays into the global, the past and present relay into the far future. As researchers, in comparison to the deep, rich, highly nuanced local histories and practices we glimpsed, we were two humans doing little more than passing through. We have attempted to acknowledge and respect this reality, while asserting that what is unfolding here is neither separate from nor unrecognizable compared to what is unfolding (strangely) elsewhere, be it the wildfires of New Mexico, the rising seas in New York City, or the urgent cultural adaptations to the post-Fukushima environmental and economic realities unfolding in Japan.

As strangers in our local, increasingly strange landscapes, how might we turn brains, bodies, and spirits toward those vastly different planetary realities – the futures – that are now arriving at our doorsteps? How do we scale to the inflow of changes that aren't in sync with the physical, cognitive, and cultural structures that our species and social groups have evolved? The physical and psychological effects of big, fast planetary change are real. This reaffirms our sense that making our physical and psychological states sturdy enough to navigate present and increasingly inevitable future change as gracefully as possible is a highly urgent individual and personal task.

The stories that we now tell ourselves along the way, as we are in the midst of big fast change, will be highly consequential. The Inhabiting Change project is our attempt to invent a meaningful way to move-with and in the changes that are already unfolding, and to prepare for those to come. Via this project, we have begun the work of attempting to cultivate an awareness of the limits our species has already crossed. From here, we will try to move meaningfully with the material realities that are right now composing near and deep futures.

What difference might the Future North project return to the changing North it encountered?

Notes

1   See http://smudgestudio.org/smudge/change.html.
2   Jane Bennett, *Vibrant Matter: A Political Ecology of Things* (Durham, NC: Duke University Press, 2010), 108.
3   From an interview conducted by smudge studio with Pétur Matthíasson, Public Relations Manager for the Icelandic Road Administraiton, Reykjavík, Iceland, May 5, 2012.
4   See www.oculs.no/projects/landscape-journeys/about/.
5   See www.oculs.no/projects/landscape-journeys/people/ and http://lateraloffice.com/landscape-journeys-2012.
6   See https://fopnews.wordpress.com/2012/04/22/streaming-landscapes/.
7   See www.vegagerdin.is/english/.
8   See http://travel.nytimes.com/2006/06/18/travel/18ring.html?pagewanted=all.
9   See www.eaee.boun.edu.tr/bulletins/v20/v20web/iceland.htm.
10 See: www.platetectonics.com/oceanfloors/iceland.asp.
11 See www.volcano.si.edu/world/volcano.cfm?vnum=1702-03=.
12 See www.bbc.com/news/world-europe-15995845.
13 Deanne Bird, Guðrún Gísladóttir, and Dale Dominey-Howes, "Residents' Views of Volcanic Risk in Southern Iceland – a Preliminary Analysis," 2008. www.vegagerdin.is/vefur2.nsf/Files/Res_views_volcanic_risk_south_Iceland/$file/Res_views_volcanic_risk_south_Iceland.pdf.
14 Guðrún Jóhannesdóttir and Guðrún Gísladóttir, "People Living Under Threat of Volcanic Hazard in Southern Iceland: Vulnerability and Risk Perception," *Natural Hazards and Earth Systems Sciences* 10 (2010): 415. www.nat-hazards-earth-syst-sci.net/10/407/2010/.
15 Ibid., 416.
16 "Tokyo Has 70% Chance of Powerful Earthquake Within Four Years," *The Guardian*, January 23, 2012, www.guardian.co.uk/world/2012/jan/23/tokyo-powerful-earthquake-four-years.

17 Yasunari Kawabata, *The Sound of the Mountain* (New York: Random House, 1996), 8.

18 John McPhee, "Cooling the Lava," in *Bedrock: Writers on the Wonders of Geology,* edited by Lauret E. Savoy, Eldridge M. Moores, and Judith E. Moores (San Antonio: Trinity University Press, 2006), 116.

19 "Records of Court Proceedings 1692–1695," the Archives of Finnmark District Magistrate No. 25, translated in *The Witchcraft Trials in Finnmark, Northern Norway*, Liv Helene Willumsen (Bergen: Varanger Museum and Skald Publisher, 2010), 377–92.

20 See http://donnawheeler.com/witching-hour-in-norway-the-steilneset-memorial/.

21 See https://fopnews.wordpress.com/2011/05/21/lofoten/.

Bibliography

Bennett, Jane. *Vibrant Matter: A Political Ecology of Things*. Durham, NC: Duke University Press, 2010.

Bird, Deanne, Guðrún Gísladóttir, and Dale Dominey-Howes. "Residents' Views of Volcanic Risk in Southern Iceland – a Preliminary Analysis." 2008. www.vegagerdin.is/vefur2.nsf/Files/Res_views_volcanic_risk_south_Iceland/$file/Res_views_volcanic_risk_south_Iceland.pdf.

Jóhannesdóttir, Guðrún and Guðrún Gísladóttir. "People Living Under Threat of Volcanic Hazard in Southern Iceland: Vulnerability and Risk Perception." *Natural Hazards and Earth Systems Sciences* 10 (2010): 407–420. www.nat-hazards-earth-syst-sci.net/10/407/2010/.

Kawabata, Yasunari. *Sound of the Mountain*. New York: Random House, 1996.

Matthíasson, Pétur. *Public Relations Manager*. Icelandic Road Administration, Interview with smudge studio, Reykjavík, Iceland, May 5, 2012.

McPhee, John. "Cooling the Lava." In *Bedrock: Writers on the Wonders of Geology*, edited by Lauret E. Savoy, Eldridge M. Moores, and Judith E. Moores San Antonio: Trinity University Press, 2006, 116.

"Records of Court Proceedings 1692–1695." The Archives of Finnmark District Magistrate No. 25. Translated in *The Witchcraft Trials in Finnmark, Northern Norway*, Liv Helene Willumsen, 377–92. Bergen: Varanger Museum and Skald Publisher, 2010.

"Tokyo Has 70% Chance of Powerful Earthquake Within Four Years'." *Guardian*. January 23, 2012. www.guardian.co.uk/world/2012/jan/23/tokyo-powerful-earthquake-four-years.

Index

Note: Page numbers in *italics* denote references to Figures.

Milton Keynes UK
Ingram Content Group UK Ltd.
UKHW050452071024
449327UK00015B/338